Physiology of the Graafian Follicle and Ovulation

The development and selection of ovarian follicles is one of the most active areas of contemporary reproductive research. Relevant experimental work extends from laboratory rodents, across a wide range of domestic species, to human clinical studies, especially as related to problems of fertility and *in vitro* fertilisation. This volume provides comprehensive coverage of the field, integrating research findings from animal and human studies and condensing the vast published literature into a meaningful and digestible physiological account that highlights the key rôle played by the oocyte in influencing all stages of follicular development.

R.H.F. HUNTER has a wealth of experience in reproductive research, having spent 40 years studying reproduction in laboratory and farm animals. His work has included numerous surgical studies of reproductive tissues that have provided him with unique insights into ovarian function and the process of ovulation. He has published extensively in his field, including the books *Physiology and Technology of Reproduction in Female Domestic Animals* (1980), *The Fallopian Tubes: Their Rôle in Fertility and Infertility* (1988) and *Sex Determination, Differentiation and Intersexuality in Placental Mammals* (1995).

Regnerus De Graaf
Delphis Medicinæ Doctor

Contents

Preface   *page* xi
Acknowledgements   xiv
List of abbreviations   xviii

I Mammalian ovaries, Graafian follicles and oocytes: selected
   historical landmarks   1
      Introduction   1
      Steps in classical antiquity   1
      Sixteenth and seventeenth century contributions   3
      Eighteenth and nineteenth century views   14
      Twentieth century highlights   16
      Prospects for the current century   19
      References   20

II Formation and structure of ovaries: elaboration of follicular
    compartments   24
      Introduction   24
      Sex determination   24
      Differentiation of an ovary; primordial germ cells   27
      Oogenesis, meiosis, growth of oocyte   33
      Formation and function of zona pellucida   38
      Follicular growth and formation of a Graafian follicle   42
      Follicular development, acquisition of follicle stimulating
         hormone dependence   48
      Transcription factors in folliculogenesis   50
      Oocyte macromolecules and maternal RNA programme   51
      Concluding remarks   51
      References   52

III   Physiology of the ovaries and maturing Graafian follicles   60
        Introduction   60
        Ovarian vasculature   61
        Ovarian and neighbouring lymphatic pathways   69
        Ovarian innervation, especially of follicles   77
        Ovarian blood flow   83
        Temperature gradients within the ovaries   90
        Why do Graafian follicles grow so large?   96
        Concluding remarks   98
        References   99

IV   Ovarian follicular-antral fluid   106
        Introduction   106
        Formation of follicular fluid   107
        Physical condition of follicular fluid   110
        Composition of follicular fluid   112
        Intra-follicular pressure, PO_2, PCO_2, pH   122
        Contribution to Fallopian tube and peritoneal fluids   124
        Involvement of follicular fluid in events of fertilisation   127
        Concluding remarks   130
        References   131

V   Endocrine potential and function of a Graafian follicle   138
        Introduction   138
        Steroids and gonadotrophins: 'two-cell' theory of oestradiol
            synthesis   139
        Steroid acute regulatory protein   147
        Ovarian proteins – inhibin, activin and follistatin   150
        Peptide growth factors   158
        Cytokines and eicosanoids   163
        Endorphins and enkephalins   169
        Involvement of nitric oxide   170
        Concluding remarks   172
        References   172

VI   Follicular recruitment, growth and development: selection – or
        atresia   186
        Introduction   186
        Recruitment of follicles, selection, dominance   187
        Waves of follicular development   188

Endocrine activity associated with dominance — 197
Models for follicular selection and dominance — 198
Follicle growth inhibitory factors — 204
Atresia of follicles and germ cells — 204
Apoptosis within ovarian follicles — 209
Concluding remarks — 215
References — 216

VII Follicular responses to the pre-ovulatory surge of gonadotrophic
hormones — 224
  Introduction — 224
  Events underlying the pre-ovulatory surge — 226
  Gonadotrophin surge attenuating factor — 229
  Resumption of meiosis — 230
  Expansion and mucification of cumulus oophorus — 244
  Extracellular matrix of follicle — 248
  Remodelling of basement membrane — 250
  Concluding remarks — 251
  References — 252

VIII The process of ovulation and shedding of an oocyte — 262
  Introduction — 262
  Spontaneous versus induced ovulation — 263
  Timing of ovulation and dimensions of follicle — 266
  Process of ovulation – general features — 268
  Process of ovulation – morphological highlights — 270
  Process of ovulation – biochemical events — 274
  Contribution of leucocytes and cytokines — 276
  Rôle of fimbriated extremity of Fallopian tube — 281
  Ischaemic model for studying ovulation — 282
  Genes involved in the ovulatory cascade — 283
  Concluding remarks — 284
  References — 286

IX Post-ovulatory fate of follicle and oocyte: contributions of
somatic cells and follicular fluid — 295
  Introduction — 295
  Collapsed follicle evolves into corpus luteum — 296
  Rôle of fimbriated infundibulum, cilia and myosalpinx — 297
  Liberated follicular cells as paracrine tissue — 301

Contribution and fate of follicular fluid 304
Regional fluid environments within the Fallopian tube(s) 306
Local ovarian influences on tubal physiology 307
Increasing progesterone secretion modifies tubal physiology 310
Progression of fertilised versus unfertilised eggs 311
Post-ovulatory ageing of oocytes 313
Concluding remarks 316
References 317

X Failure of ovulation: status of the gonads 325
Introduction 325
Cystic follicles in animals 326
Polycystic ovarian disease in women 328
Ovarian dysgenesis 331
Kallmann's syndrome 333
Ovotestis and ovulation failure 334
Ovulation failure as a response to stress 336
Leptin involvement in ovulation failure 337
Concluding remarks 338
References 339

XI Induction of ovulation in women and domestic animals 345
Introduction 345
Treatments for overcoming anovulation in women 346
Induction of multiple ovulation in women 350
Maintenance of a functional corpus luteum 355
Control of ovulation time in animals 356
Induction of superovulation in animals 357
Manipulation of prepuberal animals 360
Concluding remarks 361
References 362

XII Concluding thoughts and a current perspective 366
Introduction 366
Historical landmarks 366
Formation of ovarian follicles 368
Blood vessels, lymphatics, innervation and temperature 369
Contribution of antral fluid 370
Endocrine potential of Graafian follicle 371
ʼllicular selection, dominance, atresia 372

Responses to the gonadotrophin surge   373
Components of ovulation and shedding of the oocyte   374
Post-ovulatory fate of follicle and contents   375
Ovulatory failure and ovarian anomalies   376
Induction of ovulation and control of ovulation time   377

Index   379

The plate sections are between pp. 140 and 141, and pp. 268 and 269.

Preface

This book was written in Edinburgh between July 1999 and July 2001, with some specific additions and considerable editing between September and December 2001. However, it is only reasonable to confess that the hope of producing such a manuscript had long been on my mind. During training in Cambridge in the 1960s, and very much under the guidance of Professor E.J.C. Polge, CBE, FRS, in the laboratory and the late Mr L.E.A. Rowson, OBE, FRS, in the operating theatre, there were repeated opportunities to examine ovaries close to the time of ovulation. Indeed, on more than one occasion, the actual collapse (not rupture!) of follicles was witnessed during surgical intervention, a point referred to in the text and one which enabled development of a surgical model for observing the terminal changes in Graafian follicles of pigs. Studies employing induced ovulation, superovulation, and culture of follicles at different stages of development all prompted thoughts concerning the process of ovulation itself, although such ruminations took second place to studies of fertilisation and early embryonic development. Our focus was on events in the Fallopian tubes rather than on those in the gonads.

Only whilst holding a Carlsberg Research Professorship in the Department of Veterinary Reproduction in Copenhagen during 1997 and 1998 was there an opportunity to read about, and reflect more extensively on, follicular development and the process of ovulation. This was at a time when we were examining the possibility of temperature gradients within ovarian tissues of large domestic species. Such reading was facilitated by the excellent library in the Royal Veterinary University and by the superb collection of books in Professor Torben Greve's own clinical department. In fact, I particularly wish to acknowledge the inspiration received during my time in Copenhagen, and the generosity and assistance of many very special colleagues there.

Having made such comments about the origin and genesis of the text, there naturally follows the question as to precisely why it has been written. The response is straightforward. As a university lecturer, it has been apparent that the topics of follicular maturation and ovulation continue to fascinate undergraduates and postgraduates alike, so one objective in producing this monograph was that it should be helpful to both students and their teachers. And although, of course, books continue to be produced across the compass of reproductive physiology, one could not fail to register that development and selection of ovarian follicles is one of the most active and fashionable areas of contemporary reproductive research. Relevant experimental work extends from laboratory rodents, across a wide range of large domestic species, to human clinical studies, especially as related to problems of fertility and *in vitro* fertilisation.

Many of the leading journals are currently devoting one quarter to one third of their published material to ovarian studies, and therein resides a major problem. It is simply not possible for a single author to read, digest and incorporate more than a small proportion of such extensive material into his or her own writing nor, in reality, would a more computerised approach to the literature have overcome this limitation. No matter how much ground one attempts to cover in bibliographical searches, eventually one has to focus and produce some agreeable sentences, not a list or encyclopaedia with an indigestible number of references and even more points of minute detail. So the author readily makes this confession – I have tried to produce a readable text, not an all embracing *magnum opus*. The latter would scarcely be possible in the twenty-first century, even by the most motivated and industrious of academics with no other duties or activities.

As noted in an earlier work,

> That which you cannot as you would achieve,
> you must perforce accomplish as you may.
> W. Shakespeare, *Titus Andronicus*, Act II, sc. 1.

Within this limitation, it is hoped that there is a sufficient body of references to guide those so inclined to further material on a given topic. This should be possible from the text, aided no doubt by access to 'on-line' facilities. One point concerning the references is in a sense historical, for there is a significant overlap between much that is being published today and work that was in print 30 to 40 years ago, even though the *raison d'être* or interpretation may be slightly different. I have endeavoured to bring this out in various respects, and have a strong feeling that one could prepare a scholarly and entertaining review demonstrating that what purports to be new in ovarian studies is not original at all. The terminology may have changed, a molecular flavour may have been introduced here and there, but the nub of the idea has a long and respectable pedigree. This point won't be laboured for fear one has transgressed oneself, much to the delight of potential critics! Even so, one will undoubtedly have overlooked many key papers, not least the work of friends and colleagues, and for such omissions I apologise and ask for understanding.

The chapters are presented in what should appear as a reasonable sequence, each with some form of perspective in the introduction and highlights in the conclusion. In order that the chapters can be perused as individual entities, there is a mild overlap of certain points or themes – as will be apparent from the table in the final chapter – but nowhere is this extensive. The intention is not to be repetitive, but simply to support the flow of ideas and to offer observations in a meaningful context. The latter remarks are also relevant to the actual species covered in the text: examples have been given where they seem useful and add clarity or inspiration, but no attempt has been made at systematic comparisons between species. Bearing in mind the author's long-standing commitments within veterinary faculties, a certain bias towards large farm animals will be understandable.

Finally, a comment concerning the intended readership. Mention has already been made of undergraduates and postgraduates. More specifically, it is hoped that this will be seen as a useful text by advanced (Honours) undergraduates in biological, biomedical and animal sciences, as well as by medical and veterinary students and members of clinical departments. The book should also offer background reading for MSc and doctoral candidates (PhD and/or MD) and for post-doctoral fellows and clinicians developing projects in related reproductive spheres. And, perhaps most of all, there is the wish that busy (i.e. overworked) university lecturers may find at least some of the themes helpful to their own teaching. My previous books were generously reviewed by various well-known academics, so there remains a modest hope that the present one will be seen as a not-completely redundant contribution. This thought is written in spite of the endless flow of up-to-the-minute, multi-authored conference volumes, frequently under the name of prominent or even distinguished editors.

Edinburgh   R.H.F. Hunter
January 2002

Acknowledgements

Inspiration comes from diverse sources, not always fully appreciated during a spell of writing. In the inelegant jargon of the times, it must certainly be multi-compartmental and doubtless has its roots in listening to excellent lectures, observing significant changes in tissues at surgery and in the laboratory, reading and browsing in well-stocked libraries and, perhaps most important of all, conversing with talented colleagues and students.

Over and above these general remarks, one remains forever indebted to gifted teachers and also to a wonderfully tolerant family who understood one's need to write. The teachers almost all stem from Cambridge days and include the late Sir Joseph Hutchinson, Professor T.R.R. Mann and Mr L.E.A. Rowson and, happily still with us, Professor E.J.C. Polge, Professor R.V. Short and Professor Sir Brian Heap. All have been and/or continue to be supportive and I remain warmly grateful for such help. A further notable source of inspiration concerns post-doctoral days in Paris, where Professor Charles Thibault made the life of a young researcher especially agreeable and stimulating through example and friendship. Such stimulation and friendship continue to this day, and repeated spells in France have undoubtedly added a useful dimension to Anglo-Saxon views of the world!

Moving more specifically to preparation of the text, my debt to many close colleagues is considerable. Assistance was offered in various ways. Drafts of individual chapters were generously commented on by Drs K.P. Bland, A.A. Macdonald and N.L. Poyser in Edinburgh and by Dr B. Cook in Glasgow; and by Professors T.G. Baker and T.D. Glover in England, Professor N. Einer-Jensen in Odense, Denmark, Professor C.A. Price in Montréal and Professor B.P. Setchell in Adelaide. To all these experts, I express my very best thanks.

Preparation of the typescript both in draft and final form was once again undertaken by Mrs Frances Anderson, on this occasion at some extra inconvenience to herself having moved from Edinburgh to the Scottish Highlands. I am particularly grateful for all her help, not least her skills in employing the most modern electronic and computerised technology. For a simple pen and ink man, this means of avoiding frustration and delays with the ubiquitous screen has been a welcome relief. Others worship the computer – indeed I have been relentlessly bullied towards it by a Glaswegian colleague (now happily retired!) – but pleasure and discipline can still be found whilst composing sentences in the traditional manner. Ultimately, it is the thoughts or ideas that count in an intellectual exercise, not the means of imposing them upon the empty page.

As to illustrative material, skilled assistance came from various quarters. Mr Colin Warwick, Senior Photographer in the Faculty of Veterinary Medicine, University of Edinburgh, devoted a great deal of time to producing prints and larger photographic plates with which to support the text. My debt to him is considerable, not only for the tasks accomplished, but also for his interest and enthusiasm. Further specialist assistance with illustrative material was given by Miss Jane Sharland of the Royal Collection of Pictures at Windsor Castle, Mr Nicholas Smith of Cambridge University Library, Miss Jean Archibald of Edinburgh University Library, Mrs Niki Pollock of Glasgow University Library, together with the following colleagues: Professors D.T. Baird, T.G. Baker, N. Einer-Jensen, J.E. Fléchon, M.J.K. Harper, P. Hyttel, A. Illius, N.-O. Sjöberg, C. Thibault and the late D.G. Szöllösi. Production of the finished volume would not have been possible without this much appreciated cooperation.

The following people kindly supplied reprints, preprints, photocopies or guidance to specific information: W.R. Allen, D.G. Armstrong, B. Avery, J. Bahr, D.T. Baird, T.G. Baker, M. Barrault, K.P. Bland, R.J. Blandau, O. Bomsel, M. Brännström, I. Bruck-Bøgh, A.G. Byskov, B. Campbell, H. Charlton, B. Cook, N. Dekel, M.A. Driancourt, R.G. Edwards, N. Einer-Jensen, R. Einspanier, L. Espey, B. Fléchon, J.E. Fléchon, P.A. Fowler, S. Franks, F. Gandolfi, T.D. Glover, R. Gore-Langton, R.G. Gosden, A. Gougeon, G. Greenwald, T. Greve, C. Grøndahl, A.R. Günzel-Apel, M.J.K. Harper, R.A.P. Harrison, S.G. Hillier, M.G. Hunter, P. Hyttel, W. Jöchle, P.G. Knight, K.A. Lawson, H.J. Leese, H. Lehn-Jensen, M.C. Levasseur, F. Lopez-Gatius, M. Luck, A.A. Macdonald, E.A. McGee, A. McLaren, G. Macchiarelli, J. Martal, V. Mezzogiorno, R. Moor, P. Motta, H. Niemann, S. Nottola, N.L. Poyser, C. Price, D. Rath, F.M. Rhodes, M. Roberts, M. Schmidt, B.P. Setchell, R.V. Short, N.O. Sjöberg, F. Stewart, A. Templeton, C. Thibault, J.L. Tilly, E. Töpfer-Petersen, A. Tsafriri, G. Vajta, D. Waberski, D.C. Wathes, R. Webb and A. Zeleznik.

Permission to copy illustrations, figures or the content of tables was generously granted by the following authors: F. Arakane, A. Arici, D.G. Armstrong, D.T. Baird, T.G. Baker, A.R. Bellve, J.D. Biggers, R.J. Blandau,

M. Brännström, B.W. Brown, N.W. Bruce, O. Bukulmez, R.K. Burns, R.K. Christenson, B. Cook, M.J. Cunningham, N. Dekel, R.P. Dickey, G.A. Dissen, M.A. Driancourt, R.G. Edwards, N. Einer-Jensen, J.W. Everett, T. Fair, B. Fléchon, J.E. Fléchon, S. Franks, I.S. Fraser, R.G. Gosden, A. Gougeon, M.J.K. Harper, S.G. Hillier, D.E. Holtkamp, A.J.W. Hsueh, P. Hyttel, J.G. Koritké, E. Lenton, M.C. Levasseur, M.F. McDonald, E.A. McGee, K.P. McNatty, A.S. McNeilly, G. Macchiarelli, R.M. Moor, Y. Morita, B. Morris, R.J. Norman, S. Nottola, S.R. Ojeda, F. Olivennes, Ch. Owman, H. Peters, H.M. Picton, E.J.C. Polge, N.L. Poyser, T. Reissmann, S.R.M. Reynolds, L. Roblero, G.T. Ross, M. Rosselli, M.B. Sass, R.J. Scaramuzzi, R.H. Schwall, R.V. Short, M.J. Sinosich, N.O. Sjöberg, L.D. Staples, D.G. Szöllösi, C. Thibault, B. Thuel, J.L. Tilly, R. Webb, J.L. Yovich and F. Zachariae.

Blessing to cite tabulated data or reproduce illustrative material was sought from the following publishing houses: Academic Press, Antonio Delfino Editore, Cambridge University Press, Ellipses, Granada Publishing (Paul Elek), Karger, Longman, Oxford University Press, Raven Press, Springer-Verlag and Williams & Wilkins.

Permission was similarly sought from the Editors of the following academic journals: *Acta Endocrinologica*; *Archives d'Anatomie, d'Histologie et d'Embryologie*; *Archiv für die Physiologie*; *Archives of Histology and Cytology*; *Anatomy and Embryology*; *Biochemical and Biophysical Research Communications*; *Biology of Reproduction*; *Contributions to Embryology of the Carnegie Institution*; *Developmental Biology*; *Endocrine Reviews*; *European Journal of Obstetrics, Gynaecology and Reproductive Biology*; *Fertility and Sterility*; *Human Reproduction*; *Human Reproduction Update*; *Johns Hopkins Hospital Reports*; *Journal of Anatomy*; *Journal of Experimental Zoology*; *Molecular Reproduction and Development*; *Nature (London)*; *Oxford University Press*; *Proceedings of the Royal Society Series B*; *Recent Progress in Hormone Research*; *Reproduction, Fertility and Development*; *Reproduction* (formerly *Journal of Reproduction and Fertility*); *Reproduction, Nutrition, Development*; *Reviews of Reproduction*; and *Tissue and Cell*.

Any omissions from the above list are completely unintentional. If any there should be, then the author offers sincere and unreserved apologies for such an oversight. In the convention of a single-authored text, the responsibility for errors remains his alone.

Two further areas of acknowledgement must be placed on record. Dr Alan Crowden, Director of Science Publishing, and his colleagues at Cambridge University Press have been especially thoughtful, patient and considerate throughout this exercise. I wish to thank the printers, photographers and designers and, particularly warmly, Sandi Irvine, my Copy-Editor.

Final words concern the debt to my wife. Preparation of this work has involved diverse sacrifices on her part and that of our family, and a degree of flexibility and tolerance which one suspects few wives could have shown throughout a period of more than two years. In addition to the domestic freedom afforded, there has been a vast amount of help with checking, rechecking and editing of the various drafts, and then with the further saga of proofs and index. For this exceptional and continuing collaboration, as well as for understanding and affection, I very much wish to record my profound gratitude.

R.H.F. Hunter

List of abbreviations

ACTH	adrenocorticotrophic hormone
ANOVA	analysis of variance
bFGF	basic fibroblast growth factor (FGF-2)
BMP	bone morphogenetic protein
cAMP	cyclic $3',5'$-adenosine monophosphate (cyclic AMP)
CGRP	calcitonin gene-related peptide
CSF	(1) cytostatic factor; (2) colony stimulating factor
CTP	carboxy-terminal peptide
Cx43	connexin 43
E-cadherin	epithelial cadherin
eCG	equine chorionic gonadotrophin
EGF	epidermal growth factor
EGFR	EGF receptor
eNOS	endothelial NOS
FADD	Fas-associated protein with death domain
FasL	Fas ligand
α-FF antigen	α-follicular fluid antigen
FF-MAS	follicular fluid MAS
FGF	fibroblast growth factor
FGIF	follicle growth inhibitory factor
FLICE	FADD-like IL-1β-converting enzyme
FLIP	FLICE/caspase 8 inhibitory protein
FSH	follicle stimulating hormone
GAG	glycosaminoglycan
GCSF	granulocyte colony stimulating factor
GDF	growth differentiation factor
GMCSF	granulocyte–macrophage CSF
GnRH	gonadotrophin releasing hormone

GnSAF	gonadotrophin surge attenuating factor
GnSIF	gonadotrophin surge inhibiting factor
GRP	gastrin releasing peptide
HB-EGF	heparin-binding epidermal growth factor
hCG	human chorionic gonadotrophin
HGF	hepatocyte growth factor
hMG	human menopausal gonadotrophin
hPG	human pituitary gonadotrophin
IFN	interferon (e.g. IFN-γ)
Ig	immunoglobulin
IGF	insulin-like growth factor
IGF-1R	IGF-1 receptor
IGFBP	IGF-binding protein
IL	interleukin (e.g. IL-1)
IL-6/LIF	interleukin 6/leukaemia inhibitory factor
iNOS	inducible NOS
i.u.	international unit(s)
IVF	in vitro fertilisation
LH	luteinising hormone
LHRH	LH releasing hormone
LIF	leukaemia inhibitory factor
M	monocyte
MAP kinase	mitogen activated protein kinase
MAS	meiosis activating substance/sterol
MCP	monocyte chemoattractant protein
MCSF	macrophage CSF
MIS	(1) meiosis-inducing substance; (2) Müllerian inhibiting substance
MMP	matrix metalloprotease
MPF	metaphase promoting factor
MPGF	male pronucleus growth factor
MPS	meiosis-preventing substance
M_r	relative molecular mass
MRNA	messenger RNA
N-cadherin	neuronal cadherin
NCAM	neural cell adhesion molecule
NE	norepinephrine
NG	neutrophilic granulocyte
NGF	nerve growth factor
nNOS	neuronal NOS

NOS	nitric oxide synthase
NPY	neuropeptide Y
OHSS	ovarian hyperstimulation syndrome
PA	plasminogen activator
PAF	platelet activating factor
PAPP-A	pregnancy associated plasma protein A
$P\text{CO}_2$	partial pressure of carbon dioxide
PCOS	polycystic ovarian syndrome
PDGF	platelet-derived growth factor
PG	prostaglandin (e.g. $PGF_{2\alpha}$)
PGS	prostaglandin synthase
PKA	protein kinase A
PMSG	pregnant mare serum gonadotrophin
$P\text{O}_2$	partial pressure of oxygen
PP5	placental protein 5
rFSH	recombinant FSH
rLH	recombinant LH
ROS	reactive oxygen species
SCF	stem cell factor (Steel factor)
SDS-PAGE	sodium dodecyl sulphate–polyacrylamide gel electrophoresis
SHBG	sex hormone-binding globulin
StAR	steroid or steroidogenic acute regulatory protein
TGF	transforming growth factor
TIMPs	tissue inhibitors of metalloproteases
TNF	tumour necrosis factor
tPA/uPA	tissue PA/urokinase PA
VEGF	vascular endothelial growth factor
VIP	vasoactive intestinal peptide
ZP	zona protein

I

Mammalian ovaries, Graafian follicles and oocytes: selected historical landmarks

Introduction

To the younger readers of this volume, the inclusion of a brief history of observations on mammalian ovaries may seem quite unnecessary, perhaps even an indulgence. Many such readers will have a molecular orientation, will be seeking a balanced assessment of recent research in their own highly specialised ovarian field, and will doubtless be hoping for inspiration and fruitful new lines of enquiry. All this is readily appreciated, as is the fact that most of the younger generation will have had neither the time nor the inclination to browse in the libraries of an ancient university or medical school. Such activity is, of course, not to everyone's taste but it could be to the advantage of many, in particular to reflect on how their chosen field has developed down the decades or even centuries and to note the considerable contributions of their predecessors. Viewed on such an extended timescale, their own sophisticated researches on a new growth factor, binding protein, gene sequence or mutation may fall into a different perspective: excellent – even distinguished – work certainly, but overall only a tiny fraction of the jigsaw that constitutes an understanding of ovarian function in the year 2002. So absolutely no apology is offered for the concise chapter that follows. Rather, the wish is expressed that all those who handle this volume will spare a few moments to glance at some of the studies undertaken before they themselves came on the scene. A valuable history on discovery of the ovaries can be found in Short (1977).

Steps in classical antiquity

Although its component parts and their relationships were far from being appreciated, the reproductive system attracted the attention of ancient Greek philosophers and physicians, the female organs in particular becoming the subject of

1

classical works. Here, most attention appears to have focused on the uterus, even though the ovaries would undoubtedly have been seen by Hippocrates (460–370 B.C.). They came to be thought of as the 'female testes', but this was not the view of Aristotle (384–322 B.C.). He regarded the horned (bifurcated) uterus as representing the female gonads, and within this scheme he interpreted the *catamenia* (i.e. menstrual coagulum) elaborated by the uterus as the female contribution to generation of an embryo. Intermingling of the female fluid with the male semen or *sperma* in some remarkable manner formed the rudiment of the fetus – the heart – within the uterus. The heart then directed elaboration of a complete fetus from further menstrual blood. In this concept, there was clearly no requirement to highlight the ovaries, even though Aristotle must have examined these prominent structures during his extensive observations on many species of animal and doubtless pondered their significance.

Within this Western – essentially Mediterranean – tradition, credit for first specifically describing human ovaries is generally given to Herophilus of Chalcedon, a third century B.C. anatomist who taught and practised medicine at Alexandria. His major treatise on midwifery came to be highly esteemed and widely consulted during antiquity. Despite such acclaim, he had introduced an erroneous line of thought, stemming from poor observation and considerable imagination. Even though he described the uterus and cervix, Herophilus presumed that the ducts now termed Fallopian tubes transmitted female semen from the ovaries (the 'female testes') to the urinary bladder. To propose a generative rôle for the ovaries was clearly inspirational, and a bold step in that it disagreed with Aristotle's view of the uterus as the source of female semen. Even so, the failure of Herophilus to have followed the Fallopian tubes to their correct destination appears surprisingly careless to modern eyes, and transmission of female semen to the bladder would seem unusual in any scheme of reproduction.

Soranus of Ephesus (*circa* A.D. 100), a Greek physician who practised medicine in Rome, perpetuated the misinterpretation that ducts pass from the ovaries to the urinary bladder in his classical work on gynaecology. Having studied in Alexandria and been much influenced by the descriptions of Herophilus, Soranus was in effect re-stating that the putative female seed has no concern with procreation, being expelled from the body with the urine. The extensive writings of Soranus had a prominent influence on the Latin West, and the views assembled in his *Gynaecia* long outlived him, indeed in some quarters for many centuries (Temkin, 1956). In a context of a rôle for the ovaries, and in the light of subsequent contributions by Galen, this may come as a considerable surprise. However, clear evidence has been offered by Bodemer (1969) for the manner in which sixteenth century texts were influenced by the views of Soranus.

The Greek physician and biologist, Galen (A.D. 130–200) also followed the traditional pilgrimage across the Mediterranean to the renowned medical school at Alexandria. He studied there from A.D. 152 to A.D. 157 before progressing to Rome and, although strongly taken by the writings of Herophilus, he escaped from prevailing doctrines in proposing that the female does elaborate semen by filtration from the bloodstream within the ovaries ('female testicles'). Such semen would then be carried through the Fallopian tubes, which, he correctly noted, coursed to the uterus and not to the bladder. Despite this anatomical evidence based in part on observations of the bicornuate uterus in domestic ungulates, Galen envisaged that it was the male and female fluids intermingling within the uterus that coagulated to form a fetus. His views thus restored some prominence to the ovaries in the process of generation and in one sense gave them parity with the male gonads. The doctrines of Galen with their echoes of Aristotle remained prominent for almost fourteen centuries after his death, undoubtedly a mark of his commanding stature in the medical world. In a reproductive context, such was the state of enquiry and of philosophical endeavour towards the end of the Graeco-Roman period – a legacy of ideas and speculation that remained prominent well into the period of mediaeval thought and teaching. Although the wonderful drawings of Leonardo da Vinci (1452–1519) took shape during the latter part of this span, many were lost and those remaining were not published systematically for the benefit of the scientific world until the late nineteenth century. For the historical record, however, it is worth noting that Leonardo clearly depicted the gravid human uterus and indeed the morphology of the neighbouring ovary (Fig. I.1). Moreover, generations of medical students owe a special debt to Leonardo in that he added excitement to the topic of placentation by bestowing the cotyledonary specialisations of the cow placenta upon the materno-fetal interface in humans. How did this come about? Simply that he supplemented his painstaking dissection of human cadavers with animal specimens, not least ones obtained from the slaughterhouse.

Sixteenth and seventeenth century contributions

As is nowadays quite widely appreciated, aesthetically pleasing and surprisingly accurate drawings of the human reproductive organs were published in 1543 in the masterpiece of Andreas Vesalius (1514–1564). Born in Brussels, he studied in Louvain and Paris and then held a professorship in the medical school of his adopted Padua. The seven volumes of his work were together entitled *De Humani Corporis Fabrica Libri Septem*, although they are usually referred to simply as the *Fabrica*. Generally considered to form the basis of

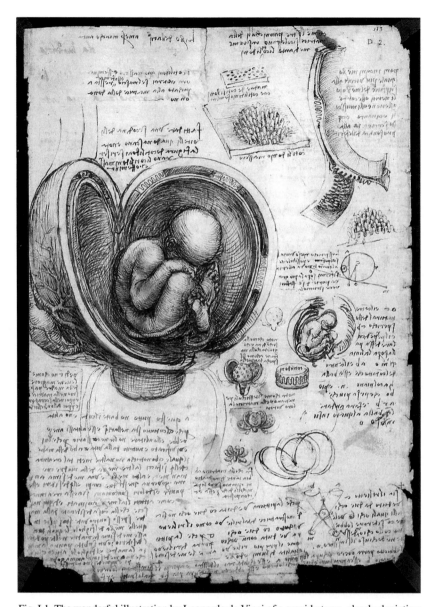

Fig. I.1. The wonderful illustration by Leonardo da Vinci of a gravid uterus, clearly depicting one of the ovaries and associated blood vessels.

As an aside, Leonardo has taken the remarkable liberty of introducing the cotyledonary placenta of a cow into a human uterus. One explanation offered for this error is that by the time the corpse was available for dissection, the contents of the uterus were in an advanced state of degeneration, so tissues were procured from a local abattoir to enable completion of the drawing. (Courtesy of 'The Royal Collection © 2001, Her Majesty Queen Elizabeth II'.)

modern systematic teaching and research in anatomy, the *Fabrica* illustrate the disposition of the ovaries and Fallopian tubes, but the interpretation of their relationship retains a strong flavour of Galen. Vesalius inferred that the ovaries and Fallopian tubes played a rôle parallel to that of the testes and associated male ducts. The tubes were thus interpreted as 'semen conveying' vessels and, with considerable artistic licence, illustrated in Book V of the *Fabrica* as ducts coiled around the 'female testes' (Fig. I.2). In fairness to Vesalius, it is worth emphasising that the existence of male and female gametes had not yet even entered the imagination, let alone the notion that the ovaries might be the source of egg cells (i.e. oocytes).

Fallopius (Gabriele Fallopio, 1523–1562), a sometime pupil of Vesalius who succeeded to the same Chair in Padua, did not take matters forward in the above regard. Although he produced the first specific description of the Fallopian tubes in his *Observationes Anatomicae* (1561), he still regarded these structures as seminal ducts for the transmission of female semen. In other words, a misinterpretation coupled with a too-willing acceptance of dogma handed down through the ages. Little could Fallopius have imagined that these 'uterine tubes' would one day be shown to provide the meeting place of the gametes, that is, the actual site of the process of fertilisation in mammals. The uterus was still regarded as the organ wherein the embryo was generated and yet – as one notes with hindsight – Fallopius provides a wonderfully eloquent description of the fimbriated extremity of the tube, the portion that actually 'captures' the egg from the ovary at ovulation and displaces it into the ostium of the Fallopian tube.

> This seminal duct (*meatus seminarius*) originates from the cornua uteri; it is thin, very narrow, of white colour and looks like a nerve. After a short distance it begins to broaden and to coil like a tendril (*capreolus*), winding in folds almost up to the end. There, having become very broad, it shows an *extremitas* of nature of skin and colour of flesh, the utmost end being very ragged and crushed, like the fringe of worn out clothes. Further, it has a great hole which is held closed by the fimbriae which lap over each other. However, if they spread out and dilate, they create a kind of opening which looks like the flaring bell of a brazen tube. Because the course of the seminal duct, from its origin up to its end, resembles the shape of this classical instrument – anyhow, whether the curves are existing or not – I named it *tuba uteri*. These uterine tubes are alike not only in men, but also in the cadavers of sheeps and cows, and all the other animals which I dissected.
>
> *(Herrlinger & Feiner, 1964)*

Similarly attracted to the Padua school was a pupil of Fallopius called Fabricius (1537–1619), who in due course also succeeded to the Chair of Anatomy. He concluded that the Fallopian tube or duct was an organ of secretion, having produced a reasonably accurate account of its function in the formation

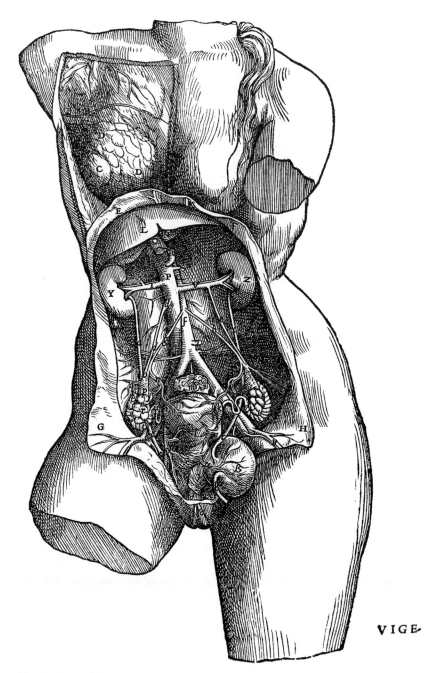

VIGE·

Fig. I.2. To specialists in mammalian reproduction, this is perhaps the most appealing plate from the *Fabrica* of Andreas Vesalius (1543). Both ovaries feature prominently, even though seemingly embraced by coiled ducts at the extremity of the Fallopian tubes. Such an arrangement would correspond to the disposition of epididymal tissues alongside the male gonad. (Courtesy of Glasgow University Library.)

of chicken eggs. As a gifted anatomist, Fabricius provided accurate descriptions of the uterus and ovaries, and likewise of the uterine and ovarian arteries and their anastomoses. Indeed, in a splendid drawing from *De Formato Foetu* (1604), depiction of the ovaries alongside the gravid uterus of a sow is sufficiently precise to enable individual corpora lutea to be distinguished (Fig. I.3). Overall, however, and despite his contributions on the chicken egg and on embryology of the chick, Fabricius did not abandon completely the theories of Galen concerning the involvement of semen and menstrual blood in mammalian generation.

As is well known in both medical and historical circles, the Englishman William Harvey (1578–1657) trained in Cambridge and then completed his formal education at the University of Padua (1602), undoubtedly still the leading medical school in Europe. He was there a sometime student of Fabricius and much interested in reproduction – a sphere quite distinct from that in which he subsequently achieved fame, i.e. circulation of the blood. Harvey examined the contents of the uterus in various species of mammal after they had mated and, largely on the basis of his studies in red and fallow deer in which no trace of conception could be found for a protracted interval after the onset of rutting, concluded that the existence of female semen was a myth and indeed that male semen was not involved in formation of the fetus. He was thus out of step with his distinguished predecessors, the more so since he proposed a new theory of generation far removed from concoctions of semen and menstrual blood that had featured since antiquity. Although Harvey in no sense suggested a specific function for the mammalian ovaries, nor did he even entertain a rôle for them in reproduction, he was sufficiently inspired by his work on the chicken embryo to propose that: *Ex ovo omnia* – 'All living things come from eggs' (Fig. I.4) in his *Exercitationes de Generatione Animalium* (1651). Actual evidence in support of this magisterial statement became available only some 200 years later in the reports of Barry (1843), Bischoff (1854) and Van Beneden (1875) on sperm penetration into the mammalian egg. Nonetheless, Harvey did think in terms of epigenesis – a gradual emergence of the differentiating embryo and fetus from the egg – rather than endorsing the concept of preformation. In error, he considered that the embryo itself was generated from a drop of blood that is elaborated within the fertile egg.

Conducting studies in Leiden, where he was Professor of Anatomy, van Horne (1668) clearly distinguished ovarian vesicles or follicles, but thought of them as eggs and imagined that the Fallopian tubes might in some manner function to transmit the vesicles to the uterus. Initially working with avian gonads (Fig. I.5), it was Regnier de Graaf (1641–1673) of Delft (see Frontispiece) who highlighted the tertiary or vesicular follicles that were later to take his name

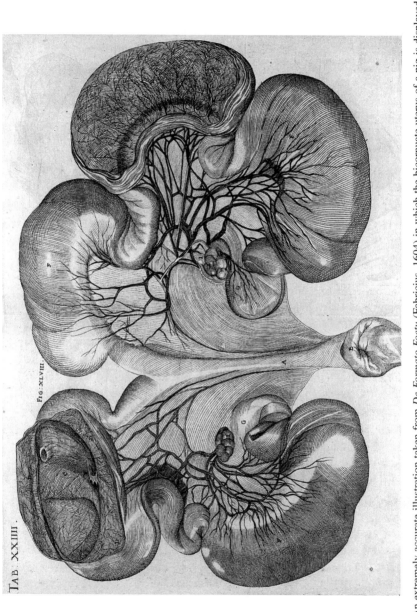

Fig. I.3. An extremely accurate illustration taken from *De Formato Foetu* (Fabricius, 1604) in which the bicornuate uterus of a pig is displayed at an advanced stage of gestation. Dominated by the presence of mature corpora lutea, the ovaries and associated blood vessels are clearly distinguishable. (Courtesy of Cambridge University Library.)

Fig. I.4. The splendid plate that appears as Frontispiece to William Harvey's *Exercitationes de Generatione Animalium* (1651). The caption states that 'All living things come from eggs', and Zeus is shown liberating live creatures from a substantial egg. (Courtesy of Edinburgh University Library.)

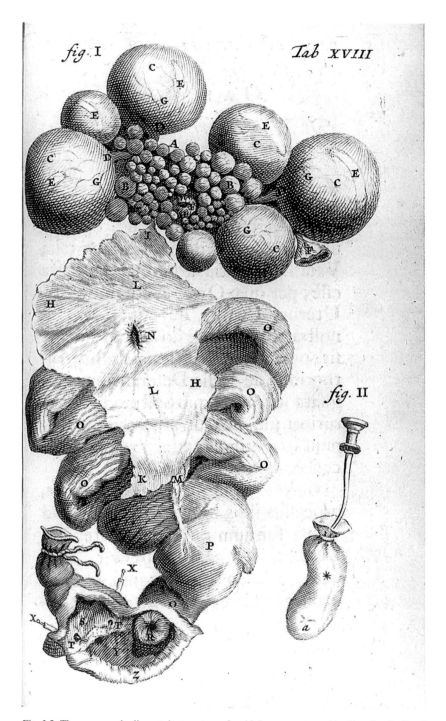

Fig. I.5. The ovary and adjacent duct system of a chicken as portrayed by Regnier de Graaf (1672). The ovarian follicles are much larger than the corresponding mammalian structures, for they become distended with yolk. (Courtesy of Edinburgh University Library.)

in mammalian ovaries. He maintained that the follicles were the source of the eggs, even though an egg transport function for the Fallopian tubes was not yet generally accepted: Harvey's (and Galen's) notion of conception – of an ovum formed *in utero* – still prevailed.

By 1672, however, de Graaf had traced the would-be passage of eggs through the tubes to the uterus and he had also emphasised that sterility invariably follows the castration of females, a fact already known to Aristotle from observations on domestic animals. Although de Graaf had an appreciation of ovarian function, his mistake was to believe that the follicles were eggs. Quite apart from his avian studies, he was here led astray by observations in the rabbit – an induced ovulator – in which the large follicles disappeared after copulation to be replaced by corpora lutea and yet were presumed to be the blastocysts noted in the uterus a few days later. As de Graaf himself fully appreciated, the problem of dimensions of the ovarian follicles in relation to the diameter of the Fallopian tubes could not be reconciled with the passage of eggs unless, in reality, it was the contents of the follicles that entered the tubes. But he had not observed actual rupture of the follicle or release of the egg, for microscopy was at its inception and de Graaf seems not to have been tempted by the primitive models then available to examine the contents of ovarian follicles. His drawings remain essentially large scale and those of human ovaries, Fallopian tubes and associated vasculature are particularly fine (Fig. I.6). As a postscript appropriate to mention at this point, de Graaf's recording of a tubal ectopic pregnancy (Fig. I.7) and his detailed discussion of this condition were used as a powerful argument in support of the proposition that the eggs from which fetuses are to be generated pass from the ovaries ('female testicles') to the uterus by way of the Fallopian tubes.

Before the close of the seventeenth century, there was a related – albeit posthumous – contribution from the famous Italian anatomist Marcello Malpighi (1628–1694). Malpighi had made his first known drawing of an ovary in 1666, a monkey ovary, and had clearly noted Graafian follicles. Apparently, he had reasoned that the actual egg must lie within the ovary and that the follicle itself was not shed from the ovary, perhaps as a result of his well-known correspondence with de Graaf. In contrast to de Graaf's original notion that the follicle itself was the egg, Malpighi proposed in 1681 that the egg was derived from the luteal tissue that was frequently present in a mature mammalian ovary (Adelmann, 1966). Working with cow ovaries, he recorded that the yellow tissue of the corpus luteum (his terminology) possibly contributed to formation of the egg, doubtless being led astray by analogies with yolk colour in the chicken egg, which he had studied intensively. To the extent that the corpus luteum arises from a Graafian follicle after ovulation, Malpighi had offered a

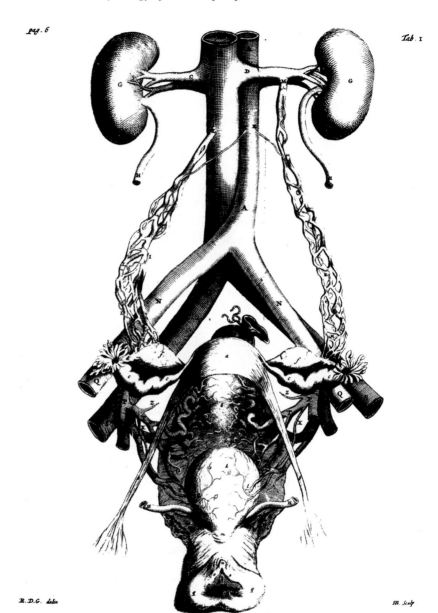

Fig. I.6. A drawing of the human uterus and Fallopian tubes as presented by Regnier de Graaf (1672). Both ovaries feature prominently, as do the major blood vessels supplying these vital organs. (Courtesy of Edinburgh University Library.)

y. 260

Tab XXI

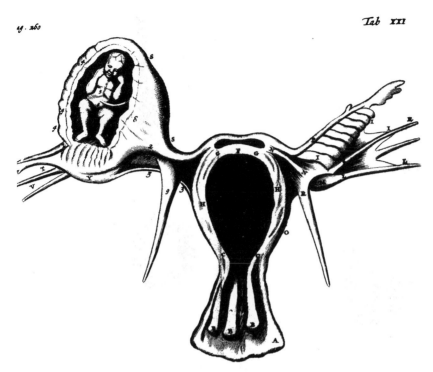

Fig. I.7. A tubal ectopic pregnancy as portrayed with some artistic licence in Regnier de Graaf (1672). Such an observation enabled de Graaf to deduce that eggs pass from the ovaries to the uterus through the Fallopian tubes. He added the somewhat macabre comment that 'as such a fetus grows, it prepares death for its mother'. (Courtesy of Edinburgh University Library.)

somewhat tenuous connection between the egg and a Graafian follicle. On the basis of only this link and the fact that luteinisation commences before ovulation, enthusiastic scholars of the Malpighi tradition have seemingly given their master undue prominence in discovering the origin of the mammalian egg. An elegant and convincing demonstration of this feature of the ovaries was almost 150 years into the future.

Soon after the two key publications of Regnier de Graaf (1668, 1672), translated by Jocelyn & Setchell (1972) and reviewed by Setchell (1974), microscopy began to shed light on the first steps of reproduction, although of course the significance of these earliest observations could not be clearly grasped. Another Dutchman, Antonie van Leeuwenhoek (1678), also of Delft, examined semen under his own rudimentary model of microscope and reported on the millions of little animalcules so observed to the Royal Society of London. This key stage was achieved in 1677 but only published in 1678 due to a delay in translation from Dutch into Latin (Leeuwenhoek, 1678). The observations were made on the semen of fish, frogs and mammals and were thought to have constituted the first factual description of spermatozoa. Leeuwenhoek (1683) went on to postulate that de Graaf's 'egg' required to be impregnated by one of the animalcules for pregnancy to occur – an amazing piece of intuition – although no such observation was made at that time. As a postscript of some interest, it was de Graaf who introduced Leeuwenhoek to the Royal Society in 1673 (B.P. Setchell, personal communication).

Eighteenth and nineteenth century views

As improved models of microscope were developed, de Graaf's views and indeed findings were eventually endorsed in 1797 by William Cruikshank, who likewise identified rabbit eggs in the Fallopian tube on the third day after the animals had mated. It should be emphasised that this step was facilitated by the prominent mucin layer that accumulates around the eggs in this species during their passage towards the uterus. By 1827, von Baer had discovered the origin of the mammalian egg, and thereby had commenced to dispel the existing confusion between eggs and follicles. This was achieved by dissecting open Graafian follicles and examining the liberated contents within their cumulus masses. von Baer recorded the triumphant moment:

> Led by curiosity . . . I opened one of the follicles and took up the minute object on the point of my knife, finding that I could see it very distinctly and that it was surrounded by mucus. When I placed it under the microscope I was utterly astonished, for I saw an ovule just as I had already seen them in the tubes, and so clearly that a blind man could hardly deny it.
>
> *(Translation from Corner, 1933)*

At this time, von Baer was working in Königsberg but he was in fact born in Estonia and trained in medicine at Tartu University. As a sequel to this key publication concerning the origin of the mammalian ovum, there was frequent reference to 'Baer's bladder' (Syritsa & Kalm, 1999).

This was therefore the first significant step in understanding cellular components of the ovaries. In fact, Prévost & Dumas (1824) studying reproduction in the dog had anticipated von Baer's discovery in the sense that they had deduced that de Graaf's follicles probably contained the eggs and that fertilisation would occur only after shedding of such putative eggs from the ovary. They further speculated that perhaps fluid from the Graafian follicles assisted in transporting eggs to the uterus, although this is now appreciated not to be so, for the eggs or embryos of most mammals so far examined remain in the Fallopian tubes for 2–3 days or more before passing through the utero-tubal junction. Follicular fluid entering the Fallopian tube at ovulation is refluxed rather promptly into the peritoneal cavity.

Bischoff (1842) suggested that contractions of the Fallopian tube and its mesenteries were necessary for passage of eggs to the uterus, and Thiry (1862) proposed that ciliary activity was responsible for the displacement of eggs from the ovaries into the tubes. Both of these advances followed Blundell's (1819) key work in which he drew attention to muscular movements of the Fallopian tube. Thus did physiological observations take tentatively to what had previously been overwhelmingly an anatomical stage. But not to be overlooked in this nineteenth century focus on eggs, their origin and displacement is the earlier classical experiment of John Hunter (1787) involving unilateral ovariectomy to deduce the contribution of individual ovaries to the long-term fertility of sows. Whereas Aristotle had a passing acquaintance with the behavioural influence of castration in sows, Hunter's work more than 2000 years later was the first exploitation of ovariectomy as a specific experimental approach to clarifying function. The study was far ahead of its time but imperfectly executed in the sense that the 'hemi-spayed' sow was slaughtered prematurely: its presumptive infertility was seemingly attributed to ovarian cysts, even though the animal periodically came into oestrus. Nonetheless, Hunter deduced that the ovary had a limiting finite power in terms of the number of offspring an animal can produce, a statement not to be faulted even today.

Surgical exploration of the reproductive organs was increasingly frequent in humans and animals during the nineteenth century, and laid the foundations of the discipline that would in due course become endocrinology of the twentieth century. The fact that menstruation was under ovarian control was gradually clarified (see Tilt, 1850; Corner, 1950), and corpora lutea slowly came to be appreciated as glands of secretion rather than as post-ovulatory scar tissue. Treatment with ovarian extracts was noted to produce morphological and be-havioural influences, and soon the uterus was seen as a target organ of ovarian activity, although not precisely in those terms. Grafting of ovaries into rabbits previously ovariectomised prevented atrophy of the uterine tissues (Knauer, 1896), indicating that the ovaries were glands of internal secretion – as they became called in due course. This step was quickly followed by the suggestion of Prenant (1898) that the corpus luteum itself was a gland of internal secretion, its product(s) entering directly into the bloodstream. Extracts of corpora lutea were soon prepared and diverse experiments undertaken to clarify the rôle of the supposed secretion.

Twentieth century highlights

Of the two major compartments of the ovaries after puberty (putting the stroma to one side), research seems to have concentrated more on the corpus luteum than on the Graafian follicle during the first part of the twentieth century. Numerous experiments were performed bearing on luteal function, in one sense culminating in the isolation of progesterone in the mid 1930s in George Corner's Rochester laboratory in the USA (Corner, 1942). As to the follicle, and after much confusion, a distinction was gradually drawn between spontaneous and induced ovulation, with some preliminary classification of species into the two categories. As may be imagined, there was considerable debate as to whether species thought to be spontaneous ovulators could become induced ovulators under appropriate conditions, a debate that continues to the present day – and in the reverse sense, too (see Chapter VIII). Bioassays were gradually developed for the secretory products of corpora lutea and Graafian follicles using selected target tissues, and such assays underwent progressive refinement during the first half of the century, eventually to be superseded by chemical, radio-chemical and even immunological assays. The sexual cycle of mammals had been defined in Heape's classical paper of 1900 and there then followed 40 years of intensive, exciting and even competitive endocrinology (see Parkes, 1962). Scientists in France, Germany, the Netherlands, Switzerland, the United Kingdom and the USA were all much involved. By now, in fact, there was vigorous participation of researchers at various North American universities, and not primarily those of the so-called Ivy League.

Not only were the ovarian steroid hormones isolated and analysed by the end of the 1930s, with synthesis of such steroids soon following, but it was appreciated by the mid 1920s that the ovaries themselves were under the control of the pituitary gland (Smith & Engle, 1927). Pursuit of the gonadotrophic hormones followed rapidly, initially by means of the skilled surgical intervention termed hypophysectomy. Separation of the lobes of the pituitary gland enabled gonadotrophic activity to be attributed to the anterior pituitary whereas the endocrine activity of the posterior pituitary, especially of the reproductive hormone oxytocin, was found to be derived from the hypothalamus. The concept of hypothalamic control of the anterior pituitary gland by means of peptide releasing factors was pursued most prominently and elegantly by G.W. Harris from the mid to late 1930s, and the involvement of the hypothalamo-hypophyseal portal system summarised in both his *Neural Control of the Pituitary Gland* (Harris, 1955) and masterly Upjohn Lecture of the Endocrine Society (Harris, 1964).

Hysterectomy in laboratory and large farm animals – but not in women or other primates – had indicated that the uterus in some manner influenced the

Table I.1. *Selected volumes concerned with ovarian function in mammals published during the twentieth century*

Marshall (1910)	*The Physiology of Reproduction*
Parkes (1929)	*The Internal Secretions of the Ovary*
Hartman (1936)	*Time of Ovulation in Women*
Pincus (1936)	*The Eggs of Mammals*
Corner (1942)	*The Hormones in Human Reproduction*
Parkes (1956)	*Marshall's Physiology of Reproduction*, 3rd edition
Villee (1961)	*Control of Ovulation*
Young (1961)	*Sex and Internal Secretions*, 3rd edition
Zuckerman & Mandl (1962)	*The Ovary*
Perry (1971)	*The Ovarian Cycle of Mammals*
Austin & Short (1972)	*Reproduction in Mammals*
Greep (1973)	*American Handbook of Physiology*
Mossman & Duke (1973)	*Comparative Morphology of the Mammalian Ovary*
Crosignani & Mishell (1976)	*Ovulation in the Human*
Zuckerman & Weir (1977)	*The Ovary*, 2nd edition
Jones (1978)	*The Vertebrate Ovary*
Midgley & Sadler (1979)	*Ovarian Follicular Development and Function*
Thibault & Levasseur (1979)	*La Fonction Ovarienne chez les Mammifères*
Edwards (1980)	*Conception in the Human Female*
Motta & Hafez (1980)	*Biology of the Ovary*
Peters & McNatty (1980)	*The Ovary. A Correlation of Structure and Function in Mammals*
Austin & Short (1982)	*Reproduction in Mammals*, 2nd edition
Lamming (1984)	*Marshall's Physiology of Reproduction*, 4th edition
Knobil & Neill (1988)	*The Physiology of Reproduction*
Hillier (1992)	*Gonadal Development and Function*
Adashi & Leung (1993)	*The Ovary*
Thibault, Levasseur & Hunter (1993)	*Reproduction in Mammals and Man*
Findlay (1994)	*Molecular Biology of the Female Reproductive System*
Knobil & Neill (1994)	*The Physiology of Reproduction*, 2nd edition
Grudzinskas & Yovich (1995)	*Gametes – The Oocyte*

lifespan of the cyclic corpus luteum. Many experimental approaches were used to unravel the humoral interaction between uterine and luteal tissues, both in cyclic and in pregnant animals. By the end of the 1960s, it was appreciated that luteotrophic (embryonic) and luteolytic (uterine) factors were at play and that it was the balance between these positive and negative factors that determined luteal lifespan. Prostaglandin $F_{2\alpha}$ ($PGF_{2\alpha}$) was proposed as the luteolytic factor by the late 1960s (Pharriss & Wyngarden, 1969), and demonstrated to be elaborated by the endometrium of a healthy, non-gravid uterus (i.e. in the absence of pyometritis) shortly before the demise of the corpus luteum (McCracken, 1971; Goding, 1974). As to the embryo and working initially with the much elongated conceptus of ungulates (Moor & Rowson, 1964; Rowson & Moor, 1966, 1967), the luteotrophic factor that acts to suppress secretion of uterine $PGF_{2\alpha}$ was shown to be a protein initially termed trophoblastin, a useful name giving some indication of its origin (Martal *et al.*, 1979). However, now that the molecular age is well and truly upon us, the luteotrophic factor – at least in ruminants – is referred to as interferon tau (τ) (Roberts *et al.*, 1999; Winkelman *et al.*, 1999). As a postscript to this paragraph, it should be noted that the ovary itself can also synthesise prostaglandins, both in Graafian follicles and in the mature corpus luteum, at least in primates. Indeed, the latter doubtless affords a means of explaining ovarian cyclicity after hysterectomy in women.

By the 1980s, the hypothalamic-posterior pituitary peptide oxytocin was also shown to be an ovarian hormone (see Flint & Sheldrick, 1982; Wathes & Swann, 1982; Wathes, 1989; Wathes & Denning-Kendall, 1992), and much involved in the events of luteolysis and seemingly more widely in ovarian physiology. And, by the mid-to-late 1980s, many other peptides representing the diverse family of growth factors were shown to be critically involved in ovarian function. Finally, in this brief perspective, reference must be made to the traffic in white blood cells, especially polymorphonuclear leucocytes, through ovarian tissues. Although an observation of long-standing, this field received a particular emphasis during the 1990s, and products of ovarian macrophages such as cytokines are now believed to make a pivotal contribution to tissue modifications close to the time of ovulation (see Brännström, 1997).

A notable feature of ovarian research and indeed of reproductive research in the twentieth century, very much during its second half, was the appearance of major studies not only from Europe and the whole of North America, but also from Japan, the newly founded Israel, Australia and New Zealand, Latin America, India and – in the past twenty years – from China and Korea. In proportion to its size and population, published work from the former Soviet Union remains disappointingly thin. By and large, reproductive research in most of Africa is on an exceedingly modest scale, and this is seemingly true also for much of South-East Asia – Burma, Thailand, Malaysia, Indonesia and Vietnam. And, as noted earlier in this chapter, whereas great endeavours came from ancient Greece and Alexandria 2000 or more years ago, modern contributions from Egypt, Greece and Turkey and indeed most of the Middle East remain slight. Despite the widely trumpeted rôle of the internet, there is no reason to suppose that this situation will change rapidly in the new century.

Good research requires not only modern technology – it also requires initiative, originality, enthusiasm, organisation and hard work, all in ample measure and set against an appropriate social backcloth.

Prospects for the current century

As is widely recognised by most of those immersed in the scientific endeavour, the rate of change, of important discovery, of technical development is not only impressive, it is fast rendering many of the more industrious participants breathless or even exhausted. There is much to be said in favour of bouts of intensive and fruitful research activity but there is also a pressing and absolute requirement for time to read widely, to ponder deeply and to discuss. Daily routines should not simply be composed of the laboratory bench, the operating theatre, the ubiquitous computer screen, and yet all too many colleagues are seemingly hypnotised by the last and strait-jacketed by the first.

Taking time to gaze into the crystal ball – not necessarily a worthy activity – the future for ovarian and ovarian-related research would seem to be full of exciting prospects. For example, the contribution of the autonomic nervous system to normal and abnormal ovarian function certainly needs to be revisited, and significant advances seem possible in clarifying the aetiology of polycystic ovarian disease, to cite just one clinical problem. Molecular studies will enable identification of increasing numbers of genes involved in ovarian function and malfunction, and their downstream products may then become open to manipulation in a precise manner. Even so, it is perhaps the related or dependent reproductive technologies that will contribute most impressively. Here, one could cite, for example, culture and growth of primordial follicles to provide a means of transplantation therapy for aplastic ovaries; isolation and proliferation of appropriate stem cells, whether from cloned or conventional *in vitro*-generated embryos, again to offer a therapeutic approach to defective gonads; cryopreservation of portions of ovarian tissue during irradiation treatment for cancer to offer the prospect of restoring ovarian function after autotransplantation; and targeted tissue 'messengers' for mounting a local attack on ovarian cancer cells. Also to be predicted would be a controlled stimulation of Graafian follicles for purposes of *in vitro* fertilisation without the risks associated with extensive superovulation. And, likewise, a simple means of recognising the time of ovulation without recourse to repeated ultrasonic scanning. Reducing the rate of atresia in ovarian follicular populations should not be beyond the bounds of possibility, thereby offering an attractive route to extending the reproductive lifespan.

Developments in all these and in numerous other reproductive spheres should make the next 20 years a period of unparalleled progress. Thereafter who knows and who would dare to suggest? Not this author. But, taking human nature into consideration and a certain predisposition for civilisations mature in years to court disaster, issues reproductive might well have a dramatically modified constitution or be getting seriously out of hand.

References

Adashi, E.Y. & Leung, P.C.K. (eds.) (1993). *The Ovary.* New York: Raven Press.

Adelmann, H.B. (1966). *Marcello Malpighi and the Evolution of Embryology.* Ithaca, NY: Cornell University Press.

Austin, C.R. & Short, R.V. (eds.) (1972). *Reproduction in Mammals.* Cambridge: Cambridge University Press.

Austin, C.R. & Short, R.V. (eds.) (1982). *Reproduction in Mammals*, 2nd edn. Cambridge: Cambridge University Press.

Baer, K.E. von (1827). *De Ovi Mammalium et Hominis Genesi.* Leipzig.

Barry, M. (1843). Spermatozoa observed within the mammiferous ovum. *Philosophical Transactions of the Royal Society, Series B*, **133**, 33.

Bischoff, T.L.W. (1842). *Entwicklungsgeschichte des Kanincheneies.* Braunschweig.

Bischoff, T.L.W. (1854). *Entwicklungsgeschichte des Rehes.* Giessen.

Blundell, J. (1819). Experiments on a few controverted points respecting the physiology of generation. *Medical and Chirurgical Society Transactions*, **10**, 245–72.

Bodemer, C.W. (1969). History of the mammalian oviduct. In *The Mammalian Oviduct*, ed. E.S.E. Hafez & R.J. Blandau, pp. 3–26. Chicago: University of Chicago Press.

Brännström, M. (1997). Intra-ovarian immune mechanisms in ovulation. In *Microscopy of Reproduction and Development: A Dynamic Approach*, ed. P.M. Motta, pp. 163–8. Rome: Antonio Delfino Editore.

Corner, G.W. (1933). The discovery of the mammalian ovum. In *Lectures on the History of Medicine, 1926–1932.* Philadelphia: Mayo Foundation Lectures.

Corner, G.W. (1942). *The Hormones in Human Reproduction.* Princeton, NJ: Princeton University Press.

Corner, G.W. (1950). The relation of the ovary to the menstrual cycle. Notes on the history of a belated discovery. *Annals of the Faculty of Medicine*, University Republica, Montevideo, **35**, 758–66.

Crosignani, P.G. & Mishell, D.R. (eds.) (1976). *Ovulation in the Human.* Proceedings of the Serono Foundation Symposium, No. 8. New York, San Francisco & London: Academic Press.

Cruikshank, W. (1797). Experiments in which, on the third day after impregnation, the ova of rabbits were found in the Fallopian tubes; and on the fourth day after impregnation in the uterus itself; with the first appearances of the foetus. *Philosophical Transactions of the Royal Society*, **87**, 197–214.

de Graaf, R. (1668). *Tractatus de Virorum Organis Generationi Inservientibus.* Leyden: Hack.

De Graaf, R. (1672). *De Mulierum Organis Generationi Inservientibus Tractatus Novus.* Leyden: Hack.

Edwards, R.G. (1980). *Conception in the Human Female.* London & New York: Academic Press.

Fabricius, H. (1604). *De Formato Foetu.* Padua.

Fallopius, G. (1561). *Observationes Anatomicae.* Venice.

Findlay, J.K. (ed.) (1994). *Molecular Biology of the Female Reproductive System.* London & San Diego: Academic Press.

Flint, A.P.F. & Sheldrick, E.L. (1982). Ovarian secretion of oxytocin is stimulated by prostaglandin. *Nature (London),* **297**, 587–8.

Goding, J.R. (1974). The demonstration that PGF$_{2\alpha}$ is the uterine luteolysin in the ewe. *Journal of Reproduction and Fertility,* **38**, 261–71.

Greep, R.O. (ed.) (1973). *American Handbook of Physiology.* Section 7, *Endocrinology* II, *Female Reproductive System,* Part II. Washington, DC: American Physiological Society.

Grudzinskas, J.G. & Yovich, J.L. (eds.) (1995). *Gametes – the Oocyte.* Cambridge Reviews in Human Reproduction. Cambridge: Cambridge University Press.

Harris, G.W. (1955). *Neural Control of the Pituitary Gland.* London: Edward Arnold.

Harris, G.W. (1964). Sex hormones, brain development and brain function. *Endocrinology,* **75**, 627–48.

Hartman, C.G. (1936). *Time of Ovulation in Women: A Study on the Fertile Period in the Menstrual Cycle.* Baltimore, MD: The Williams & Wilkins Company.

Harvey, W. (1651). *Exercitationes de Generatione Animalium.* London.

Heape, W. (1900). The 'sexual season' of mammals and the relation of the 'pro-oestrum' to menstruation. *Quarterly Journal of Microscopical Science,* **44**, 1–70.

Herrlinger, R. & Feiner, E. (1964). Why did Vesalius not discover the Fallopian tubes? *Medical History,* **8**, 335–41.

Hillier, S.G. (ed.) (1992). *Gonadal Development and Function.* Serono Symposia Publications, vol. 94. New York: Raven Press.

Hunter, J. (1787). An experiment to determine the effect of extirpating one ovarium upon the number of young produced. *Philosophical Transactions of the Royal Society, London,* **77**, 233–9.

Jocelyn, H.D. & Setchell, B.P. (1972). Regnier de Graaf on the human reproductive organs. *Journal of Reproduction and Fertility,* Supplement, **17**, 1–222.

Jones, R.E. (ed.) (1978). *The Vertebrate Ovary.* New York & London: Plenum Press.

Knauer, E. (1896). Einige Versuche über Ovarientransplantation bei Kaninchen. *Zentralblatt für Gynaekologie,* **20**, 524–8.

Knobil, E. & Neill, J. (eds.) (1988). *The Physiology of Reproduction.* New York: Raven Press.

Knobil, E. & Neill, J. (eds.) (1994). *The Physiology of Reproduction,* 2nd edn. New York: Raven Press.

Lamming, G.E. (ed.) (1984). *Marshall's Physiology of Reproduction,* 4th edn, vol. 1. Edinburgh & London: Churchill Livingstone.

Leeuwenhoek, A. van (1678). Observationes de Anthonii Leeuwenhoek, de natis è semine genitali animalculis. *Philosophical Transactions of the Royal Society,* **12**, 451–2.

Leeuwenhoek, A. van (1683). An abstract of a letter from Mr Anthony Leeuwenhoek of Delft about generation by an Animalcule of the male seed. *Philosophical Transactions of the Royal Society*, **13**, 347–55.

Marshall, F.H.A. (1910). *The Physiology of Reproduction*. London: Longmans Green.

Martal, J., Lacroix, M.-C., Loudes, C., Saunier, M. & Wintenberger-Torrès, S. (1979). Trophoblastin, an antiluteolytic protein present in early pregnancy in sheep. *Journal of Reproduction and Fertility*, **56**, 63–73.

McCracken, J.A. (1971). Prostaglandin $F_{2\alpha}$ and corpus luteum regression. *Annals of the New York Academy of Sciences*, **180**, 456–72.

Midgley, A.R. & Sadler, W.A. (eds.) (1979). *Ovarian Follicular Development and Function*. New York: Raven Press.

Moor, R.M. & Rowson, L.E.A. (1964). Influence of the embryo and uterus on luteal function in the sheep. *Nature (London)*, **201**, 522–3.

Mossman, H.W. & Duke, K.L. (1973). *Comparative Morphology of the Mammalian Ovary*. Madison: University of Wisconsin Press.

Motta, P.M. & Hafez, E.S.E. (eds.) (1980). *Biology of the Ovary*. The Hague: Nijhoff.

Parkes, A.S. (1929). *The Internal Secretions of the Ovary*. London: Longmans.

Parkes, A.S. (ed.) (1956). *Marshall's Physiology of Reproduction*, 3rd edn. London: Longmans, Green & Co.

Parkes, A.S. (1962). Prospect and retrospect in the physiology of reproduction. *British Medical Journal*, **ii**, 71–5.

Perry, J.S. (1971). *The Ovarian Cycle of Mammals*. Edinburgh: Oliver & Boyd.

Peters, H. & McNatty, K.P. (1980). *The Ovary. A Correlation of Structure and Function in Mammals*. Berkeley: University of California Press.

Pharriss, B.B. & Wyngarden, L.J. (1969). The effect of prostaglandin $F_{2\alpha}$ on the progestagen content of ovaries from pseudopregnant rats. *Proceedings of the Society for Experimental Biology and Medicine*, **130**, 92–4.

Pincus, G. (1936). *The Eggs of Mammals*. New York: Macmillan.

Prenant, A. (1898). La valeur morphologique du corps jaune. Son action physiologique et thérapeutique possible. *Revue Générale des Sciences Pures et Appliquées*, **9**, 646–50.

Prévost, J.L. & Dumas, J.A.B. (1824). Troisième mémoire de la génération dans les mammifères, et des premiers indices du développement de l'embryon. *Annales des Sciences Naturelles*, **3**, 113–38.

Roberts, R.M., Ealy, A.D., Alexenko, A.P., Han, C.S. & Ezashi, T. (1999). Trophoblast interferons. *Placenta*, **20**, 259–64.

Rowson, L.E.A. & Moor, R.M. (1966). Development of the sheep conceptus during the first fourteen days. *Journal of Anatomy*, **100**, 777–85.

Rowson, L.E.A. & Moor, R.M. (1967). The influence of embryonic tissue homogenate infused into the uterus, on the lifespan of the corpus luteum of the sheep. *Journal of Reproduction and Fertility*, **13**, 511–16.

Setchell, B.P. (1974). The contributions of Regnier de Graaf to reproductive biology. *European Journal of Obstetrics, Gynaecology and Reproductive Biology*, **4**, 1–13.

Short, R.V. (1977). The discovery of the ovaries. In *The Ovary*, 2nd edn, vol. 1, ed. S. Zuckerman & B.J. Weir, pp. 1–39. New York: Academic Press.

Smith, P.E. & Engle, E.T. (1927). Experimental evidence regarding the rôle of the anterior pituitary in the development and regulation of the genital system. *American Journal of Anatomy*, **40**, 159–217.

Syritsa, A. & Kalm, S. (1999). The history of the ovum and its development. *Italian Journal of Anatomy and Embryology*, **104** (Supplement No. 1), 685.

Temkin, O. (1956). *Soranus' Gynecology*. Baltimore, MD: Johns Hopkins Press.

Thibault, C. & Levasseur, M.C. (1979). *La Fonction Ovarienne chez Les Mammifères*. Paris & New York: Masson.

Thibault, C., Levasseur, M.C. & Hunter, R.H.F. (1993). *Reproduction in Mammals and Man*. Paris: Ellipses.

Thiry, L. (1862). Über das Vorkommen eines Flimmerepithelium auf dem Bauchfell des weiblichen Frosches. *Göttinger Nachrichten* **171**.

Tilt, E.J. (1850). *On Diseases of Menstruation and Ovarian Inflammation, in Connexion with Sterility, Pelvic Tumours, and Affections of the Womb*. London.

Van Beneden, E. (1875). La maturation de l'oeuf, la fécondation et les premières phases du développement embryonnaire des mammifères d'après des recherches faites chez le lapin. *Bulletin de l'Académie Royale de Belgique, Classe Science*, **40**, 686–9.

van Horne, J. (1668). *Suarum circa Partes Generationis in Utroque Sexu Observationum Prodromus*. Leyden.

Vesalius, A. (1543). *De Humani Corporis Fabrica Libri Septem*. Basel.

Villee, C.A. (ed.) (1961). *Control of Ovulation*. Oxford, New York & London: Pergamon Press.

Wathes, D.C. (1989). Oxytocin and vasopressin in the gonads. *Oxford Reviews of Reproductive Biology*, **11**, 225–83.

Wathes, D.C. & Denning-Kendall, P.A. (1992). Control of synthesis and secretion of ovarian oxytocin in ruminants. *Journal of Reproduction and Fertility*, Supplement, **45**, 39–52.

Wathes, D.C. & Swann, R.W. (1982). Is oxytocin an ovarian hormone? *Nature (London)*, **297**, 225–7.

Winkelman, G.L., Roberts, R.M., Peterson, J.A., Alexenko, A.P. & Ealy, A.D. (1999). Identification of the expressed forms of ovine interferon-tau in the preimplantation conceptus: sequence relationships and comparative biological activities. *Biology of Reproduction*, **61**, 1592–1600.

Young, W.C. (ed.) (1961). *Sex and Internal Secretions*, 3rd edn. Baltimore, MD: The Williams & Wilkins Company.

Zuckerman, S. & Mandl, A.M. (eds.) (1962). *The Ovary*. London & New York: Academic Press.

Zuckerman, S. & Weir, B.J. (1977). *The Ovary*, 2nd edn. New York, San Francisco & London: Academic Press.

II
Formation and structure of ovaries: elaboration of follicular compartments

Introduction

A major function of ovarian tissue is to be the site of multiplication of the germ cell line during prenatal development and then a storage site for primary oocytes throughout postnatal life until reserves of these unique cells are exhausted. Although not all primordial germ cells reach their target tissues in the presumptive gonads upon migration from the yolk sac, the small proportion that may be lost, for example to the embryonic adrenal glands (Zamboni & Upadhyay, 1983), never achieves the possibility of being released into the reproductive tract or even the peritoneal cavity. Indeed, this is true of all primordial germ cells in extra-gonadal sites. Shedding of an oocyte into the genital duct at ovulation can occur only upon formation and growth of a Graafian follicle. Such vesicular or antral follicles are structures peculiar to the ovaries and represent the terminal stage of a process commonly referred to as follicular growth. Whereas the endocrine potential of Graafian follicles is discussed extensively in this monograph, mature follicles also represent a highly specialised physico-mechanical system – a vehicle – for liberating oocytes at a precise time into a reproductive tract that may offer the opportunity for fertilisation and further development. Formation of such a vehicle will be discussed in a section on follicular growth. First, however, it is necessary to consider the major topics of sex determination and initial differentiation of an ovary.

Sex determination

Most treatments of this major developmental decision tend to focus overwhelmingly on testis determination, usually leaving the formation of an ovary as a secondary or residual description, a mere consequence of an absence of maleness. Nonetheless, a rôle for ovary-determining genes was strongly promoted throughout the volume of Hunter (1995), a view that has since become more fashionable and which will be elaborated on in due course, since molecular studies are now bringing this orientation to the forefront. To the extent that such studies have demonstrated changes in gene expression occurring very early in XX genital ridge development (Francavilla & Zamboni, 1985), then a conclusion must be that differentiation of an ovary is not simply permissive and passive; rather, a specific gene pathway must be involved.

Irrespective of this particular emphasis, the rule for eutherian mammals remains that it is the genetic constitution of the fertilising spermatozoon that is the primary determinant of sex in the resulting zygote. Upon delivering a haploid complement of chromosomes into the vitellus at the time of sperm–egg fusion, X-chromosome bearing spermatozoa act to generate females whereas Y-chromosome bearing spermatozoa prompt formation of males. Such a genetic mechanism has been taken to indicate evolutionary progress from the lower vertebrates in which environmental influences exercise such a large and often decisive rôle in directing the path of sexual differentiation. As to the sex chromosomes themselves, an XY constitution represents the heterogametic sex in mammals and a Y chromosome prescribes testis formation. In fact, maleness is triggered by a Y chromosome irrespective of the number of X chromosomes that may be found in the embryo in anomalous circumstances. Even when the somatic sex chromosome constitution has been determined as XXY, XXYY or XXXY, such animals will still differentiate as males. They will not be functional males, however, since spermatogenesis is impaired in the presence of two X chromosomes. Inactivation of the X chromosome during spermatogenesis is seemingly essential for fertility, an underlying interpretation being that some X-coded gene product is preventing development of male germ cells (McLaren & Monk, 1981).

In the absence of a Y chromosome and initiation of testis formation, then the female condition may be expressed. A long-standing point of view has therefore been that the lack of a male-determining programme from a gene or genes on the Y chromosome rather than the incisive imposition of a female-determining programme permits expression of a female or ovary-determining pathway. The female sex-determining programme would thus be the constitutive or default programme that normally can be utilised only when not suppressed or over-ridden by a precocious testis-determining programme. The suggested male dominance and male precocity led to the search for a male-determining gene or genes, and such activity was intensified from 1959 onwards when it first became clear that a testis-determining factor or gene was located on the Y chromosome (Ford *et al.*, 1959; Jacobs & Strong, 1959; Welshons & Russell, 1959). Thirty years of research with progressively more sophisticated chromosome banding and then molecular techniques culminated in identification of *SRY* in man, *Sry* in other mammals, as a principal testis-determining gene (Gubbay *et al.*, 1990; Sinclair *et al.*, 1990; Koopman *et al.*, 1991). *SRY* is thought to initiate the differentiation of Sertoli cells from one of the somatic cell lineages in the indifferent gonad, the embryonic genital ridge. Even so, as judged by elegant techniques of microinjecting gene constructs into the pronuclei of female mice embryos (sex established retrospectively) and the resultant phenomenon of sex reversal, *Sry* is not in itself sufficient to initiate the process of spermatogenesis (Koopman *et al.*, 1991). Sex-reversed female mice were not functional males, even though adorned with penis and scrotum. Genes downstream from *Sry*, for example *Sox-9*, *Wt* and *Amh*, and certainly others, are clearly of critical importance in elaborating a functional testis with seminiferous epithelium able to produce gametes (see Figure 2 in Hunter, 1996).

   Such intensive research activity focusing on maleness tended to overlook the simple fact that there must also be specific gene programmes for generating ovarian tissue. Despite a strong plea in Hunter (1995) for more emphasis on ovary-determining genes, it is only recently that such crucial genes have begun to be identified and to be given prominence. Two genes of relevance to the expression of an ovarian pathway are *Wnt-4* and *Dax-1*. *Wnt-4* expression has been observed in the developing kidney and in the mesonephros, tissues arising in close proximity to the formative gonads. Studies in mice indicated *Wnt-4* expression in the mesenchyme of the indifferent gonad in both sexes, together with a second *Wnt* family member, *Wnt-6* (Vainio *et al.*, 1999). As sex-specific differentiation of mouse gonads becomes apparent at around 11.5 days *post coitum*, *Wnt-4* expression is maintained in the female gonad but downregulated in the male. Such sexually dimorphic expression continues whilst the sexes become morphologically distinct (Vainio *et al.*, 1999). Of additional interest, *Wnt-4* expression is found either partly or completely within somatic cell lineages, the evidence here being its expression in germ cell deficient ovaries of *Steel* (*Sl*) mutant embryos (Vainio *et al.*, 1999). During normal development, ovary-specific expression of *Wnt-4* appears to suppress Leydig cell proliferation.

   Despite this emphasis on somatic cells, preliminary findings point to an additional rôle of *Wnt-4* in supporting oocyte development, since *Wnt-4* mutant mice ovaries have fewer than 10% of the oocytes found in wild-type or heterozygous siblings (Vainio *et al.*, 1999). There would seem to be a link here with the finding that *Wnt-4* mutant females are masculinised, loss of oocytes paving the way to formation of testicular tissue. The manner whereby *Wnt-4* may be involved in maintaining development of the female germ line is not yet understood (Vainio *et al.*, 1999).

   Turning to the gene *Dax-1*, also implicated in the female pathway of sex determination, evidence here is derived from instances of male-to-female sex reversal. *Dax-1* seemingly antagonises or represses the action of *Sry*. To this extent, it may be more correctly interpreted as an anti-testis gene rather than as one acting positively to induce formation of ovarian tissues. *Dax-1* may be responsible for dosage-sensitive sex reversal in humans in instances in which XY individuals carrying duplications of part of the short arm of the X chromosome, Xp21, develop as females (Bardoni *et al.*, 1994). The duplicated region seemingly contains a gene or genes which compromise testis formation when acting at double dosage. In normal circumstances, *Dax-1* begins to be expressed at the same time as *Sry* in the genital ridge of both XY and XX embryos, but is downregulated as testis development proceeds in XY animals. In XX animals, by contrast, it persists throughout formation of the ovaries, such prolonged expression being incompatible with testis development (Swain *et al.*, 1996). In an experimental mouse model, introduction of a *Dax-1* construct into pronucleate eggs as a transgene with appropriate dosage adjustments could induce complete sex reversal in XY animals (Swain *et al.*, 1998). Overall, what became clear from these skilful studies is that precise levels and timing of gene expression are critical in the formation of functional gonads. This remark brings the discussion briefly to the topic of intersexes.

Females represent the homogametic sex in mammals and the embryonic gonads almost always develop as ovaries in the presence of two X chromosomes. The caveat in the preceding sentence takes account of circumstances in which the sex chromosome constitution is karyotyped as XX and yet one or both of the gonads may have formed an ovotestis or even a testis-like structure. Such instances of intersexuality in supposedly genetic females have been discussed in detail (Hunter, 1995). Different forms of genetic error in XX embryos can impose molecular programmes leading to the formation of testicular tissue (Fig. II.1). A key factor may be the relatively belated development of ovarian tissue, rendering the embryonic gonads (genital ridges) vulnerable to developmental errors and susceptible to influences of the precocious testicular programme. It should be emphasised that germ cells have not been detected in the testicular tissue of an ovotestis or testis-like structure in XX animals examined after puberty, and even the Sertoli cells are of questionable competence in such animals (Hunter, Baker & Cook, 1982; Hunter, Cook & Baker, 1985; Hunter, 1996). However, there are extensive and fully functional interstitial cells of Leydig.

Differentiation of an ovary; primordial germ cells

Fully compatible with the earlier description that the sex chromosome constitution of the fertilising spermatozoon determines the direction of sexual development is the fact that the embryonic gonad has the potential for differentiation

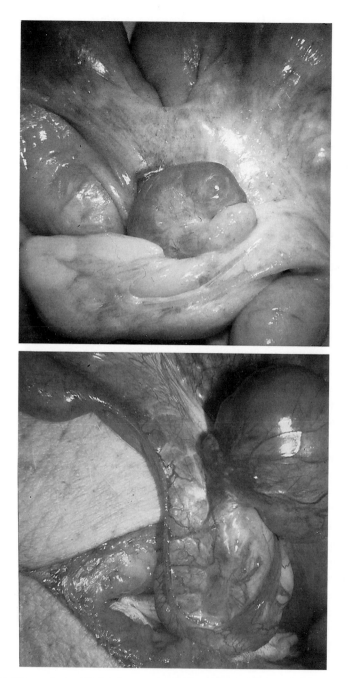

Fig. II.1. Wayward gonadal development in the domestic pig illustrating the formation of both ovarian and testicular tissue in close proximity. This anomaly was revealed surgically in animals with an XX sex chromosome complement from the inbred herd of the University of Edinburgh.

as either a testis or an ovary. The gonadal anlagen are initially indistinguishable morphologically between the two sexes. If a male programme is not imposed, i.e. in the absence of *Sry* and its protein product that leads to differentiation of Sertoli cells (Burgoyne, 1988; Koopman *et al.*, 1990, 1991), then the somatic cells of the genital ridge – a supporting precursor cell lineage – differentiate as follicle (pre-granulosa) cells in female embryos. The currently accepted dogma is that both granulosa and Sertoli cells are derived from the same mesonephric precursor cells (Upadhyay & Zamboni, 1982; Byskov & Høyer, 1988), but the gene or genes that act to induce differentiation of granulosa or, more strictly, pre-granulosa cells is still not known. Nor is it fully understood to what extent the somatic cells of the differentiating ovary are derived from the separate lineages of coelomic epithelium, mesenchymal cells and mesonephros. However, the principal contribution to the ovarian cell mass is thought to be derived from the mesonephros (Byskov & Høyer, 1988), a view not promoted in the earlier reviews of Witschi (1951) and Burns (1961), nor in the essay of Ullmann (1989) dealing with marsupials. What is fully appreciated and not controversial is that gonads are uniquely derived from migration and movement of germinal and somatic cells, and give the first clear evidence of sexual dimorphism in specific tissues.

The extra-gonadal origin of the primordial germ cells is not in doubt. These cells are the embryonic precursors of the gametes and, together with the gametes, constitute the germ cell lineage. They can first be identified in the epiblast early in embryonic life (Ginsberg, Snow & McLaren, 1990; Buehr, 1997). Elegant genetic and molecular approaches involving chimera analysis in mice have highlighted a rôle for bone morphogenetic protein 4 (*Bmp*4) in the induction of primordial germ cell precursors (Lawson *et al.*, 1999). Moreover, the extent of *Bmp*4 expression influences the size of the founding population of primordial germ cells in the epiblast, the secreted protein signal being derived from the trophectoderm lineage. Cell–cell interactions between neighbouring cell layers are thought to be critical in these events, and the possibility of more than one inductive signal still remains.

Frequently raised, the question is still outstanding as to why the primordial germ cells commence to proliferate as a discrete group so far from the presumptive gonads and are thereby subjected to a potentially hazardous phase of migration (for reviews, see Hunter, 1995; Buehr, 1997; Donovan, 1999). No matter what the full response to this question turns out to be, migration of the diploid germ cells into the genital tissues may provide or impose subtle organisational programmes upon the gonadal primordia in a way that would not have been possible if the germ cells had arisen *de novo* within the differentiating genital tissues. Migration of the germ cells from an extra-gonadal site may also permit a degree of flexibility in gonadal organisation that would not have existed if the germ cells had been elaborated in the primordia themselves (Hunter, 1995).

Multiplication of primordial germ cells commences during their migration to the differentiating genital ridges, a progression that may be facilitated by the filamentous-like processes seen linking mouse primordial germ cells (Gomperts *et al.*, 1994). Even though the classical studies of Witschi (1948, 1951) found that the number of primordial germ cells varied between human embryos, a species-specific number of mitoses is said to occur during migration along the hind gut and up the supporting mesentery to the genital ridges (Picton & Gosden, 1999). Upon arrival, they become concentrated in a region determined by the genetic sex of the embryo and proliferate to form the population of cells that will enter into meiosis – promptly in females, belatedly in males. If the gonad is to differentiate as an ovary, the proliferating germ cells become concentrated in the cortex to give rise to oogonia. Development of the cortex proceeds in contrast to progressive involution of the medulla, a phase facilitated by specific tissue movements. Whereas tissues of the genital ridges commence differentiation under the influence of their own genetic constitution, there can be little doubt – as suggested above – that the colonising germ cells impose major organisational influences (Hunter, 1995).

During formation of an embryonic ovary, the germ cells may become enclosed within elongated cords (Pflüger's cords) that have proliferated from the coelomic epithelium and which have retained connections with the mesonephros. Alternatively, germ cells may be disposed in a less organised way within the somatic tissues although still connected to the mesonephros during early differentiation and growth of an ovary. Modelling of ovarian tissues may be especially prompted by the cords of mesonephric cells colonising the central part of the presumptive gonad and displacing germ cells towards the periphery – the region of the cortex (Fig. II.2). Germ cells not displaced from the involuting medulla are destined to degenerate, presumably due to a specific apoptotic programme (see Chapter VI), whereas those in the surrounding cortex continue to proliferate. Mesenchymatous cells commence to form a thin tunica albuginea. Once cortical dominance is established and nests of germ cells have formed to become distinguishable as oogonia, the embryonic ovaries further differentiate and develop during oogenesis so that formation of primary oocytes is essentially complete by the time of birth. Not only has there been a massive multiplication of germ cells, but a significant loss in the germinal population is noted even before birth (Fig. II.3).

Sterile mouse mutants have permitted characterisation of some of the genes involved in primordial germ cell development. Two such genes are *Dominant*

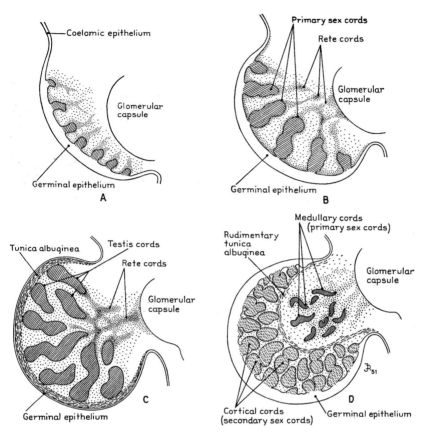

Fig. II.2. The portrayal by Burns (1961) of the main features of gonadal differentiation in lower vertebrate (amniote) embryos. These are the views of Witschi that highlight the origin of the medullary and cortical components of the organ and some principal steps in the organisation of a testis or an ovary.

White Spotting (*W*) and *Steel* (*Sl*). These genes and their mutations influence not only germ cell development but also differentiation of the haemopoietic cell lineages. The *W* locus encodes a receptor tyrosine kinase (c-Kit) whereas the *Sl* locus encodes the ligand for c-Kit; the latter has a confusing variety of names including Kit-ligand, stem cell factor (SCF) and Steel factor, and is a transmembrane growth factor. The *W/Sl* signal transduction pathway is essential for survival of primordial germ cells in mice. Whilst c-Kit receptor is expressed in the primordial germ cells, Steel factor (Kit-ligand) is expressed in the surrounding somatic cells. Mutations of *W* that inactivate or delete the catalytic domain of the c-Kit receptor have severe deficiencies in the primordial germ cells; if viable, the animals are usually sterile. Mutations

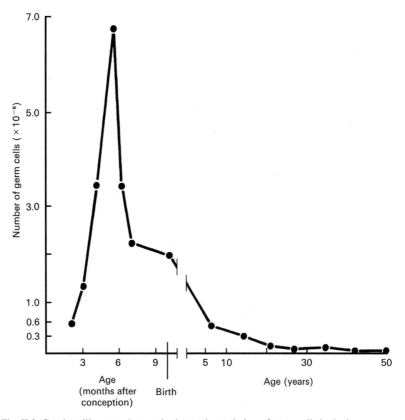

Fig. II.3. Graph to illustrate changes in the total population of germ cells in the human ovary from the stage of early embryonic development through that of the fetal peak and so to the slow decline during the reproductive lifespan. (After Baker, 1982; courtesy of Professor T.G. Baker.)

of the *Sl* locus have similar influences. Steel factor appears to be needed in a membrane-bound form for the long-term survival of primordial germ cells. In the absence of signalling via the c-Kit receptor, primordial germ cells succumb to programmed cell death or apoptosis (see De Felici, 2001; and Chapter VI).

As suggested in the reviews by Donovan (1999) and Donovan *et al.* (2001), the c-Kit/SCF signal transduction pathway may provide a sensitive means of controlling both migration and survival of primordial germ cells in mice. However, other molecules such as peptide growth factors are undoubtedly involved in proliferation and survival, and the experimental evidence was reviewed by Hunter (1995), Buehr (1997) and Donovan (1999). Prominent in the proliferation–survival axis is an influence of the interleukin 6/leukaemia inhibitory factor (IL-6/LIF). Both SCF and LIF suppress apoptosis among mouse primordial germ cells in culture (Pesce *et al.*, 1993), but the physiological significance of this observation remains unclear. Nonetheless, in the presence of these two factors and basic fibroblast growth factor, primordial germ cells can be effectively immortalised in culture and thus made available for experimental use as pluripotent cells that can re-enter the germ cell lineage *in vivo* – when introduced into a blastocyst.

Oogenesis, meiosis, growth of oocyte

After enormous proliferation in the stock of oogonia by waves of mitosis, and their initial arrangement in cord-like fashion in the embryonic gonad, these cords gradually become less organised, although filamentous processes maintain contact as intercellular bridges between many of the oogonia. However, such processes gradually break down too, as cells of mesenchymal origin proceed to envelop each oogonium. Mitotic multiplication enables numbers to reach a peak during mid-to-late fetal life (see Fig. II.3), and this is followed by the first phase of meiosis and indeed by extensive atresia of naked germ cells before the time of birth (for reviews, see Baker, 1982; Picton & Gosden, 1999; Strömstedt & Byskov, 1999). Despite the impact of atresia, a fully grown fetus will – depending on the species – contain tens or hundreds of thousands of oocytes, but the extent to which the potential for growth and development is equally distributed amongst the population of oocytes in the neonate remains a difficult and unresolved question. Certainly, only a minute proportion of the original stock is ever released at ovulation and a smaller proportion still has an opportunity to be fertilised. To some degree, these biological constraints have been overcome by so-called reproductive technologies – techniques for recovering oocytes directly from the ovaries before subjecting them to *in vitro* maturation and *in vitro* fertilisation. But such techniques for alleviating infertility still liberate only a small fraction of the reserve oocytes, and do not really address the question concerning the potential competence of all the oocytes in fetal ovaries.

Expressed in another way, why should so many oocytes be formed when so few will ever be shed from the ovaries? Does this simply represent an evolutionary vestige, a carryover from aquatic ancestors, or does it involve cellular and molecular mechanisms that permit pre-ovulatory selection of competent oocytes, or perhaps a combination of both? As of writing, the balance of evidence would seem to favour the first possibility, that of an evolutionary vestige, but one that has since become linked to – and exploited for purposes of – ovarian endocrine activity residing in developing follicles.

Concerning meiotic maturation, germ cells progress through the leptotene and zygotene stages to become arrested at late pachytene or diplotene of the first meiotic prophase. The large vesicular resting nucleus of the oocyte is referred to as a dictyate nucleus or germinal vesicle that usually contains a conspicuous nucleolus. The fate of most such germ cells is degeneration but a small proportion will progress through the steps of ovarian follicle formation. Only when encompassed by a well-developed antral follicle does liberation of an oocyte from the ovary at ovulation become a possibility. And only in the final stages of pre-ovulatory maturation of a follicle is meiotic arrest in the primary oocyte overcome to permit completion of the first meiotic division and formation of a secondary oocyte with first polar body by the time of ovulation (Fig. II.4). Members of the canine family appear exceptional in this regard (Evans & Cole, 1931; Mahi-Brown, 1991), for the first meiotic division is completed after ovulation in the confines of the Fallopian tube rather than in the Graafian follicle (see Chapters VII and IX).

A primary oocyte can grow and undergo further maturation involving organisation of cytoplasmic organelles only if it becomes surrounded by somatic cells. As already suggested, naked oocytes succumb to apoptosis. Growth occurs in part due to inherent synthetic capability of the oocyte but also because material is incorporated from surrounding pre-granulosa cells. The latter process is not impeded by the acellular coat that forms around the oocyte, the zona pellucida (see below), because interdigitations between the follicular cell processes and the egg surface microvilli develop specialised gap junctions that facilitate transfer of nutrients and substrate. The oocytes acquire considerable cytoplasmic (vitelline) reserves, of which yolk-containing vesicles and other lipid-like deposits are especially prominent in the eggs of some farm animal species. Synthesis of specific proteins is also substantial during the phase of growth (Schultz & Wassarman, 1977). Maximum diameters achieved by the end of the growth phase would be approximately 80 μm for a mouse oocyte and 120 μm for a human oocyte; these dimensions correspond to a three-fold difference in volume (Gosden & Bownes, 1995).

At a molecular level, rapid growth of an oocyte during the pre-antral stage is thought to depend on at least three trophic factors, these being epidermal growth factor (EGF), fibroblast growth factor (FGF) and c-Kit together with its ligand. There is sound evidence that the c-Kit–Kit-ligand interaction is the most important of these, for Packer *et al.* (1994) were able to demonstrate a stepwise dosage influence of Kit-ligand on growth of the oocyte, and incubation of oocytes on monolayers of granulosa cells enhanced the amount of mRNA coding for Kit-ligand. Conversely, blocking the c-Kit–Kit-ligand interaction prevented pre-antral growth of oocyte and follicle (Yoshida *et al.*, 1997). In a systematic

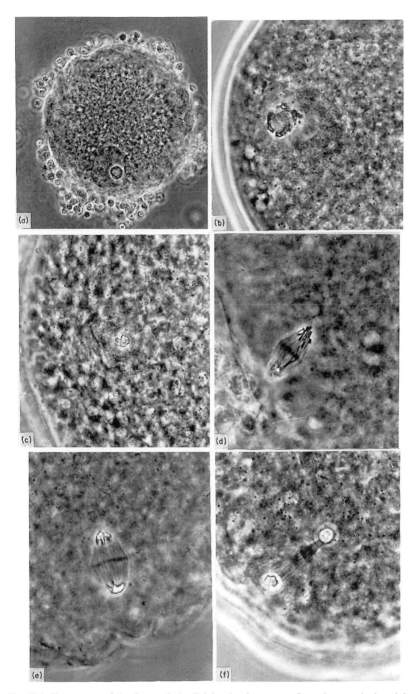

Fig. II.4. Key events of the first meiotic division in pig oocytes from the germinal vesicle stage (a and b) through prometaphase (c), anaphase (d), telophase (e) and abstriction of the first polar body (f). (After Hunter, 1980, but adapted from Hunter & Polge, 1966.)

35

review of the topic, Driancourt & Thuel (1998) emphasised the feedback loop whereby the oocyte stimulates the production of Kit-ligand and Kit-ligand stimulates growth of the oocyte and, in particular, its critical importance at this stage of development. Both EGF and FGF may also assist growth at this stage by means of paracrine signalling between the oocyte and granulosa cells. These molecules have been detected in oocytes, and appropriate receptors are present on the corresponding granulosa cells. Indeed these growth factors may indirectly enhance the amount of Kit-ligand produced by the proliferating granulosa cells, further increasing growth of the oocyte (Driancourt & Thuel, 1998).

The growth phase of an oocyte is critical in terms of its impact on any zygote that in due course may be formed, not least since mitochondria and their associated DNA are largely donated by the female gamete. (A fertilising spermatozoon has negligible cytoplasm, although it does possess a significant residue of cytoplasmic organelles.) Not only do mitochondria increase in abundance during the growth phase, but there are characteristic changes in their morphology and distribution that extend into early embryonic life. These are seen as a transition from elongated organelles with transverse cristae into somewhat spherical and vacuolated structures with fewer and concentrically arched cristae. Similarly, units of the Golgi apparatus enlarge and appear as more flattened organelles distributed in the cortex and towards the periphery of an oocyte. Molecules stored in an ovulated oocyte play a key rôle in the early development of a zygote, since the embryonic programme of macromolecular synthesis is only gradually unfolded and activated during progression of the early cleavage stages along the Fallopian tube (see Sawicki, Magnuson & Epstein, 1981; Flach *et al.*, 1982).

Most of the growth of mouse oocytes is noted by the time these are encompassed by two layers of somatic cells, and the growth phase is essentially terminated by the start of follicular antrum formation at some 2–3 weeks. Cessation of expression of Kit-ligand in the somatic cells may be a key factor in such arrest (Driancourt & Thuel, 1998). This stands in marked contrast to humans and domestic farm animals in which species a period of 4–6 months may be required for oocytes to reach terminal volume (Picton & Gosden, 1999). Associated with the synthetic activity are abundant nuclear pores sited within an undulating membrane around the germinal vesicle. These pores reflect the passage of transcripts and possibly of substrate during the extended phase of cytoplasmic synthetic activity. This membrane commences dissolution only during pre-ovulatory resumption of meiosis (or atresia), having first shown an even more pronounced degree of folding.

The oocyte's strategy of synthesising and storing macromolecules is with clear objectives in view. First, this cell of remarkable dimensions needs to build up reserves as part of the differentiation process, not least to permit interactions with adjoining somatic cells. Second, it needs to acquire and store a developmental programme to be passed on to the developing zygote for use before the embryonic genome commences to regulate synthetic activity at an early stage of cleavage. High levels of RNA synthesis are thus a hallmark of growing oocytes, although this activity characteristically dwindles and more or less ceases by the stage of pre-ovulatory resumption of meiosis should this critical event ever be reached. The stored RNA and proteins can be viewed as a maternal legacy of vital importance to the next generation, such macromolecules contributing to the newly formed embryo in both a programming (informational) and structural manner. In the latter context, actin filaments can be demonstrated as a prominent component of the cytoskeleton. Maternal transcripts are progressively degraded as the embryonic genome is progressively activated (for a review, see Gosden & Bownes, 1995).

Transcription and translation are most active in oocytes contained within growing follicles as compared with those in primordial follicles (see below). Indeed, the synthetic activity continues during follicular development until an oocyte is fully grown or succumbs to atresia. The rate of RNA synthesis has been quantified with respect to dimensions of the mouse oocyte and characterised in terms of classes of RNA (for summaries, see Gosden & Bownes, 1995; Picton & Gosden, 1999). By the time an oocyte has achieved three quarters of mature size, the total content of RNA has increased 300-fold, of which approximately two-thirds is ribosomal RNA and a quarter transfer RNA. Most of the RNA is stable and long lived, being demonstrable for up to 3 weeks. Although the absolute rate of protein synthesis declines, the overall accumulation of macromolecules in an oocyte is impressive: several hundred species of polypeptide have been noted, with relative proportions changing during growth of the oocyte.

Almost all germ cells in fully differentiated ovaries are primary oocytes or degenerating oogonia. Factors regulating the onset of meiosis are the subject of intensive research, yet they remain incompletely understood at the level of both the follicle and the oocyte. A current interpretation concerns the interplay of two somatic cell proteins – meiosis-inducing substance (MIS) and meiosis-preventing substance (MPS) – the outcome of which determines progression into meiotic prophase (Byskov, 1986). This analytical approach has its limitations and these were discussed by Upadhyay & Zamboni (1982) and Zamboni & Upadhyay (1983). Both X chromosomes require to be active before meiosis begins in female germ cells. And, whereas germ cells are not required for the formation and organisation of a testis (Merchant, 1975), an ovary fails to develop in the absence of germ cells. Maintenance of ovarian follicles also strictly depends on the presence of germ cells (Jost & Magre, 1988). These observations would suggest that ovarian determination is dependent upon gene action within the germ cells (Burgoyne, 1989). As already indicated, meiosis is resumed only as a preliminary to the events of ovulation after a pre-ovulatory surge of gonadotrophic hormones (see Chapter VII). However, an apparent resumption of meiosis is frequently a feature of atresia. Nuclear maturation from diplotene of the first meiotic prophase can be obtained in culture if oocytes are liberated towards the end of their intra-follicular growth phase (see Pincus & Enzmann, 1935; Chang, 1955; Edwards, 1962, 1965).

These comments concerning nuclear maturation must not divert attention from the requirement also for cytoplasmic maturation. Although the latter may be revealed by gel electrophoresis of macromolecules contained in oocytes of different size and from different categories of follicle (Moor & Warnes, 1978), a more impressive demonstration of cytoplasmic incompetence in oocytes liberated prematurely from Graafian follicles can be seen in sperm–egg interactions *in vivo* or *in vitro*. Induction of premature ovulation leads to polyspermic penetration of the vitellus and a failure of sperm heads to undergo chromatin decondensation and swelling in the cytoplasm (Polge & Dziuk, 1965; Hunter, Cook & Baker, 1976). Attempted fertilisation *in vitro* of *in vitro*-matured rabbit oocytes similarly led to failure of a sperm head in the cytoplasm to evolve into a pronucleus, leading to a suggested deficiency in a male pronucleus growth factor (Thibault & Gérard, 1970, 1973). As to polyspermic penetration of the vitellus, this reflects an inadequate production and migration of cortical granules in the cytoplasm to a position just beneath the vitelline membrane (Szöllösi, 1967; Fléchon, 1970), a location essential for fusion, exocytosis and establishment of a block to polyspermy upon egg activation by a fertilising spermatozoon.

As to the overall population of oocytes, *de novo* formation of these cells does not and cannot occur in ovaries depleted of germ cells as a consequence, for example, of X-irradiation (Baker & Neal, 1969, 1977). When fully depleted of oocytes, the female is sterile, although advances in reproductive technology may permit alleviation of this previously irreversible condition (Gosden, 2000).

Formation and function of zona pellucida

Before moving on to discuss development of ovarian follicles, there is a further crucial aspect concerning the oocyte. It involves progressive interactions between somatic cells and growing oocytes that lead to formation of an acellular structure, the zona pellucida. This glycoprotein-rich 'membrane' or coat around the oocyte can first be distinguished at the stage of secondary follicle when secretion appears as islands of fibrillar material (Fig. II.5). Formation of a complete and uniform zona pellucida is seen before the stage of Graafian follicle is reached (for reviews, see Wassarman, 1988; Dunbar, Prasad & Timmons, 1991; Wassarman & Albertini, 1994).

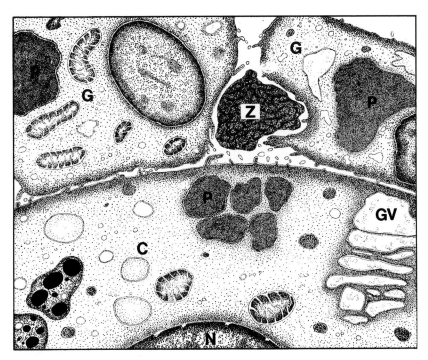

Fig. II.5. Partially diagrammatic electron micrograph depicting formation of the zona pellucida (z) as islands of fibrillar material. Although stated to be derived uniquely from oocyte secretions in the mouse, the possibility of a somatic cell contribution must be considered in other species. GV represents Golgi vesicles and C represents the ooplasm. G, granulosa; P, 'precursor-like' material; N, nucleus. (After Austin & Short, 1972; courtesy of Professor T.G. Baker.)

The origin of the zona pellucida lies primarily in secretion from growing oocytes. Areas of uniform fine fibrillar substance begin to appear close to the oocyte surface and develop into a dense mesh of interconnected filaments. In due course, these completely encompass the oocyte, thereby appearing to separate it from follicular cells. However, this is not so. The plasma membrane or oolemma has become increasingly undulating with growth of the oocyte, leading eventually to a very extensive microvillous surface. Such microvilli project into the zona pellucida, although only a small distance. Junctional contacts are established with the corona cell foot processes referred to in the next section (Fig. II.6). Gap junctions result in a syncytium by means of which the follicle cells and oocyte are metabolically coupled, an engagement required for growth and development and perhaps for resumption of meiosis. Although derived principally from synthetic activity of the oocyte itself, notably from glycoproteins generated in the Golgi apparatus of the cortex, there is also evidence for a subsequent contribution to the zona pellucida from adjacent follicular cells.

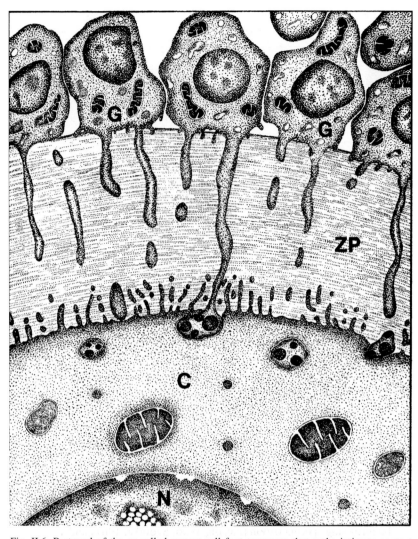

Fig. II.6. Portrayal of the so-called corona cell foot processes that make intimate contact with the plasma membrane around the oocyte, very often to the extent of indenting that membrane. The junctional contacts (gap junctions or adhesion regions) provide a means of dialogue between the germ and somatic cells and enable passage of suitable substrate molecules. G, granulosa; ZP, zona pellucida; C, ooplasm; N, nucleus. (Courtesy of Professor T.G. Baker.)

In one sense, this should not cause surprise for follicular cell secretions can certainly be detected on the surface of the zona pellucida by the pre-ovulatory stage of follicular maturation (see Szöllösi & Hunter, 1978).

At a molecular level, the zona pellucida consists of two to four well-conserved proteins, as revealed in mouse, rabbit, pig, monkey and human oocytes. The coding sequences of the human and mouse genes are 74% identical (Wassarman, 1988). Zona proteins (ZP) are heavily glycosylated in mouse oocytes, are designated as ZP1, ZP2 and ZP3, and ranged according to molecular weight (200, 120 and 83 kDa, respectively). A functional zona depends on all three proteins, since the zona matrix comprises polymeric filaments of ZP2 and ZP3 that are cross-linked non-covalently by ZP1 dimers. In other words, each glycoprotein is an important structural component of the zona pellucida (Wassarman, 1988). Mouse *Zp1*, *Zp2* and *Zp3* genes are present as single copies on chromosomes 19, 7 and 5, respectively. With the exception of *Zp2*, expression of these genes is thought not to occur in primordial follicles but commences only when the oocyte begins to grow and thus the zona to form. Such expression is essentially restricted to the period of 2 weeks needed for the oocyte to reach its maximum volume, and has dwindled to no more than 5% of peak values in those exceptional oocytes that achieve the stage of ovulation. Comparable to other long-lived RNAs in the maternal legacy, they are rather promptly deadenylated and degraded as transcription ceases during resumption of meiosis. ZP1, ZP2 and ZP3 are coordinately expressed and may have common transcriptional regulatory elements (Wassarman, 1988; Wassarman & Albertini, 1994; Wassarman, Liu & Litscher, 1997).

The functions of the zona pellucida are diverse and crucial, and include:

1. Recognition of male gametes due to the species-specificity of sperm-binding sites on the zona pellucida (Wassarman & Albertini, 1994).
2. An involvement in inducing the sperm acrosome reaction (Szöllösi & Hunter, 1978; Yanagimachi, 1988, 1994), although other factors at the site of fertilisation in the Fallopian tube – especially unique fluid constituents – should not be overlooked in prompting this membrane vesiculation reaction (Hunter, 1988, 1997).
3. Establishment of a defence or block against multiple sperm penetration of the vitellus (polyspermy) due to activation of cortical granule release by the fertilising spermatozoon (Austin, 1956; Szöllösi, 1967) and a consequent downgrading of zona sperm receptors, at least in the mouse egg (Wassarman, 1988; Wassarman & Albertini 1994), if not in those of many other mammals; such species differences have been largely overlooked.
4. Support, protection and intimate positioning of blastomeres during early development before intercellular junctions have conferred a degree of stability (Mintz, 1962; Edwards, 1964; Moor & Cragle, 1971).
5. Maintenance of fluid in the perivitelline space as a primary developmental microenvironment and facilitation of molecular exchange with Fallopian tube luminal fluids (Hunter, 1988, 1994; Leese, 1988). Such exchange is frequently said to be limited to molecules smaller than 160–170 kDa

(Austin & Lovelock, 1958; Austin, 1961; Picton & Gosden, 1999), but this supposed restriction is worth further investigation. It may change before and after fertilisation, and with the stage of early embryonic development. Secretions of the Fallopian tube, especially of unique glycoproteins such as oviductin (Kan, St-Jacques & Bleau, 1988, 1989; Wagh & Lippes, 1993) could have an important influence on zona permeability and might contribute to the phenomenon of 'zona hardening'.

Follicular growth and formation of a Graafian follicle

In order for the fetal ovary to become functional in later life, there is a requirement for (1) oogonia to enter prophase of the first meiotic division and become primary oocytes, (2) follicles to be elaborated from the relatively simple primary structure to the well-organised tertiary (Graafian) follicle, and (3) the somatic cell layers of the follicle to develop an ability to synthesise and secrete diverse steroid and peptide hormones, and to respond to endocrine cues from the hypophysis. Interactions between theca and granulosa cells (see below) and between germ and somatic cells are of critical importance for this endocrine rôle. Nor should an incisive molecular influence of oocytes in the differentiation of the supporting cell lineage as ovarian follicular cells rather than as testicular Sertoli cells be overlooked. The time at which follicles begin to form in different species depends on when oocytes reach the diplotene stage of the first meiotic prophase, and may occur during fetal life or soon after birth. However, in contrast to early organisation of the testis, ovarian follicles tend to form postnatally (Jost & Magre, 1988), especially in larger species with longer lifespans, although studies, for example in pigs (Casida, 1935) and calves (Erickson, 1966), would challenge such a point of view.

Folliculogenesis, the formation and development of follicles (Webb *et al.*, 1999), covers the stages from the initial organised contact of oocytes with incipient granulosa cells (pre-granulosa cells) until a primary oocyte is supported by a cumulus cell mass within a Graafian follicle (Fig. II.7). In our own species, entry of primordial follicles into the growth phase represents a continuous process that extends from before birth until the menopause (Baker, 1972; Gougeon, 1996). To keep matters in perspective, it should be emphasised that only a small proportion of the initial complement of oocytes ever becomes arranged within a Graafian follicle. Most oocytes and indeed follicles are lost due to atresia (Sturgis, 1961; and see Chapter VI). Such loss can occur at any of the stages to be described in the paragraphs that follow.

A masterly and comprehensive review of follicular development and its control across diverse species has been published by Greenwald & Roy (1994),

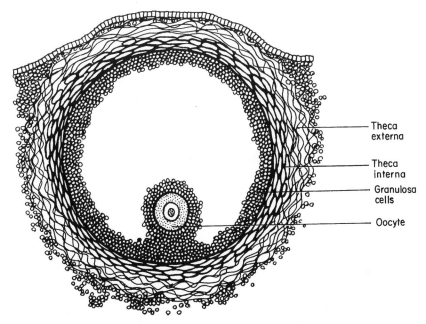

Fig. II.7. Semi-diagrammatic representation of a Graafian follicle to illustrate the theca externa, theca interna and mural granulosa. An oocyte is supported within the antrum by a protrusion of granulosa cells termed the cumulus oophorus. (After Hunter, 1980.)

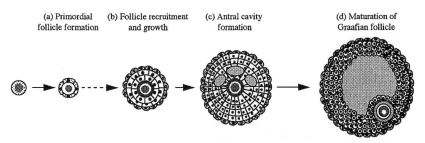

Fig. II.8. Semi-diagrammatic illustration of formation of a Graafian follicle showing the progression from primordial follicle through the stage of antral cavity formation to that of tertiary or Graafian follicle with a multi-layered granulosa cell compartment. (Adapted from Picton & Gosden, 1999; courtesy of Dr H. Picton.)

building on the earlier ones of Greenwald (1978) and Hirshfield (1991). Termed a primordial follicle, the first stage of development consists of an oocyte that has prompted so-called pre-granulosa cells to assume a characteristic flattened arrangement around its limiting membrane (Fig. II.8). Whilst the nature of a putative attractant produced by the oocyte at this preliminary stage of follicular formation has yet to be revealed, specific molecular influences of the far larger

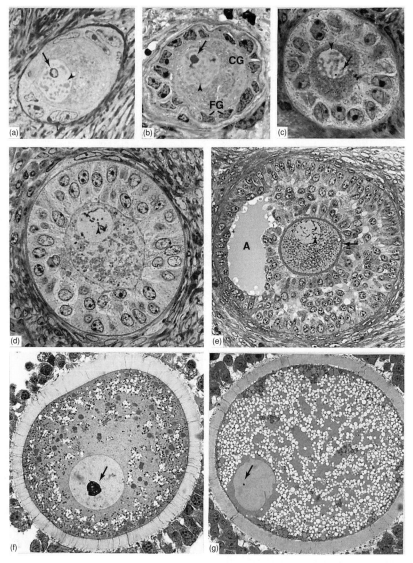

Fig. II.9. (a) Light micrograph showing a resting primordial follicle. Note the flattened granulosa cells, the oocyte containing an eccentrically located nucleus displaying patches of condensed chromatin (arrowhead) and a nucleolus (arrow). 717× (b) Light micrograph showing an activated primordial follicle. Note the flattened (FG) and cuboidal granulosa cells (CG), the oocyte containing an eccentrically located nucleus displaying patches of condensed chromatin (arrowhead) and a nucleolus (arrow). 666× (c) Light micrograph showing a primary follicle. Note the cuboidal granulosa cells, the oocyte containing an eccentrically located nucleus displaying patches of condensed chromatin (arrowhead) and a nucleolus (arrow). 640× (d) Light micrograph showing a secondary follicle. Note the bilayer of

germ cell upon somatic cells in its vicinity could reasonably be anticipated. In fact, it may be helpful to think of germ cells acting in this way to influence or even to programme their own destiny. Nonetheless, specific factors that initiate development of primordial follicles are not known. To the critical question as to what drives folliculogenesis, there could be diverse answers depending on the precise stage under consideration and the level of growth response anticipated. Even so, a critical rôle for peptide factors is assumed, with paracrine signalling from the oocyte in terms of EGF and FGF being a strong possibility.

In most mammals so far examined, primordial follicles are located in the outermost portion of the ovary, the cortex, but this is not so in equids, owing to the unusual morphology of the ovary. In such species, cortical tissue is confined to one area and is largely surrounded by medullary tissue. This specialisation defines the site of ovulation. Follicles can release their contents only through the ovulation groove or fossa, a region adjacent to the cortical tissue.

In the mouse, approximately five pre-granulosa cells can be identified. Once follicular growth is initiated, these expand clonally during 3 weeks to form the mature epithelium of >50 000 cells in a Graafian follicle (Boland & Gosden, 1994).

Pre-granulosa cells making contact with an oocyte gradually become cuboidal as envelopment of the oocyte proceeds, and this stage with a mixture of flattened and cuboidal cells is sometimes referred to as an intermediary follicle (Fig. II.8). A complete single layer of cuboidal granulosa cells surrounding an oocyte and resting on a newly formed basement membrane characterises a primary follicle (Figs. II.8 and II.9). Proliferation of additional layers of granulosa cells signals the formation of a secondary follicle. Recruitment of theca cells from the stroma is already distinguishable to the outside of the basement membrane, and these begin to receive a capillary network. Theca interna cells differentiate and soon demonstrate organelles characteristic of steroid-secreting cells, notably villiform cristae in the mitochondria. Cells of the theca externa remain more spindle shaped and merge with the surrounding stroma. Once again, some form of molecular prompting of division and organisation of neighbouring somatic cells by the encompassed oocyte would be anticipated. Growth of the oocyte itself is depicted in Fig. II.10.

Caption for fig. II.9. (*cont.*) cuboidal granulosa cells, the oocyte containing an eccentrically located nucleus displaying patches of condensed chromatin (arrowhead). 480× (e) Light micrograph showing an early tertiary follicle. Note the multiple layers of granulosa cells, the antral cavity (A), the oocyte (arrow) containing an eccentrically located nucleus displaying patches of condensed chromatin (arrowhead). 237× (f) Light micrograph of an oocyte <110 μm. Note the eccentrically located, round nucleus containing a nucleolus (arrow). 300× (g) Light micrograph of an oocyte >110 μm. Note the peripherally located nucleus containing a nucleolus (arrow). 320× (After Fair *et al.*, 1997; courtesy of Professor P. Hyttel.)

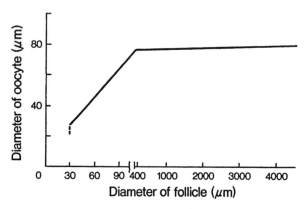

Fig. II.10. A modified version of the classical graph of Green & Zuckerman (1951) illustrating growth of the oocyte as a function of diameter of the follicle, based upon analysis of human ovarian tissue. (Courtesy of Professor T.G. Baker.)

By the time the investment of granulosa cells has become multi-layered, and theca cells have become prominent owing to further cell division, pools or islets of fluid will have commenced to form within the granulosa intercellular spaces. Gradual coalescence of the accumulated pools leads to the appearance of a cavity or antrum containing so-called follicular fluid. Such antral or vesicular follicles represent the stage of tertiary or Graafian follicles, structures with a high degree of specialisation in their various cellular compartments. By this stage, the oocyte is no longer located centrally within the follicle but rather is suspended towards one side in a protrusion of granulosa cells, the cumulus oophorus (see Fig. II.7). Organisation and proliferation of such cumulus cells are strongly influenced by the oocyte (Buccione, Schröder & Eppig, 1990).

Viewing the structure of a Graafian follicle from the exterior towards the centre, the protruding surface of a mature follicle is limited by the so-called germinal epithelium – a single layer of flattened or cuboidal cells. As is widely appreciated, this is not germinal at all, the epithelium being continuous with the coelomic or visceral peritoneum. Below the germinal epithelium are the respective layers of theca externa and theca interna. Smooth muscle (contractile) elements can be distinguished in the layers of the theca externa (Thomson, 1919; Guttmacher & Guttmacher, 1921) and likewise an important degree of innervation (see Chapter III). Both layers of the theca become well vascularised. There is also a prominent lymphatic network (see Chapter III). The basement membrane separating the theca interna from granulosa cells prevents invasion of the capillary bed except in the final hours before ovulation, following the surge of gonadotrophic hormones and gradual dissolution of this limiting structure.

Avoiding vascularisation of the granulosa cells appears to be associated with preventing their luteinisation and a switch from oestradiol to progesterone secretion. There might also be consequences for the oocyte. Nonetheless, diverse theca–granulosa cell interactions do underlie development and maturation of Graafian follicles.

The layers of granulosa cells arranged inwards from the basement membrane show morphological and functional specialisations. Most conspicuous of these is the cloud-shaped mass of cells projecting into the follicular antrum that supports and envelops the oocyte. The innermost layer of cumulus cells arranged around the surface of the zona pellucida is termed the corona radiata. In fact, such corona cells not only have an intimate interface with the surface of the zona but, as noted above, themselves project long cytoplasmic processes through the substance of the zona to make specific contact with the plasma membrane of the oocyte, the oolemma. The bulbous tips of these processes are termed corona cell foot processes (see Fig. II.6), and facilitate nutritional exchange between the somatic cells and oocyte by means of highly specialised gap junctions (Szöllösi, 1975; Anderson & Albertini, 1976).

The mural granulosa, that portion closest to the basement membrane, constitutes the other major subdivision of the granulosa cell mass. Whereas the corona cells and a large proportion of the cumulus cell investment are expelled from a Graafian follicle at ovulation, together with the bulk of the accumulated follicular fluid, much of the mural granulosa remains *in situ* and provides a contribution in the form of granulosa lutein cells to the developing corpus luteum.

In general terms, the smaller and less elaborate follicles, i.e. primordial, intermediary and primary stages, constitute the reserves or store from which more advanced follicles continue to emerge and develop throughout the reproductive lifespan. Overall, the stages of primordial and primary follicle constitute >95% of the follicles contained in an ovary. Further development and incorporation into reproductive events is termed recruitment, usually involving a group or cohort of follicles. This may be followed by progression into waves of development (see Chapter VI). However, only a small proportion of primary follicles ever achieves the Graafian stage, for the vast majority become atretic and undergo prompt degeneration as a consequence of apoptosis. Thus the fate of primary follicles is either one of development or, more probably, the degenerative condition of atresia that may occur at any subsequent stage (see Chapter VI).

Key reviews of follicular growth and development in the human ovary have been presented by Gougeon (1993, 1996, 1997, 2000) and Driancourt *et al.* (2001). The proportion of atretic and resting follicles has been estimated to be about 50% at birth, and this proportion then decreases until approximately 30 years of age. After the age of 30, loss of resting follicles is due mainly to entrance of follicles into the growth phase. Growth of follicles occurs at different rates, and can be usefully considered as three phases.

1. Pre-antral growth phase, beginning with the evolution of primordial follicles into primary follicles and ending with the formation of pre-antral follicles, a transition that may require several months in humans.
2. Tonic growth phase, seen as the development of pre-antral follicles into small antral follicles. This transition is estimated to require approximately 2 months in humans. It ends when such small antral follicles are approximately 2–5 mm in diameter.
3. Exponential growth phase, when small antral follicles can then undergo exponential growth with development into large, mature ovulatory follicles. This phase in humans lasts approximately 21 days, beginning in the mid-luteal phase of one cycle and ending with ovulation of the follicle in the next cycle. In the exponential phase, several early antral follicles are recruited to begin final maturation during the luteal phase of the menstrual cycle. A process of selection then occurs during the following follicular phase whereby a single follicle becomes dominant by the middle of the follicular phase and is then destined to ovulate. The number of dominant follicles is species specific.

If a follicle is successful in avoiding the pitfall of atresia, its physiological impact upon surviving the hazards thus far is impressive. As a complex and sophisticated secretory unit, a mature Graafian follicle exerts potent and diverse influences in an autocrine, paracrine and endocrine manner. Indeed, the endocrine activity of a mature Graafian follicle, especially in terms of its synthesis and secretion of steroid hormones, integrates the function of the gonad with that of the female genital tract and with responses generated in the pituitary gland, hypothalamus and higher nervous centres. The physiological context is, of course, one of female receptivity to male advances (oestrus and/or lordosis in farm and laboratory species), coitus, ovulation and the possibility of fertilisation if functional gametes interact in a timely manner.

Follicular development, acquisition of follicle stimulating hormone dependence

Viewed in one light, development begins when follicles leave the pool of resting structures and show growth, cell division and differentiation. These processes are influenced overall by secretion of gonadotrophic hormones from the anterior lobe of the pituitary gland. Intra-ovarian factors also play a key rôle, acting by both autocrine and paracrine routes to modulate and perhaps amplify local signals involved in follicular development. However, there is strong evidence that primordial follicles can be formed in the absence of specific gonadotrophic stimulation (see Findlay & Drummond, 1996), not least since they can be generated when gonadotrophin receptor activity is undetectable (Richards *et al.*, 1987). As alluded to above, an organising rôle for the oocyte is therefore strongly suggested and likewise by means of locally available peptide factors.

Despite evidence from hypophysectomy experiments (Dufour, Cahill & Mauléon, 1979), a model with very low to zero gonadotrophin titres, and from the comments above, transformation of the flattened pre-granulosa cells of the primordial follicle into cuboidal cells of a primary follicle may involve basal levels of gonadotrophin support (see Ross, 1974). Signals from the oocyte would also seem to be needed for this transformation, since conversion of primordial into primary follicles does not occur until oocytes achieve a diameter of at least 20 μm.

The tonic growth phase involves low levels of gonadotrophins whereas the exponential growth phase is heavily dependent on gonadotrophins. These remarks prompt consideration of follicle stimulating hormone (FSH) receptors in ovarian follicles. The gonadotrophin-driven influence on folliculogenesis is by specific binding of these hormones to receptors on the surface of follicular cells. Gonadotrophin receptors are G protein-coupled receptors that traverse the plasma membrane and activate adenylyl cyclase and increase cyclic $3',5'$-adenosine monophosphate (cAMP) as second messenger in appropriate follicular cells. As would be anticipated, FSH binds specifically to FSH receptors on granulosa cells (see below and Chapters V and VI).

FSH receptors have been detected in human ovaries from the stage of secondary follicle onwards. Furthermore, the density of FSH receptors on granulosa cells is higher in antral than in pre-antral follicles, and this could provide one explanation for the increasing responsiveness of later follicular development to gonadotrophic stimulation as compared with earlier stages. In the mouse ovary, expression of the FSH receptors was absent in primordial stages and not consistently found until after the primary stage (Oktay, Briggs & Gosden, 1997). Significant expression of both FSH and luteinising hormone (LH) receptors has been detected during the period from the primary to the mid-to-late secondary stages of mouse follicle development (O'Shaughnessy, McClelland & McBride, 1997). As for domestic farm animals, FSH receptors can be detected on the granulosa cells of growing follicles from bovine ovaries (Wandji, Pelletier & Sirard, 1992). Shortly after formation of an antrum, follicles become dependent on FSH (Scaramuzzi *et al.*, 1993).

The relevance of FSH receptors to follicular development has been emphasised by clinical observations. Women with premature ovarian failure due to hypergonadotrophic ovarian dysgenesis lack secondary sexual development despite elevated serum FSH concentrations. Molecular studies have revealed a point mutation in their FSHR gene resulting in a complete absence of second messenger activity – cAMP generation – in granulosa cells, thereby effectively preventing ovarian stimulation. Such ovaries reveal predominantly primordial follicles, but some primary follicles and an occasional pre-antral follicle can be noted. Overall, however, the ovaries of such women are conspicuously underdeveloped, highlighting the relevance of a functional FSH receptor.

Transcription factors in folliculogenesis

Despite a fairly long-standing appreciation of a variety of intra-ovarian factors that can influence growth and differentiation of follicles (Toneta & DiZerega, 1989; Baird & Smith, 1993), control mechanisms involved in initiation of follicular growth are still not clearly understood. It is not even certain whether recruitment depends primarily on follicular stimulation rather than on alleviation of inhibition (Gosden, 1998); of course, both processes might be occurring simultaneously. There is the suggestion, for example, that activin derived from secondary follicles could be causing small pre-antral follicles to remain dormant (Mizunuma *et al.*, 1999), although the manner in which such an influence is overcome by individual follicles has yet to be specified.

Primary and primordial follicles express a variety of molecules that contribute to growth and differentiation, and a full list of these would be impressive for its seeming functional diversity. It includes growth differentiation factor 9 (GDF-9), insulin-like growth factor (IGF), activin and activin receptor II, transforming growth factor β (TGF-β) and the Wilm's tumour gene (*WT*1). Neither GDF-9 nor Kit-ligand (see above) are currently considered candidates for a putative triggering factor, the evidence being that gene knockouts in mice for GDF-9, Steel[t] and Steel[panda] block growth of follicles at the primary stage but not at the primordial stage (Huang *et al.*, 1993; Dong *et al.*, 1996).

One recent study on the involvement of WT1 found that this protein was a marker for immature follicles in both mammals and birds. Acting as a transcriptional repressor influencing growth factor and growth factor receptor genes, WT1 appeared to have an important rôle in the slow growth of early (immature) follicles (Chun *et al.*, 1999). In brief, it may act to suppress factors that stimulate proliferation of granulosa cells in immature follicles.

Oocyte macromolecules and maternal RNA programme

At a molecular level, the proto-oncogene c-*mos* was one of the first genes proposed to be involved in meiosis. In the mouse, c-*mos* mRNA is expressed in growing and fully grown oocytes, but the Mos protein is found only in fully grown oocytes during maturation and at the second meiotic arrest awaiting fertilisation; it is not present after fertilisation (Strömstedt & Byskov, 1999). Mos is a serine-threonine protein kinase representing a component of the cytostatic factor (CSF). Mice deficient in c-*mos* have conspicuously reduced fertility and, rather than arresting at second meiotic metaphase, the oocytes frequently undergo parthenogenesis. In summary, c-*mos* appears essential for the checkpoint at metaphase II arrest (Strömstedt & Byskov, 1999).

Involvement of the genes *W* (*Dominant White Spotting*) and *Sl* (*Steel*) and of *IL-6/LIF* has already been discussed in a context of primordial germ cells, but subsequent influences in oocytes of various species need further examination (see De Felici, 2001; Donovan *et al.*, 2001). Some valuable detail on the synthesis and uptake of macromolecules is given in the review by Gosden & Bownes (1995).

A more recent molecular finding is that mouse oocytes themselves express FSH and LH receptors, quite distinct from expression in the surrounding somatic cells (Patsoula *et al.*, 2001). Such receptors are currently viewed as products of the maternal genome, which would be the interpretation also for such products in the newly formed zygote, since embryonic gene transcription commences at the two-cell stage in this species (Flach *et al.*, 1982). The presence of FSH and LH mRNA transcripts in mouse oocytes raises key questions concerning a direct involvement of gonadotrophins in promoting different aspects of oocyte maturation. At least in part, such a potential route of stimulation might underlie the resumption of meiosis in denuded oocytes *in vitro* when gonadotrophins are added to the culture medium.

Concluding remarks

Key themes in this chapter have included the transition from primordial germ cell to competent oocyte, factors prompting recruitment of follicles from the primordial pool, and diverse relationships between an oocyte and its surrounding somatic cells. There was consideration also of the manner in which a developing follicle becomes dependent on pituitary gonadotrophic support, and the arrest of meiosis at the diplotene stage of the first prophase with resumption only shortly before ovulation. Completion of meiosis occurs promptly after ovulation upon activation by a fertilising spermatozoon. Most phases of meiosis are therefore accomplished close to the time of ovulation, emphasising the dominance of the diplotene (dictyate) phase and raising questions as to a protracted germinal influence on the developing follicle.

Unresolved remain the questions of why so many oocytes are generated in a mammalian ovary when so few will ever be used fruitfully, and the extent to which selection processes are at work in diverting such a high proportion of oocytes along a pathway to atresia. Expressed in another way, to what degree is such selection based upon the potential quality and competence of an oocyte?

Some readers will have noted the absence of any meaningful remarks concerning the contribution of the vascular network and nervous system in the early development of ovarian follicles. This follows in a systematic way in Chapter III, but perhaps the present chapter can conclude by entertaining possible paracrine contributions of angiogenic factors (e.g. vascular endothelial growth factor (VEGF)) and nerve growth factors to the early stages of folliculogenesis. Since such factors are certainly involved in the later stages of follicular development, they might also be contributing to progression from the primordial stage. And, of course, molecules emanating from the oocytes themselves could well play a key rôle even in the earliest stages of folliculogenesis.

References

Anderson, E. & Albertini, D.F. (1976). Gap junctions between the oocyte and companion follicle cells in the mammalian ovary. *Journal of Cell Biology*, **71**, 680–6.

Austin, C.R. (1956). Cortical granules in hamster eggs. *Experimental Cell Research*, **10**, 533–40.

Austin, C.R. (1961). *The Mammalian Egg*. Oxford: Blackwell Scientific Publications.

Austin, C.R. & Lovelock, J.E. (1958). Permeability of rabbit, rat and hamster egg membranes. *Experimental Cell Research*, **15**, 260–1.

Austin, C.R. & Short, R.V. (eds.) (1972). *Reproduction in Mammals*, 1st edn. Cambridge: Cambridge University Press.

Baird, D.T. & Smith, K.B. (1993). Inhibin and related peptides in the regulation of reproduction. *Oxford Reviews of Reproductive Biology*, **15**, 191–232.

Baker, T.G. (1972). Primordial germ cells. In *Reproduction in Mammals*, 1st edn, ed. C.R. Austin & R.V. Short, pp. 1–13. Cambridge: Cambridge University Press.

Baker, T.G. (1982). Oogenesis and ovulation. In *Reproduction in Mammals*, 2nd edn, vol. 1, ed. C.R. Austin & R.V. Short, pp. 17–45. Cambridge: Cambridge University Press.

Baker, T.G. & Neal, P. (1969). The effects of X-irradiation on mammalian oocytes in organ culture. *Biophysika*, **6**, 39–45.

Baker, T.G. & Neal, P. (1977). Action of ionizing radiations on the mammalian ovary. In *The Ovary*, 2nd edn, ed. S. Zuckerman & B.J. Weir, pp. 1–58. New York, San Francisco & London: Academic Press.

Bardoni, B., Zanaria, E., Guioli, S., *et al.* (1994). A dosage sensitive locus at chromosome Xp21 is involved in male to female sex reversal. *Nature Genetics*, **7**, 497–501.

Boland, N.I. & Gosden, R.G. (1994). Clonal analysis of chimaeric mouse ovaries using DNA *in situ* hybridisation. *Journal of Reproduction and Fertility*, **100**, 203–10.

Buccione, R., Schröder, A.C. & Eppig, J.J. (1990). Interactions between somatic cells and germ cells throughout mammalian oogenesis. *Biology of Reproduction*, **43**, 543–7.

Buehr, M. (1997). The primordial germ cells of mammals: some current perspectives. *Experimental Cell Research*, **232**, 194–207.

Burgoyne, P.S. (1988). Role of mammalian Y chromosome in sex determination. *Philosophical Transactions of the Royal Society of London, Series B*, **322**, 63–72.

Burgoyne, P.S. (1989). Thumbs down for zinc finger? *Nature (London)*, **342**, 860–2.

Burns, R.K. (1961). Role of hormones in the differentiation of sex. In *Sex and Internal Secretions*, 3rd edn, ed. W.C. Young, pp. 76–158. Baltimore, MD: Williams & Wilkins.

Byskov, A.G. (1986). Differentiation of mammalian embryonic gonad. *Physiological Reviews*, **66**, 71–117.

Byskov, A.G. & Høyer, P.E. (1988). Embryology of mammalian gonads and ducts. In *The Physiology of Reproduction*, ed. E. Knobil & J.D. Neill, pp. 265–302. New York: Raven Press.

Casida, L.E. (1935). Prepuberal development of the pig ovary and its relation to stimulation with gonadotrophic hormones. *Anatomical Record*, **61**, 389–96.

Chang, M.C. (1955). The maturation of rabbit oocytes in culture and their maturation, activation, fertilisation and subsequent development in the Fallopian tubes. *Journal of Experimental Zoology*, **128**, 379–405.

Chun, S.Y., McGee, E.A., Hsu, S.Y., *et al.* (1999). Restricted expression of WT1 messenger ribonucleic acid in immature ovarian follicles: uniformity in mammalian and avian species and maintenance during reproductive senescence. *Biology of Reproduction*, **60**, 365–73.

De Felici, M. (2001). Twenty years of research on primordial germ cells. *International Journal of Developmental Biology*, **45**, 519–22.

Dong, J., Albertini, D., Nishimori, K., Kumar, T., Lu, N. & Matzuk, M. (1996). Growth differentiation factor-9 is required during early ovarian folliculogenesis. *Nature (London)*, **383**, 531–5.

Donovan, P.J. (1999). Primordial germ cells. In *Encyclopedia of Reproduction*, vol. 3, ed. E. Knobil & J.D. Neill, pp. 1064–72. San Diego & London: Academic Press.

Donovan, P.J., De Miguel, M.P., Hirano, M.P., Parsons, M.S. & Lincoln, A.J. (2001). Germ cell biology – from generation to generation. *International Journal of Developmental Biology*, **45**, 523–31.

Driancourt, M.A. & Thuel, B. (1998). Control of oocyte growth and maturation by follicular cells and molecules present in follicular fluid. A review. *Reproduction, Nutrition, Development*, **38**, 345–62.

Driancourt, M.A., Gougeon, A., Monniaux, D., Royère, D. & Thibault, C. (2001). Folliculogenèse et ovulation. In *La Reproduction chez les Mammifères et l'Homme*, ed. C. Thibault & M.C. Levasseur, pp. 316–47. Paris: Ellipses.

Dufour, J., Cahill, L.P. & Mauléon, P. (1979). Short and long term effects of hypophysectomy and unilateral ovariectomy on ovarian follicular populations in sheep. *Journal of Reproduction and Fertility*, **57**, 301–9.

Dunbar, B.S., Prasad, S.V. & Timmons, T.M. (1991). Comparative structure and function of mammalian zonae pellucidae. In *A Comparative Overview of Mammalian Fertilisation*, ed. B.S. Dunbar & M.G. O'Rand, pp. 97–114. New York: Plenum Press.

Edwards, R.G. (1962). Meiosis in ovarian oocytes of adult mammals. *Nature (London)*, **196**, 446–50.

Edwards, R.G. (1964). Cleavage of one and two-celled rabbit eggs *in vitro* after removal of the zona pellucida. *Journal of Reproduction and Fertility*, **7**, 413–15.

Edwards, R.G. (1965). Maturation *in vitro* of mouse, sheep, cow, pig, rhesus monkey and human ovarian oocytes. *Nature (London)*, **208**, 349–51.

Erickson, B.H. (1966). Development and senescence of the postnatal bovine ovary. *Journal of Animal Science*, **25**, 800–5.

Evans, H.M. & Cole, H.H. (1931). An introduction to the study of the oestrous cycle in the dog. *Memoirs of the University of California*, **9**, 65–119.

Fair, T., Hulshof, S.C.J., Hyttel, P., Greve, T. & Boland, M. (1997). Oocyte ultrastructure in bovine primordial to early tertiary follicles. *Anatomy and Embryology*, **195**, 327–36.

Findlay, J.K. & Drummond, A.E. (1996). Control of follicular growth. In *The Ovary: Regulation, Dysfunction and Treatment*, ed. M. Filicori & C. Flamigni, pp. 13–21. Excerpta Medica International Congress Series, No. 1106. Amsterdam: Elsevier.

Flach, G., Johnson, M.H., Braude, P.R., Taylor, R.A.S. & Bolton, V.N. (1982). The transition from maternal to embryonic control in the 2-cell mouse embryo. *European Molecular Biology Organisation*, **1**, 681–6.

Fléchon, J.E. (1970). Nature glycoprotéique des granules corticaux de l'oeuf de lapine. *Journal de Microscopie*, **9**, 221–42.

Ford, C.E., Jones, K.W., Polani, P.E., de Almeida, J.C. & Briggs, J.H. (1959). A sex-chromosome anomaly in a case of gonadal dysgenesis (Turner's syndrome). *Lancet*, **i**, 711–13.

Francavilla, S. & Zamboni, L. (1985). Differentiation of mouse ectopic germinal cells in intra- and perigonadal locations. *Journal of Experimental Zoology*, **233**, 101–9.

Ginsburg, M., Snow, M.H.L. & McLaren, A. (1990). Primordial germ cells in the mouse embryo during gastrulation. *Development*, **110**, 521–8.

Gomperts, M., Garcia-Castro, M., Wylie, C. & Heasman, J. (1994). Interactions between primordial germ cells play a role in their migration in mouse embryos. *Development*, **120**, 135–41.

Gosden, R.G. (1998). Biology and technology of primordial follicle development. In *Gametes: Development and Function*, ed. A. Lauria & F. Gandolfi, pp. 71–83. Rome: Serono Symposia.

Gosden, R.G. (2000). Low temperature storage and grafting of human ovarian tissue. *Molecular and Cellular Endocrinology*, **163**, 125–9.

Gosden, R.G. & Bownes, M. (1995). Molecular and cellular aspects of oocyte development. In *Gametes – the Oocyte*, ed. J.G. Grudzinskas & J.L. Yovich, pp. 23–53. Cambridge: Cambridge University Press.

Gougeon, A. (1993). Dynamics of human follicular growth. A morphologic perspective. In *The Ovary*, ed. E.Y. Adashi & P.C.K. Leung, pp. 21–39. New York: Raven Press.

Gougeon, A. (1996). Regulation of ovarian follicular development in primates: facts and hypotheses. *Endocrine Reviews*, **17**, 121–55.

Gougeon, A. (1997). Kinetics of human ovarian follicular development. In *Microscopy of Reproduction and Development: A Dynamic Approach*, ed. P.M. Motta, pp. 67–77. Rome: Antonio Delfino Editore.

Gougeon, A. (2000). Évolution du capital folliculaire au cours du vieillissement ovarian. *Reproduction Humaine et Hormones*, **13**, 213–23.

Green, S.H. & Zuckerman, S. (1951). Quantitative aspects of the growth of the human ovum and follicle. *Journal of Anatomy*, **85**, 373–5.

Greenwald, G.S. (1978). Follicular activity in the mammalian ovary. In *The Vertebrate Ovary. Comparative Biology and Evolution*, ed. R.E. Jones, pp. 639–89. New York & London: Plenum Press.

Greenwald, G.S. & Roy, S.K. (1994). Follicular development and its control. In *The Physiology of Reproduction*, 2nd edn, ed. E. Knobil & J.D. Neill, pp. 629–724. New York: Raven Press.

Gubbay, J., Collignon, J., Koopman, P., *et al.* (1990). A gene mapping to the sex-determining region of the mouse Y chromosome is a member of a novel family of embryonically expressed genes. *Nature (London)*, **346**, 245–50.

Guttmacher, M.S. & Guttmacher, A.F. (1921). Morphological and physiological studies on the musculature of the mature Graafian follicle of the sow. *Johns Hopkins Hospital Bulletin*, **32**, 394–9.

Hirshfield, A.N. (1991). Development of follicles in the mammalian ovary. *International Review of Cytology*, **124**, 43–101.

Huang, E., Manova, K., Packer, A., Sanchez, S., Bachvarova, R. & Besmer, P. (1993). The murine Steel panda mutation affects *kit* ligand expression and growth of early ovarian follicles. *Developmental Biology*, **157**, 100–9.

Hunter, R.H.F. (1980). *Physiology and Technology of Reproduction in Female Domestic Animals*. London & New York: Academic Press.

Hunter, R.H.F. (1988). *The Fallopian Tubes: Their Role in Fertility and Infertility*. Berlin & Heidelberg: Springer-Verlag.

Hunter, R.H.F. (1994). Modulation of gamete and embryonic microenvironments by oviduct glycoproteins. *Molecular Reproduction and Development*, **39**, 176–181.

Hunter, R.H.F. (1995). *Sex Determination, Differentiation and Intersexuality in Placental Mammals*. Cambridge: Cambridge University Press.

Hunter, R.H.F. (1996). Aetiology of intersexuality in female (XX) pigs, with novel molecular interpretations. *Molecular Reproduction and Development*, **45**, 392–402.

Hunter, R.H.F. (1997). Sperm dynamics in the female genital tract: interactions with Fallopian tube microenvironments. In *Microscopy of Reproduction and Development: A Dynamic Approach*, ed. P.M. Motta, pp. 35–45. Rome: Antonio Delfino Editore.

Hunter, R.H.F., Baker, T.G. & Cook, B. (1982). Morphology, histology and steroid hormones of the gonads in intersex pigs. *Journal of Reproduction and Fertility*, **64**, 217–22.

Hunter, R.H.F., Cook, B. & Baker, T.G. (1976). Dissociation of response to injected gonadotrophin between the Graafian follicle and oocyte in pigs. *Nature (London)*, **260**, 156–8.

Hunter, R.H.F., Cook, B. & Baker, T.G. (1985). Intersexuality in five pigs, with particular reference to oestrous cycles, the ovotestis, steroid hormone secretion and potential fertility. *Journal of Endocrinology*, **106**, 233–42.

Hunter, R.H.F. & Polge, C. (1966). Maturation of follicular oocytes in the pig after injection of human chorionic gonadotrophin. *Journal of Reproduction and Fertility*, **12**, 525–31.

Jacobs, P.A. & Strong, J.A. (1959). A case of human intersexuality having a possible XXY sex-determining mechanism. *Nature (London)*, **183**, 302–3.

Jost, A. & Magre, S. (1988). Control mechanisms of testicular differentiation. *Philosophical Transactions of the Royal Society of London, Series B*, **322**, 55–61.

Kan, F.W.K., St-Jacques, S. & Bleau, G. (1988). Immunoelectron microscopic localisation of an oviductal antigen in hamster zona pellucida by use of a monoclonal antibody. *Journal of Histochemistry and Cytochemistry*, **36**, 1441–7.

Kan, F.W.K., St Jacques, S. & Bleau, G. (1989). Immunocytochemical evidence for the transfer of an oviductal antigen to the zona pellucida of hamster ova after ovulation. *Biology of Reproduction*, **40**, 585–98.

Koopman, P., Gubbay, J., Vivian, N., Goodfellow, P. & Lovell-Badge, R. (1991). Male development of chromosomally female mice transgenic for *Sry*. *Nature (London)*, **351**, 117–21.

Koopman, P., Münsterberg, A., Capel, B., Vivian, N. & Lovell-Badge, R. (1990). Expression of a candidate sex-determining gene during mouse testis differentiation. *Nature (London)*, **348**, 450–2.

Lawson, K.A., Dunn, N.R., Roelen, B.A.J., *et al.* (1999). *Bmp*4 is required for the generation of primordial germ cells in the mouse embryo. *Genes & Development*, **13**, 424–36.

Leese, H.J. (1988). The formation and function of oviduct fluid. *Journal of Reproduction and Fertility*, **82**, 843–56.

Mahi-Brown, C.A. (1991). Fertilisation in dogs. In *A Comparative Overview of Mammalian Fertilisation*, ed. B.S. Dunbar & M.G. O'Rand, pp. 281–97. New York: Plenum Press.

McLaren, A. & Monk, M. (1981). X-chromosome activity in the germ cells of sex-reversed mouse embryos. *Journal of Reproduction and Fertility*, **63**, 533–7.

Merchant, H. (1975). Rat gonadal and ovarian organogenesis with and without germ cells. An ultrastructural study. *Developmental Biology*, **44**, 1–21.

Mintz, B. (1962). Experimental study of the developing mammalian egg: removal of the zona pellucida. *Science*, **138**, 594–5.

Mizunuma, H., Liu, X., Andoh, K., *et al.* (1999). Activin from secondary follicles causes small preantral follicles to remain dormant at the resting stage. *Endocrinology*, **140**, 37–42.

Moor, R.M. & Cragle, R.G. (1971). The sheep egg: enzymatic removal of the zona pellucida and culture of eggs in vitro. *Journal of Reproduction and Fertility*, **27**, 401–9.

Moor, R.M. & Warnes, G.M. (1978). Regulation of oocyte maturation in mammals. In *Control of Ovulation*, ed. D.B. Crichton, G.R. Foxcroft, N.B. Haynes & G.E. Lamming, pp. 159–76. London: Butterworth.

Oktay, K., Briggs, D. & Gosden, R.G. (1997). Ontogeny of follicle-stimulating hormone receptor gene expression in isolated human ovarian follicles. *Journal of Clinical Endocrinology and Metabolism*, **82**, 3748–51.

O'Shaughnessy, P.J., McClelland, D. & McBride, M.W. (1997). Regulation of luteinising hormone-receptor and follicle-stimulating hormone receptor mRNA levels during development in the neonatal mouse ovary. *Biology of Reproduction*, **57**, 602–8.

Packer, A.I., Hsu, Y.C., Besmer, P. & Bachvarova, R.F. (1994). The ligand of the c-kit receptor promotes oocyte growth. *Developmental Biology*, **161**, 194–205.

Patsoula, E., Loutradis, D., Drakakis, P., Kallianidis, K., Bletsa, R. & Michalas, S. (2001). Expression of mRNA for the LH and FSH receptors in mouse oocytes and preimplantation embryos. *Reproduction*, **121**, 455–61.

Pesce, M., Farrace, M.G., Piacentini, M., Dolci, S. & De Felici, M. (1993). Stem cell factor and leukemia inhibitory factor promote primordial germ cell survival by suppressing programmed cell death (apoptosis). *Development*, **118**, 1089–94.

Picton, H.M. & Gosden, R.G. (1999). Oogenesis, in mammals. In *Encyclopedia of Reproduction*, vol. 3, ed. E. Knobil & J.D. Neill, pp. 488–97. San Diego & London: Academic Press.

Pincus, G. & Enzmann, E.V. (1935). The comparative behaviour of mammalian eggs *in vivo* and *in vitro*. I. The activation of ovarian eggs. *Journal of Experimental Medicine*, **62**, 665–75.

Polge, C. & Dziuk, P. (1965). Recovery of immature eggs penetrated by spermatozoa following induced ovulation in the pig. *Journal of Reproduction and Fertility*, **9**, 357–8.

Richards, J.S., Jahnsen, T., Hedin, L., *et al.* (1987). Ovarian follicular development: from physiology to molecular biology. *Recent Progress in Hormone Research*, **43**, 231–70.

Ross, G.T. (1974). Gonadotropins and preantral follicular maturation in women. *Fertility and Sterility*, **25**, 522–43.

Sawicki, J.A., Magnuson, T. & Epstein, C.J. (1981). Evidence for expression of the paternal genome in the two-cell mouse embryo. *Nature (London)*, **294**, 450–451.

Scaramuzzi, R.J., Adams, N.R., Baird, D.T., *et al.* (1993). A model for follicle selection and the determination of ovulation rate in the ewe. *Reproduction, Fertility and Development*, **5**, 459–78.

Schultz, R.M. & Wassarman, P.M. (1977). Biochemical studies of mammalian oogenesis: protein synthesis during oocyte growth and meiotic maturation in the mouse. *Journal of Cell Science*, **24**, 167–94.

Sinclair, A.H., Berta, P., Palmer, M.S., *et al.* (1990). A gene from the human sex-determining region encodes a protein with homology to a conserved DNA binding motif. *Nature (London)*, **346**, 240–4.

Strömstedt, M. & Byskov, A.G. (1999). Oocyte, mammalian. In *Encyclopedia of Reproduction*, vol. 3, ed. E. Knobil & J.D. Neill, pp. 468–80. San Diego & London: Academic Press.

Sturgis, S.H. (1961). Factors influencing ovulation and atresia of ovarian follicles. In *Control of Ovulation*, ed. C.A. Villee, pp. 213–18. Oxford & New York: Symposium Publications Division, Pergamon Press.

Swain, A., Zanaria, E., Hacker, A., Lovell-Badge, R. & Camerino, G. (1996). Mouse *Dax-1* expression is consistent with a role in sex determination as well as in adrenal and hypothalamus function. *Nature Genetics*, **12**, 404–9.

Swain, A., Narvaez, V., Burgoyne, P., Camerino, G. & Lovell-Badge, R. (1998). *Dax*1 antagonises *Sry* action in mammalian sex determination. *Nature (London)*, **391**, 761–7.

Szöllösi, D.G. (1967). Development of cortical granules and the cortical reaction in rat and hamster eggs. *Anatomical Record*, **159**, 431–46.

Szöllösi, D. (1975). Ultrastructural aspects of oocyte maturation and fertilisation in mammals. In *La Fécondation*, ed. C. Thibault, pp. 13–35. Paris: Masson.

Szöllösi, D. & Hunter, R.H.F. (1978). The nature and occurrence of the acrosome reaction in spermatozoa of the domestic pig, *Sus scrofa*. *Journal of Anatomy*, **127**, 33–41.

Thibault, C. & Gérard, M. (1970). Facteur cytoplasmique nécessaire à la formation du pronucleus mâle dans l'ovocyte de lapine. *Compte Rendu de l'Académie des Science, Paris*, **270**, 2025–6.

Thibault, C. & Gérard, M. (1973). Cytoplasmic and nuclear maturation of rabbit oocytes *in vitro*. *Annales de Biologie Animale, Biochimie et Biophysique*, **13**, 145–56.

Thomson, A. (1919). The ripe human Graafian follicle, together with some suggestions as to its mode of rupture. *Journal of Anatomy*, **54**, 1–40.

Toneta, S.A. & DiZerega, G.S. (1989). Intragonadal regulation of follicular maturation. *Endocrine Reviews*, **10**, 205–29.

Ullmann, S.L. (1989). Ovary development in bandicoots: sexual differentiation to follicle formation. *Journal of Anatomy*, **165**, 45–60.

Upadhyay, S. & Zamboni, L. (1982). Ectopic germ cells: natural model for the study of germ cell sexual differentiation. *Proceedings of the National Academy of Sciences, USA*, **79**, 6584–8.

Vainio, S., Hekkilä, M., Kispert, A., Chin, N. & McMahon, A.P. (1999). Female development in mammals is regulated by *Wnt-4* signalling. *Nature (London)*, **397**, 405–9.

Wagh, P.V. & Lippes, J. (1993). Human oviductal fluid proteins. V. Identification of human oviductin-1 as alpha-fetoprotein. *Fertility and Sterility*, **59**, 148–56.

Wandji, S.A., Pelletier, G. & Sirard, M.A. (1992). Ontogeny and cellular localisation of ^{125}I-labelled insulin-like growth factor-1, ^{125}I-labelled follicle stimulating hormone, and ^{125}I-labelled human chorionic gonadotropin binding sites in ovaries from bovine fetuses and neonatal calves. *Biology of Reproduction*, **47**, 814–22.

Wassarman, P.M. (1988). Zona pellucida glycoproteins. *Annual Review of Biochemistry*, **57**, 415–42.

Wassarman, P.M. & Albertini, D.F. (1994). The mammalian ovum. In *The Physiology of Reproduction*, ed. E. Knobil & J.D. Neill, pp. 79–122. New York: Raven Press.

Wassarman, P.M., Liu, C. & Litscher, E.S. (1997). Building the mammalian egg zona pellucida: an organelle that regulates fertilisation. In *Microscopy of Reproduction and Development: A Dynamic Approach*, ed. P.M. Motta, pp. 169–76. Rome: Antonio Delfino Editore.

Webb, R., Gosden, R.G., Telfer, E.E. & Moor, R.M. (1999). Factors affecting folliculogenesis in ruminants. *Animal Science*, **68**, 257–284.

Welshons, W.J. & Russell, L.B. (1959). The Y-chromosome as the bearer of male determining factors in the mouse. *Proceedings of the National Academy of Sciences, USA*, **45**, 560–6.

Witschi, E. (1948). Migration of the germ cells of human embryos from the yolk sac to the primitive gonadal folds. *Contributions to Embryology of the Carnegie Institution*, **32**, 67–80.

Witschi, E. (1951). Embryogenesis of the adrenal and the reproductive glands. *Recent Progress in Hormone Research*, **6**, 1–23.

Yanagimachi, R. (1988). Mammalian fertilisation. In *The Physiology of Reproduction*, vol. 1, ed. E. Knobil & J.D. Neill, pp. 135–85. New York: Raven Press.

Yanagimachi, R. (1994). Mammalian fertilisation. In *The Physiology of Reproduction*, 2nd edn, ed. E. Knobil & J.D. Neill, pp. 189–317. New York: Raven Press.

Yoshida, H., Takakura, N., Kataoka, H., Kunisada, T., Okamura, H. & Mishikawa, S.I. (1997). Stepwise requirement of c-kit tyrosine kinase in mouse ovarian follicle development. *Developmental Biology*, **184**, 122–37.

Zamboni, L. & Upadhyay, S. (1983). Germ cell differentiation in mouse adrenal glands. *Journal of Experimental Zoology*, **228**, 173–93.

III

Physiology of the ovaries and maturing Graafian follicles

Introduction

Although the microvasculature and lymphatic networks of ovarian follicles have attracted attention as possible routes of inter-follicular communication, further interest could profitably focus on the pattern of innervation of developing follicles. Viewed from the perspective of a general physiologist, there has always been a suspicion that innervation of gonadal tissues would contribute importantly to vital functions, if perhaps less obviously than pituitary trophic hormones. In terms of an overall integration of ovarian activity, and not least the key processes of follicular recruitment, development and ovulation, the autonomic nervous system could certainly be involved in modulating such essential reproductive events. Nervous and endocrine programming of ovarian tissues would thus be functioning in a complementary fashion, as in various other organ systems. However, a perennial objection to this interpretation comes from transplantation evidence. Detachment and relocation of an ovary involving severing tissues in the mesovarium or below, as in the transplantation model of Goding, McCracken & Baird (1967), did not curtail ovarian function. More recent studies involving grafting of frozen–thawed ovarian tissue in mice (Gosden, 2000; Oktay, Newton & Gosden, 2000) and sheep (Gosden *et al.*, 1994; Baird *et al.*, 1999) make essentially the same point. A widespread conclusion has therefore been that innervation is not required for function of ovarian tissues, at least in the short term. This apparently logical interpretation has prevented many reproductive biologists from reflecting more imaginatively on a potential contribution of ovarian innervation to physiological events. Such reservations are responded to later in this chapter before moving on to the exciting topic of intra-ovarian temperature gradients – another field in which a rôle for the autonomic nervous system could be invoked.

After reviewing diverse components of ovarian physiology, attention finally turns to consideration of dimensions. In particular, the question is posed as to why a Graafian follicle grows so large relative to the size of its oocyte. To highlight the nature of this conundrum, why does a human pre-ovulatory follicle need to achieve a diameter of 20–22 mm or an equine follicle a diameter of 45–50 mm if the oocyte to be released is only some 100–120 μm in diameter? In seeking a meaningful response, one is prompted to consider the process of follicular maturation from various angles.

Ovarian vasculature

The ovarian blood supply becomes distinguishable in the fetus at an early stage of sexual differentiation when blood vessels from the mesonephric region gradually colonise the developing cortex, having entered the medulla through the hilus. Capillary networks are noted in the cortical connective tissue during establishment of the primordial follicles, and become progressively more embracing with the appearance of secondary follicles. Even though blood vessels were seen in close proximity to primordial follicles, specific vascularisation of such follicles was not found in women (dos Santos-Ferreira *et al.*, 1974) nor in pigs (Andersen, 1926), rats (Bassett, 1943) or rabbits (Macchiarelli *et al.*, 1991) do such follicles show an organised capillary network. By the time of birth, the ovarian vasculature has assumed its definitive form, although further growth and development will of course occur before puberty. Puberty heralds a marked increase in ovarian angiogenesis, not least due to the local and systemic influence of steroid hormone secretion and especially due to a local influence of peptide growth factors (Redmer & Reynolds, 1996).

Although there are subtle regional variations, the overall pattern of blood supply to and from the ovaries is reasonably similar between species. The ovarian artery, which stems directly from the aorta in women, enters the mesovarium through the infundibulo-pelvic (lumbo-ovarian) ligament as the main source of blood supply to the ovary (Fig. III.1). Thereafter, the principal artery courses through the entire length of the broad ligament to anastomose with the uterine artery (Figs. III.2 and III.3), which can thus be considered as the secondary blood supply of the ovary (Reynolds, 1973; Einer-Jensen, Cicinelli & Galantino, 2000). The relative contribution of the individual arterial supplies depends on physiological circumstances, especially the stage of the menstrual or oestrous cycle or the different stages of pregnancy (Gillet *et al.*, 1968; Gillet, 1971; Wehrenberg *et al.*, 1977); it also depends on species. For example, the ovarian artery is relatively small in rhesus monkey, most of the blood supply

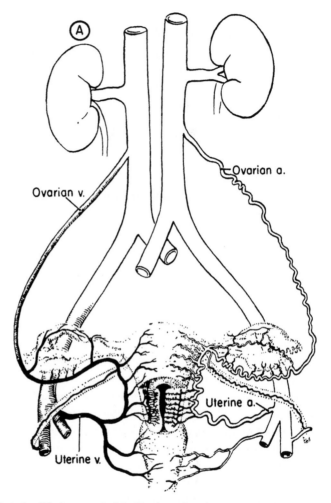

Fig. III.1. A simplified portrayal of the blood supply to human ovaries from both the ovarian and uterine arteries. v, vein; a, artery. (Adapted from Reynolds, 1973.)

to the gonad being said to come from the uterine artery, except apparently in late pregnancy (Wehrenberg *et al.*, 1977; Edwards, 1980); this would be worth further study. More specifically, 95% (range 79–100%) of the ovarian blood supply in rhesus monkey comes from the uterine artery during menstrual cycles and early pregnancy. This distribution changes significantly in late pregnancy, decreasing to a mean of 9% with compensation from the ovarian artery, as judged by a labelled microsphere technique (Wehrenberg *et al.*, 1977).

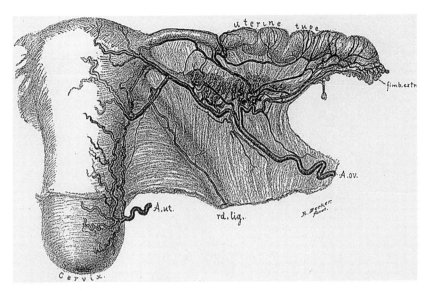

Fig. III.2. Illustration of the ovarian and uterine arteries and their relationship with the vascular arcade of the Fallopian tube (labelled uterine tube) in a newborn child. The spiral ovarian vessels are clearly distinguishable. (Adapted from Clark, 1900.)

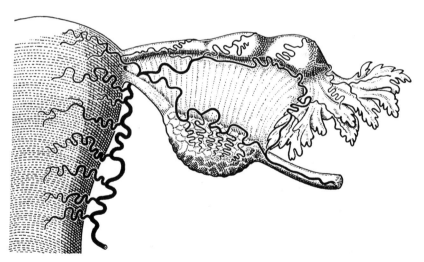

Fig. III.3. A finely drawn if slightly generalised portrayal of the ovarian and uterine blood supply, showing the anastomosis between uterine and ovarian arteries and also arterial supplies to the Fallopian tube. (Adapted from various sources, but principally from Koritké, Gillet & Pierti, 1967.)

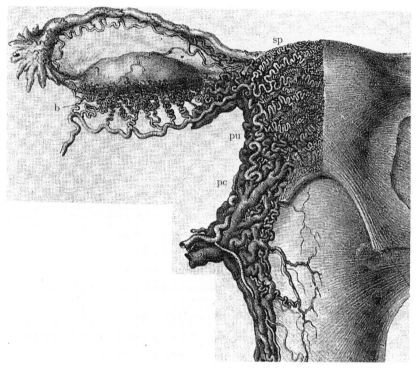

Fig. III.4. An excellent representation of the ovarian arteries and adjoining vessels in human tissue. A contribution from the uterine artery is clearly distinguishable, and the arterial supply to the Fallopian tube is well shown. (Adapted from Calza, 1807.)

In rabbits, the ovarian and uterine arterial systems remain more distinct and separate than in other species so far examined. However, in cattle, guinea-pigs, rats and hamsters, there are prominent anastomoses between the uterine and ovarian vascular systems. And, in the guinea-pig at least, the ovary is entirely supplied with arterial blood via the ovarian artery throughout the oestrous cycle and during early pregnancy (Chaichareon, Rankin & Ginther, 1976). In sheep, the ovarian branches of the ovarian artery are distinct, if remarkably tortuous (Lee & O'Shea, 1976), whilst functional arterio-venous anastomoses have been noted in the ovary itself (Mattner, Brown & Hales, 1981).

The main ovarian artery follows a relatively winding path alongside the organ in women, and small spiral arteries – commonly some 6–10 – penetrate into the ovary through the hilus (Figs. III.3 and III.4). They themselves do not anastomose with neighbours but branch extensively to form a plexus in the medulla. Spiral arterioles arise in turn from these medullary vessels to colonise the cortex, enabling dense networks of capillaries to form around growing follicles and within the theca layers. A fully meaningful interpretation for the existence of ovarian spiral arteries is uncertain, but they are thought to facilitate appropriate blood flow under conditions of changing ovarian size and structure – the remodelling of ovarian tissues that occurs throughout the menstrual cycle and during early pregnancy. A spiral artery and arterioles may also help to decrease blood pressure (pulse pressure) from the relatively high level in the ovarian artery, frictional considerations along the wall of the lengthened spiral being significant for achieving such a putative reduction (Delson, Lubin & Reynolds, 1949; Reynolds, 1973).

The venous return from the ovary is seen in part as a pampiniform plexus – several highly-coiled vessels within the broad ligament – close to where the veins emerge from the hilus (Fig. III.5). Drainage is also via the uterine vein, since a common utero-ovarian venous trunk can be demonstrated in many mammals, although seemingly not in the rabbit (Reynolds, 1973; Macchiarelli, Nottola & Motta, 1997). Asymmetry in the disposition of the ovarian veins is found in the sense that an ovarian vein enters the vena cava directly on the right-hand side of the body whereas, on the left, it enters the renal vein before joining the vena cava.

A fascinating aside in some of the literature from the nineteenth and early years of the twentieth century, reviewed by Mezzogiorno, Fusaroli & Cavallucci (1967), is the suggestion that the ovary could be considered as an erectile organ. This would be on the grounds of its engorged pampiniform plexus of blood vessels in the hilus. This interpretation was reached in part on the basis of injecting the blood vessels in autopsy specimens and noting the increase in volume of the ovarian tissues. Smooth muscle elements around the ovarian veins would act to impede ovarian drainage and were thereby suggested to contribute to the process of ovulation – 'Celle-ci pourrait contribuer à provoquer la rupture des follicules murs' (Prénant & Bouin, 1911). The idea was very much abroad at that time that ovulation really did represent the rupture of a Graafian follicle, being so to speak an explosive process.

The ovarian stroma shows an extensively branched capillary network (Fig. III.6). In the presence of growing follicles, a network surrounds each spherical structure in the form of a basket or crown (Andersen, 1926; Gillet *et al.*, 1968; Macchiarelli *et al.*, 1993). As already noted in Chapter II, the granulosa cell layers of a developing follicle are not vascularised until shortly before ovulation whereas cells of the theca layers have a prominent capillary network that drains into a corresponding venous system. Such theca capillaries become especially dense and conspicuous in the hours leading up to ovulation. Vascular permeability undergoes critical changes at this final stage (Cavender & Murdoch, 1988).

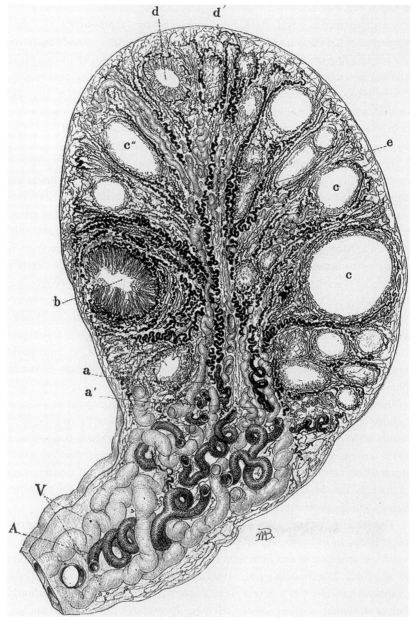

Fig. III.5. A superb drawing of a human ovary in section highlighting the arterial supply to Graafian follicles at various stages of development and showing, in particular, the pampiniform arrangement of artery (A) and veins (V) in the ovarian pedicle. (After Clark, 1900.)

Fig. III.6. To show the wreath or network of capillaries around individual Graafian follicles in the rabbit that may permit a counter-current exchange of follicular molecules. (Courtesy of Professors G. Macchiarelli and S. Nottola.)

Concerning the microcirculation in the ovarian cortex, the coiled arteries branch into numerous thin, straight or slightly undulating arterioles that supply developing follicles. The venous drainage closely follows the arterial supply. In the outer medulla and hilus, a thin venous meshwork closely envelops the arterial coils, thereby establishing numerous arterio-venous contacts. These may represent morphological devices for a local recirculation of steroid and peptide hormones, i.e. a counter-current feedback mechanism (Bendz *et al.*, 1982; Macchiarelli *et al.*, 1997). There were the further suggestions that a counter-current mechanism for hormone exchange could enable local communication between different compartments of the ovary (Bendz *et al.*, 1982) and that such a mechanism might offer a vascular means of local modulation in the control of recruitment and selection of follicles (Macchiarelli *et al.*, 1997). Perhaps the overall point is that local vasodynamics should be considered as an integral part of any model of ovarian function, reference here being to (1) changes in morphology of blood vessels, (2) changes in blood flow, and (3) changes in local transfer activity.

Using the rabbit as an experimental model, vascular baskets of different size and architecture could be correlated with follicles in various stages of development in the ovarian cortex. Small (primary) follicles showed simple thread-like capillaries whereas larger (secondary and antral) follicles showed a progressive increase in the number, size and tortuosity of round-meshed capillaries adapted to growing follicles. Interstitial stromal capillaries could be directly involved in the initial development of recruited follicles (Macchiarelli *et al.*, 1991, 1993).

In a much earlier, now classical, report dealing with pig ovaries, Andersen (1926) noted that the very youngest follicles have no special vascular system of their own: in the words of Bassett (1943), writing of rat ovaries, just a random association with inconspicuous capillaries. However, soon after formation of the follicular antrum, a one-layered wreath of blood capillaries could be distinguished. Once the theca interna has differentiated from the theca externa, a double wreath of blood vessels is apparent. The outer wreath consists of arterioles and venules in the theca externa, with an inner wreath – a capillary network – prominent in the theca interna. It is from this inner wreath of capillaries that the granulosa cell layer commences to be colonised in Graafian follicles on the verge of ovulation. This description of follicular wreaths in the pig was matched almost exactly in the subsequent description of Bassett (1943) for the rat ovary.

In the pre-ovulatory follicle itself, the blood supply is directed mainly to the wreath of vessels that lies along the inner border of the theca interna, which enables delivery of nutrients and substrate to the steroidogenically active cells of the theca and by diffusion to the avascular granulosa layer and to the oocyte

(Macchiarelli *et al.*, 1993). The ovarian microvasculature responds to oestrogen administration by prompt vasodilation – hyperaemia – attributed to local release of histamine-like substances (see Chapter VIII).

Ovarian and neighbouring lymphatic pathways

The lymphatic vessels of human ovaries received detailed description in the works of Poirier & Charpy (1901) and Polano (1903), and were covered briefly in the review by Andersen (1926). In the earlier work, there was some controversy as to the extent of the lymphatic bed around individual Graafian follicles, its relationship with follicular blood vessels, and the course of lymphatic drainage. Such matters have long since been resolved. In essence, Poirier & Charpy (1901) considered antral follicles to be surrounded by a sinus of lymphatics from which vessels drained into the lymphatics of the hilus and so into six or eight trunks alongside the veins. Some of these were noted to join the lymphatics of the Fallopian tube and uterine fundus to drain into lymph nodes anterior to the aorta. By contrast, Polano (1903) had reservations as to whether lymphatics completely surrounded antral follicles, whether they coursed alongside blood vessels and opened into lymph spaces, and whether they joined the lymphatics of the Fallopian tube and uterus. As far as the present discussion is concerned, Polano's most serious reservation centred on the existence of lymphatic pathways in the theca interna. Nonetheless, his macroscopic description is instructive and was composed – it should be recalled – a full century ago.

> The course of the lymph vessels is radiating, flowing toward the hilum, which they leave by larger valved vessels along the free edge of the ovarian–pelvic ligament. Anastomosis with the uterine or tubal lymph vessels is only exceptionally found. The distribution in the interior of the ovary shows that the cortical layer has in general less vessels than the medulla. Moreover in the parenchymatous zone wherever follicles or their end products are, one finds a deviation of the radiating vessels to form a circle.

Turning to more recent descriptions, Jdanov (1960, 1969) confirmed that human antral follicles were surrounded by a double network of lymphatic capillaries, an inner layer of small loops and an outer layer of larger loops of capillaries. Examining ovaries taken from mature women, Jdanov noted that primary follicles were not yet braided by networks of lymphatic capillaries, although the cortex was permeated by such networks. Development of small Graafian follicles elicited growth of small lymph networks in the cortex, and these served to establish a primitive system of lymphatic vessels in the ovary. The lymphatic capillaries of the external network were connected to neighbouring lymphatic capillaries and with the vessels of the cortex. Such lymphatic channels tended to follow the course of blood vessels. Andersen (1926, 1927) was perhaps the first to suggest a cyclical variation in the activity of the ovarian and neighbouring tubal lymphatic system, with vessels appearing largest in the follicular phase, a point taken up by Reiffenstuhl (1964) and Jdanov (1969). Cyclical variation received substantial emphasis in the review by Reynolds (1973), who commented that changes in the microcirculation were related to ovarian endocrine changes.

The lymphatic network of and below the hilus represents the primary drainage pathway for ovarian lymph. Based upon a close reading of Mossman & Duke (1973), ovarian lymphatics were said to drain fluid away from vascularised beds of connective tissue and from the theca of large follicles (Edwards, 1980). In passing, Edwards (1980) proposed that lymphatics might also be involved in the formation of an antrum in the expanding follicle. Quite apart from an obvious rôle in transporting fluid, the lymphatic system has attracted special interest because of its potential for transmitting endocrine information between neighbouring tissues and organs. A lymphatic capillary plexus drains along the ovarian vessels to the middle lumbar nodes and via uterine anastomoses to the sacral nodes (Fig. III.7), and the latter may facilitate the 'transmission of compounds' from the uterus to the ovary (Mossman & Duke, 1973; Edwards, 1980) if there were a retrograde flow of lymph or further transfer to an artery. Bearing in mind the usual $3-3^1/_2$ days' residence of a human embryo in the Fallopian tubes, it is further worth noting that the tubal lymph vessels anastomose with those of the ovary in the sub-ovarian plexus. And, of course, inter-follicular communication within the same ovary may utilise lymphatic pathways for highly sensitive local modulation.

Andersen's (1926) work on the domestic pig noted a lymphatic network of larger and smaller vessels, all of them with valves, draining the ovarian stroma more or less evenly. Maturing follicles had many large lymph vessels forming a double wreath in the thecal layers and joining lymphatics from the stroma to produce an extensively anastomosing network of large and irregular vessels (Fig. III.8), eventually draining to the hilus. Lymphatics were found around arteries throughout the ovary (Fig. III.9), but, wherever a lymph vessel ran close to a follicle, it was flattened against it, appearing elliptical or even biconcave in cross-section. As to follicles themselves, lymphatic vessels were initially found to lag behind blood vessels in their growth (perhaps due to technical difficulties of demonstration), although they followed essentially the same course. Lymph vessels accompanied and surrounded most of the arterioles of the developing theca externa, and only once this stage was achieved did lymphatics begin

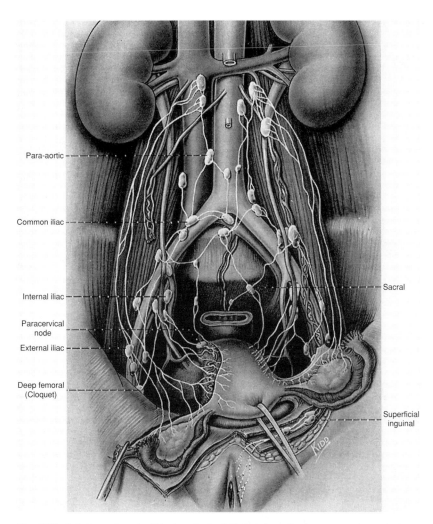

Fig. III.7. Artistic portrayal of the lymphatic drainage from the human ovaries and uterus coursing towards diverse nodes including the iliac, sacral, para-cervical and para-aortic. The internal iliac nodes are also termed hypogastric. (Adapted from Edwards, 1980.)

to form in the theca interna. Valves were never seen in any of the lymphatic branches in the theca interna. Graafian follicles possessed a complete wreath of lymphatics in the theca interna, an irregular network of diamond-shaped meshes connected to the lymphatics of the theca externa, which, in turn, anastomosed with lymphatics of the stroma.

A detailed study by Morris & Sass (1966) in sheep found close parallels with pig ovarian lymphatics as described by Andersen (1926). Stromal lymphatic

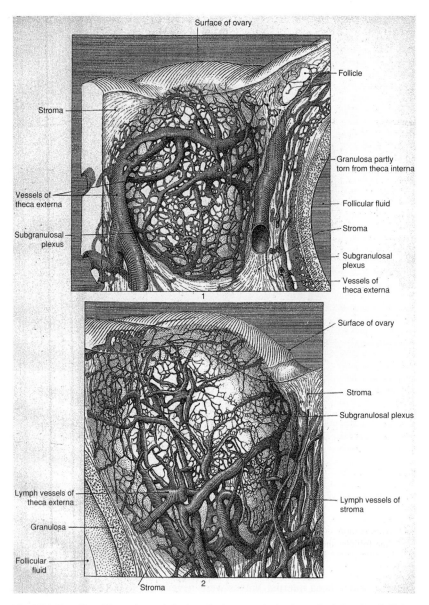

Fig. III.8. Excellent illustrations of the lymphatic networks in the wall of Graafian follicles prepared from tissues of domestic pigs. An impression can be gained from the lower picture of lymphatic capillaries developing as angiogenesis proceeds apace in the mature follicle. (After Andersen, 1926.)

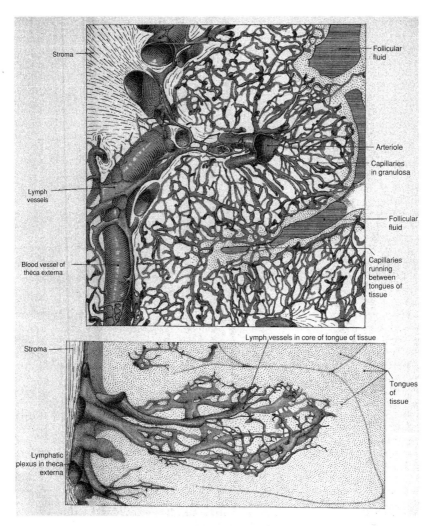

Fig. III.9. A more detailed illustration of the lymphatic vessels draining the wall of a mature Graafian follicle, and showing the intimate relationship with blood vessels in the wall of the follicle. In the lower picture, there is a suggestion of proliferation of lymphatic capillaries. (After Andersen, 1926.)

vessels were prominent in mature ovaries, as were wreaths of vessels in the theca layers of Graafian follicles (Fig. III.10). The basket-like network did not penetrate through the basement membrane to the granulosa cells but remained in close proximity to capillaries in the theca layers. In transverse section, lymphatics around larger follicles often appeared flattened or slit-like. This distortion

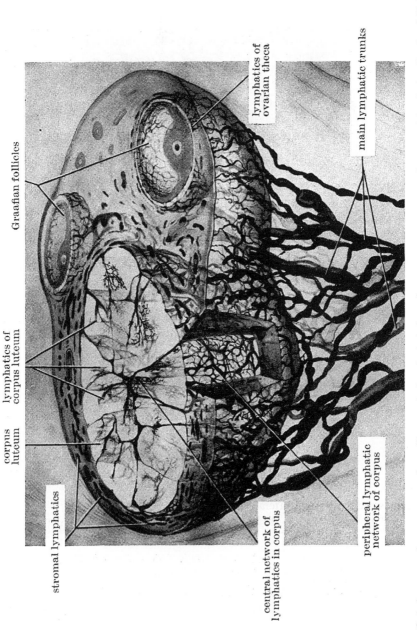

Graafian follicles

lymphatics of
ovarian theea

main lymphatic trunks

corpus
luteum

lymphatics of
corpus luteum

stromal lymphatics

central network of
lymphatics in corpus

peripheral lymphatic
network of corpus

Fig. III.10. A sheep ovary in transverse section to illustrate the lymphatic drainage from stroma, mature corpus luteum and developing Graafian follicles. Lymph vessels drain the thecal layers of a follicle. The principal lymphatic trunks are in close association with the ovarian blood supply. (After Morris & Sass, 1966.)

Fallopian tube

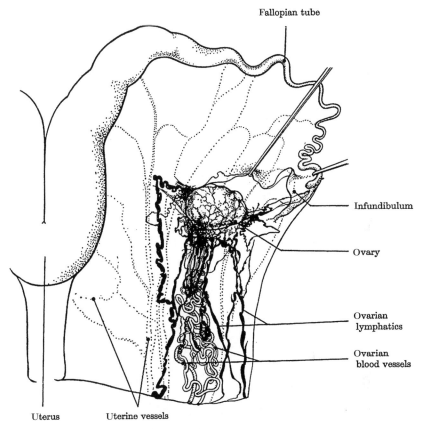

Infundibulum

Ovary

Ovarian
lymphatics

Ovarian
blood vessels

Uterus   Uterine vessels

Fig. III.11. Portrayal of the sheep reproductive tract, ovary and supporting mesenteries, largely mesometrium, to show the principal lymphatic vessels draining the ovary and the associated blood vessels. (After Morris & Sass, 1966.)

was presumed to be due to pressure exerted by expanding follicles, but might have represented a technical artefact.

Morris & Sass (1966) offered greater anatomical detail than hitherto on the principal lymphatic trunks draining the ovary (Fig. III.11), and also presented details as to the rate and direction of lymphatic flow and steroid hormone content. The lymphatics that drain the ovary are mainly associated with the utero-ovarian vascular pedicle, and closely applied to the veins of the ovarian plexus (Lindner, Sass & Morris, 1964). Typically, there were four or five major lymph vessels sited just below the ovary in the sub-ovarian plexus, and seemingly anastomosing freely near the hilus. These sources of ovarian drainage were joined by lymphatics from the proximal half of the Fallopian tube, again

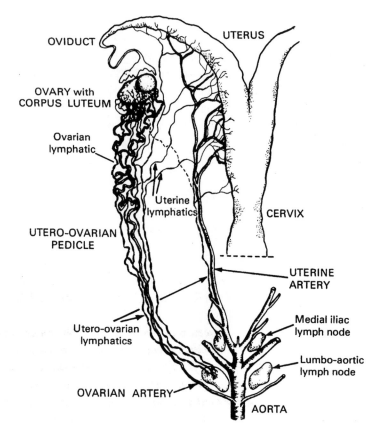

Fig. III.12. Further portrayal of sheep reproductive system to show the relationship between ovarian lymphatic drainage and that of the genital tract together with the corresponding major lymph nodes. (Adapted from Staples *et al.*, 1982.)

perhaps providing a route for local modulation of ovarian events by an embryo. On the other hand, there was no evidence for retrograde lymph flow from the uterus to the ovaries (Morris & Sass, 1966; Staples, Fleet & Heap, 1982; Fig. III.12), but embryonic programming of ovarian physiology could perhaps involve a counter-current transfer from a lymph vessel into the ovarian arterial system similar to that shown for $PGF_{2\alpha}$ to cause luteolysis (see Heap, Fleet & Hamon, 1985). If so, unilateral influences would need to be considered.

Ovarian lymphatics have been examined in other mammalian species, but such further detail does not change an overall picture of the follicular wall in particular. For example, once rabbit follicles have achieved a developmental stage of two or more layers of granulosa cells, then relatively large lymphatic vessels commence to be distinguishable in the theca layers (Täher, 1964). Thus it would seem that development of a follicular system of lymphatics follows that of the theca capillary bed, although appearance of lymphatic capillaries invariably lags slightly behind proliferation of the blood vascular system.

Readers seeking a broader account of the lymphatic system as it bears on ovarian physiology should consult Wislocki & Dempsey (1939), Burr & Davies (1951), Reiffenstuhl (1964), Mossman & Duke (1973), Reynolds (1973) and Ellinwood, Nett & Niswender (1978).

Ovarian innervation, especially of follicles

Much work during the past 30 years dealing with ovarian innervation by both sympathetic and sensory neurons has come from the Scandinavian school of Owman, Sjöberg, Walles and colleagues. A brief summary can be found in Sjöberg, Bodelsson & Stjernquist (1997), and a useful coverage of earlier studies bearing on human tissues is provided by Edwards (1980). Inter-species comparisons of ovarian innervation have been presented by Keller (1966) and Harrison & Weir (1977), and patterns of innervation in several species of mammal were also noted by Owman & Sjöberg (1966), Reynolds (1973) and Burden (1978). Modern concepts have been summarised by Mayerhofer *et al.* (1997) and Dissen & Ojeda (1999).

The ovary has a source of innervation distinct from that of the adjoining reproductive tract. Whereas the smooth muscle coats of the Fallopian tube, uterus and vagina receive a mixture of short adrenergic, cholinergic and peptidergic nerves from the utero-vaginal ganglion formations and from fibres originating in the inferior mesenteric ganglion (Fig. III.13), the ovary is supplied primarily from the aortico-renal ganglion complex (Owman & Sjöberg, 1966; Owman *et al.*, 1992). Extrinsic innervation is principally by sympathetic and sensory fibres, with a small parasympathetic component. More specifically, ovarian innervation is by fibres from the tenth and eleventh thoracic segments of the spinal cord, and also from the lumbo-sacral region (Fig. III.13). Preganglionic parasympathetic fibres from the lumbo-sacral region enter the ovary through the hilus, coursing alongside the utero-tubal artery close to where it anastomoses with the ovarian vessels. Preganglionic fibres from the spinal cord and postganglionic fibres from the ovarian and celiac plexuses are sympathetic. A few nerve bundles are found in the ovary while small fibres in the parenchyma consist mainly of non-myelinated adrenergic fibres. Nerves in the major veins of the hilus that end sub-epithelially were thought to control blood flow (Keller, 1966), although this function is generally ascribed to arterioles.

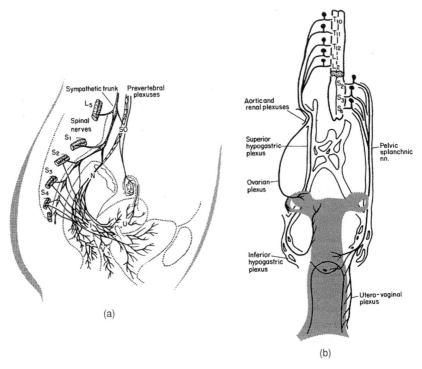

Fig. III.13. Semi-diagrammatic portrayal of innervation of the female reproductive organs, with some detail discernible for the ovary. nn, nerves. (Adapted from Edwards, 1980.)

In the review by Dissen & Ojeda (1999), a contrast was drawn between innervation of the rat ovary and that of the human, but differences could be brought out between various species. Of the two major nerves to the rat ovary, the ovarian plexus nerve runs along the ovarian artery whereas the superior ovarian nerve is associated with the suspensory ligament. The parasympathetic innervation of the rat ovary is from the vagus (Burden *et al.*, 1981). The second nerve supply to the human ovary, by contrast, is the hypogastric nerve, subdivided into several bundles that reach the ovary via the broad ligament. As to the parasympathetic innervation, cell bodies that project to the human ovary are located in sacral segments 2–4 of the spinal cord. Axons from these neurons run in the pelvic nerves and form synapses in the utero-vaginal ganglion.

Various lines of evidence have suggested the existence of a bidirectional neural connection between the hypothalamus and the ovary (Fig. III.14), again usefully summarised by Dissen & Ojeda (1999). For example, electrical stimulation of the anterior hypothalamic region of rats increased secretion of ovarian steroid hormones in the absence of pituitary and adrenal glands. Conversely,

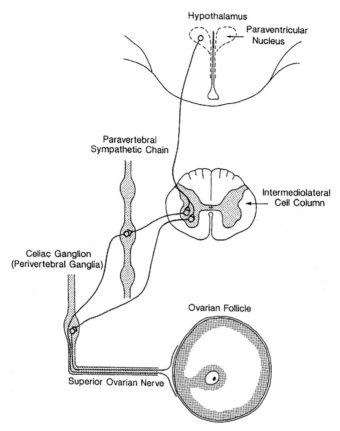

Fig. III.14. Semi-diagrammatic representation of the sympathetic innervation of a mature Graafian follicle, showing proposed pathways connecting with oxytocin-containing neurons in the paraventricular nucleus of the hypothalamus. (After Dissen & Ojeda, 1999.)

electrical lesions of the same region produced ipsilateral changes in the concentration of vasoactive intestinal peptide (VIP) and luteinising hormone-releasing hormone (LHRH) receptors. In the other sense, an influence of the ovary on hypothalamic function via an afferent route was suggested by unilateral changes in LHRH content of the hypothalamus following unilateral ovariectomy. Moreover, the existence of a hypothalamic–ovarian neural connection has been demonstrated using a viral transneuronal tracing method (Lee *et al.*, 1996). Indeed, the labelling evidence indicated that the most important groups of hypothalamic neurons that project to the ovary are those that produce oxytocin. Oxytocin neuronal activity is enhanced during suckling, opening the prospect that a direct trans-synaptic mechanism between the central nervous system and ovary could be closely involved in the widespread phenomenon of lactational anoestrus (Dissen & Ojeda, 1999). In this phase of lactation, Graafian follicles fail to develop to maturity.

In the neonatal ovary, there is a fine plexus of nerves immediately beneath the germinal epithelium. In the adult ovary, the stroma may be randomly innervated, but the initial distribution of nerve fibres suggests a relationship between their presence and the development of follicles (Schultea, Dees & Ojeda, 1992; Mayerhofer *et al.*, 1997). The ovary itself contains adrenergic and cholinergic nerve terminals, principally the former. Some innervate blood vessels, especially the larger vessels, but most are non-vascular and are located in the theca externa (Owman & Sjöberg, 1966; Owman, Rosengren & Sjöberg, 1967). Networks of autonomic fibres are present around growing follicles, particularly in the wall of Graafian follicles, where they innervate cells with contractile elements. Such cells are demonstrable in the theca externa by silver staining and electron microscopy and contain microfilaments together with the contractile proteins actin and myosin (Burden, 1972; Amsterdam, Lindner & Gröschel-Stewart, 1977; Walles *et al.*, 1978). True synapses between the adrenergic nerves and smooth muscle cells were also demonstrated, and Owman & Sjöberg (1966) and Owman *et al.* (1975) suggested that nerve fibres to follicles (Fig. III.15) could be concerned with maturation of the tissues and ovulation.

In the above context, Bahr, Kao & Nalbandov (1974) proposed that adrenergic fibres in the theca might provide direct secretory motor stimulus and/or influence the state of blood vessels around the follicles. As to nerves terminating in the stroma, Neilson *et al.* (1970) and Fink & Schofield (1971) suggested that these might inhibit secretion by the interstitial tissue and thereby could influence the rate of atresia, the selection of follicles for ovulation, and the output of steroid hormones. A modern point of view would be that the extrinsic ovarian nerves are not only involved in the regulation of blood flow but also participate directly in the control of steroidogenesis and follicular development (for a review, see Dissen & Ojeda, 1999). It is appreciated that neurotransmitters contained in ovarian nerves provoke release of steroids via specific recognition molecules on follicular steroidogenic cells.

Not only do appropriate ovarian tissues reveal adrenergic and cholinergic innervation, but peptidergic nerves have also been found in both vascular and non-vascular tissues (see Sjöberg *et al.*, 1997). Coexisting with the classic neurotransmitters, the precise function of VIP, neuropeptide Y (NPY), gastrin releasing peptide (GRP), calcitonin gene-related peptide (CGRP) and other neuropeptides in the ovary is not understood. Nonetheless, there is at least circumstantial evidence that the neuromuscular complex in the wall of Graafian

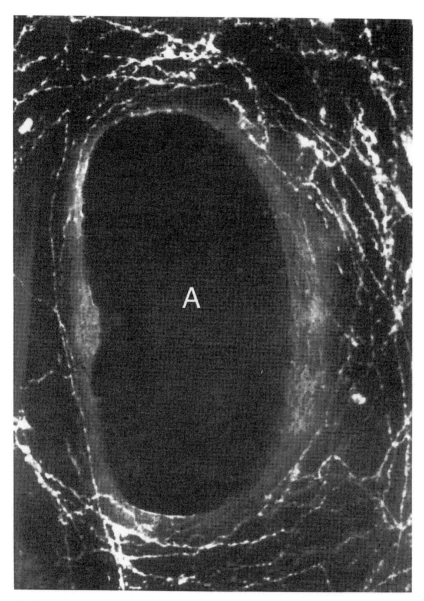

Fig. III.15. Immuno-fluorescence preparation of a mature Graafian follicle to show the dense networks of autonomic nerves in the theca externa. 'A' represents the antrum. (Adapted from Sjöberg *et al.*, 1997; after Owman *et al.*, 1975.)

follicles is involved in the process of ovulation. Contractile elements of the theca externa may not so much act to induce ovulation itself, for Graafian follicles become soft and flaccid in a number of species shortly before ovulation (see Plate 1), as to displace the cumulus–oocyte complex and associated follicular fluid at the time of ovulation. After the ovulatory collapse of the follicle, contraction of smooth muscle cells may assist in stemming haemorrhage from torn vessels in the follicle wall. Such contraction might also contribute to remodelling of the follicular surface around an initially smaller developing corpus luteum.

As to steroidogenesis, norepinephrine (NE, noradrenaline) and VIP stimulate the production of ovarian steroid hormones. Whereas both enhance the secretion of progesterone and androgens, only VIP also stimulates oestradiol production (Dissen & Ojeda, 1999). In addition, however, NE facilitates the stimulating influence of gonadotrophins on ovarian steroidogenesis. Moreover, both NE and VIP promote the biochemical differentiation of newly formed follicles by inducing the formation of FSH receptors, enabling the small follicle to respond to pituitary gonadotrophins (see Mayerhofer *et al.*, 1997). Catecholamines influence ovarian steroidogenesis through activation of β_2-adrenergic receptors, stimulating androgen secretion from theca cells and pre-ovulatory progesterone secretion from granulosa cells. Catecholaminergic sympathetic nerves may amplify the effects of circulating gonadotrophins on ovarian steroidogenesis. VIP stimulates steroid hormone secretion by enhancing the synthesis of all three components of the cholesterol side chain cleavage enzyme complex, as well as by stimulating aromatase activity (for reviews, see Greenwald & Roy, 1994; Driancourt *et al.*, 2001).

Direct evidence for a facilitatory rôle of NE and VIP in follicular development has also been provided. In essence, the experimental approach acted to increase the ovarian content of mRNA encoding the FSH receptor, such mRNA being localised to follicles by *in situ* hybridisation and such receptors being demonstrated to be functional. An overall viewpoint would be that acting via neurotransmitters coupled to the cAMP-generating system, ovarian nerves contribute to the differentiation process by which newly formed primary follicles acquire FSH receptors and responsiveness to FSH (Mayerhofer *et al.*, 1997; Dissen & Ojeda, 1999). In other words, neurotransmitters working via the cAMP-generating system facilitate initial differentiation of granulosa cells and promote the acquisition of gonadotrophin dependency.

The introduction to this chapter refers to a potential involvement of the nervous system in ovarian physiology. Despite the demonstration of function after transplantation of ovarian tissues, this experimental evidence does not preclude a critical involvement of the sensory and autonomic nervous systems in the processes underlying follicular development and indeed in integrating physio-logical events in and between the two ovaries. Of immediate relevance here, ablation of the ovarian sympathetic innervation during neonatal life in the rat, either by preventing development of such innervation by administration of anti-bodies to nerve growth factor (NGF) or by destruction of the sympathetic nerves by treatment with guanethidine, (1) delayed follicular development, (2) reduced ovarian steroid secretion, and (3) prompted irregularities of the oestrous cycle (Dissen & Ojeda, 1999). In particular, the ovaries of sympathectomised rats showed a significant reduction in large antral follicles as compared with nor-mally innervated ovaries of control animals.

As to a vital rôle of ovarian innervation, how should the transplantation evi-dence that ovarian function is not curtailed be interpreted? One line of thought, in reality of reservation, is that to accept the transplantation evidence at face value is to overlook diverse possibilities after the surgical intervention. In particular, these might include: (1) compensatory mechanisms that may have developed after ovarian transplantation, such as increased sensitivity to circulating cate-cholamines (Dissen & Ojeda, 1999); (2) vascular innervation of ovarian tissue assuming a far more important rôle after transection and transplantation than hitherto suspected; (3) specific re-innervation after the transection procedures, the extent and rate of which have yet to be determined. A long-standing pre-sumption is that the rate of innervation would be slow, a presumption that could fruitfully be tested by close examination.

To put matters in perspective, the transplantation evidence is derived from a limited sample of animals, most of which have been used for acute endocrine studies rather than subjected to mating or insemination (with the transplanted organ in a suitable site). If a population of animals (e.g. >100) bearing appro-priately transplanted ovaries were compared for reproductive efficiency with a population of control animals, it would be surprising if the fecundity of the former were to match that of the latter. Innervation of an individual Graafian follicle may not be essential for the process of ovulation but could, as suggested, contribute to the physiological programming of this vital event. It might also influence the diverse functions and destiny of such a complex structure.

Ovarian blood flow

The rate of flow in the ovarian blood vessels varies according to the stage of the oestrous or menstrual cycle, being influenced in this context principally by the prevailing ratio of oestradiol to progesterone secretion. Blood flow per unit weight of ovarian tissue is high relative to other organs (Reynolds, 1973), and notably so at the stage of maximum endocrine activity (Ellinwood *et al.*,

1978; Einer-Jensen & McCracken, 1981). Gonadotrophic hormones, especially LH, stimulate a marked increase in ovarian blood flow. The pre-ovulatory LH surge and resultant secretion of progesterone stimulate blood flow, with the corpus luteum being a specific focus, at least until the onset of luteal regression (see Rathmacher & Anderson, 1968). Prompted by gonadotrophins, angiogenic factors are critically involved in extending the vascular bed. Luteolysins such as $PGF_{2\alpha}$ act to reduce blood flow to the corpus luteum by means of vaso-constriction and redistribution within the ovarian microcirculation (Novy & Cook, 1973). Relevant to the overall theme of this book, the flow of blood seems not to be directed preferentially towards a pre-ovulatory follicle, although confusion in the literature exists on this point, doubtless a reflection of the techniques employed. However, a dilatation of blood vessels in the wall of pre-ovulatory follicles is induced by LH, which should decrease resistance and enable a specifically enhanced perfusion through the expanding vascular network.

One of the earliest studies on blood flow to the ovaries was made in rats (Zondek, Sulman & Black, 1945). In this work, an increased ovarian blood flow was noted at the time of ovulation, with hyperaemia of follicles being reported as one response to the influence of gonadotrophic hormones. Since then, it has been appreciated from many studies that LH or human chorionic gonadotrophin (hCG) prompt marked increases in total ovarian blood flow (Niswender *et al.*, 1976) and act specifically to induce vasodilation (Janson, 1975). Proximal causes for the increased ovarian blood flow close to the time of ovulation may include histamine (see Reynolds, 1973) and nitric oxide (see Chapters V and VIII), but an involvement of the autonomic nervous system must also be considered in terms of catecholamines.

An extensive series of measurements exist for ovarian blood flow in sheep, but it should be noted straight away that techniques employing both anaesthesia and venous cannulation will themselves influence the rate of flow. Mattner & Thorburn (1969) measured ovarian blood flow in sheep by cannulating the principal veins in both ovaries and tying off the subsidiary vessels. There was a sustained high level of flow during the luteal phase in both the ovulating and non-ovulating ovaries. However, a major difference was that the ovulating ovary had rates of flow approximately double those of the contralateral ovary from the fourth to the twelfth day of the oestrous cycle; that is, during the mid-luteal phase. Peaks of flow were noted at oestrus and towards the end of the luteal phase, both associated with secretion of oestradiol-17β.

When reviewing such material, it is important to distinguish between total ovarian blood flow (including the activity of shunts; Mattner *et al.*, 1981) and capillary blood flow; the two may be very different.

Table III.1. *Mean ovarian blood flow (±SEM) in sheep as estimated by a radioactive microsphere technique, and the corresponding values for the peripheral plasma concentration of progesterone at three stages of the oestrous cycle*

	Stage of oestrous cycle (day)		
	14	15	16
Number of animals	8	6	10
Peripheral plasma progesterone (ng/ml)**	3.3 ± 0.5	3.2 ± 0.7	0.2 ± 0.06
Blood flow (ml/min per 100 g)			
CL**	1122 ± 169	708 ± 203	116 ± 19
Stroma	157 ± 23	258 ± 52	176 ± 24
Follicles*	637 ± 81	742 ± 102	1096 ± 147

CL, corpus luteum; ±, standard error of the mean (SEM).
*Variation significant, $P < 0.05$. **Variation significant, $P < 0.01$.
Adapted from Bruce & Moor, 1975.

Bruce & Moor (1975) used the distribution of radioactive microspheres to measure blood flow through capillaries <15 μm in diameter in anaesthetised sheep. In general agreement with the preceding results, ovarian follicles were found to receive a high rate of blood flow per gram of tissue, in fact comparable with that of a functional corpus luteum (Table III.1). The rate of follicular blood flow increased towards oestrus whereas luteal blood flow declined in parallel with the declining concentration of plasma progesterone (Table III.2). Larger follicles, including the one destined for ovulation, had significantly lower relative rates of blood flow than smaller follicles, although the highest absolute flow occurred in the largest follicle (Bruce & Moor, 1976). Somewhat surprisingly, no difference in flow was observed between normal and atretic follicles of comparable size. The full significance of high rates of follicular blood flow remains uncertain, but is probably associated with provision of substrate and with accumulation of fluid in the antrum of an expanding follicle.

As an alternative to invasive techniques involving anaesthesia, Brown, Emery & Mattner (1980) used Doppler ultrasonic transducers implanted around blood vessels to monitor ovarian arterial blood flow (total flow) in conscious sheep. The velocity of the arterial blood supply to the ovary with a corpus luteum, but not to the contralateral ovary without a corpus luteum, was markedly increased during the luteal phase of the oestrous cycle (Fig. III.16). This was in good agreement with an earlier study in conscious sheep using essentially the same technique (Niswender *et al.*, 1975). No specific conclusions could

Table III.2. *Ovarian blood flow in sheep as estimated by a radioactive microsphere technique during the 72 hours preceding ovulation. The table shows the mean ($\pm$SEM) tissue weight (mg), rate of blood flow through capillaries 15 μm in diameter (ml/min per 100 g), and the percentage distribution of blood perfusing different ovarian tissues*

	Ovary with corpus luteum			Ovary without corpus luteum		
	Day 14	Day 15	Day 16	Day 14	Day 15	Day 16
Number of animals	7	6	10	7	6	10
Total ovary						
Weight	1380 ± 160	1200 ± 120	1100 ± 120	680 ± 120	560 ± 40	740 ± 80
Blood flow	587 ± 63	478 ± 141	149 ± 25**	181 ± 50	286 ± 98	147 ± 27
Corpus luteum						
Weight	620 ± 50	500 ± 40	340 ± 60**			
Blood flow	1122 ± 169	708 ± 203	116 ± 19**			
% of total ovarian flow	84	65	28			
Stroma						
Weight	730 ± 140	670 ± 100	730 ± 80	650 ± 120	530 ± 40	720 ± 80
Blood flow	157 ± 33	266 ± 76	160 ± 37	157 ± 35	249 ± 79	121 ± 24
% of total ovarian flow	14	32	58	86	86	75
Follicles (non-atretic)						
Number per ovary	7.1 ± 1.9	5.8 ± 1.6	7.3 ± 1.1	8.7 ± 1.8	6.8 ± 2.1	7.4 ± 2.4
Weight	2.8 ± 0.4	3.2 ± 0.6	3.5 ± 0.9	3.2 ± 0.7	5.1 ± 1.7	3.4 ± 0.5
Blood flow	668 ± 135	776 ± 173	1069 ± 247	604 ± 116	692 ± 142	1101 ± 171*
Follicles (grossly atretic)						
Number per ovary	2.3 ± 0.7	2.8 ± 0.8	3.1 ± 0.6	3.6 ± 0.8	2.8 ± 1.3	2.9 ± 0.4
Weight	2.9 ± 1.2	4.1 ± 2.2	1.9 ± 0.7	1.9 ± 0.6	2.9 ± 1.1	2.2 ± 0.7
Blood flow	447 ± 115	878 ± 228	1083 ± 329	741 ± 179	762 ± 208	1062 ± 317
Follicles (combined)						
% of total ovarian flow	2	3	14	14	14	25

Variation between days significant; * $P < 0.05$; ** $P < 0.01$.
After Bruce & Moor, 1976.

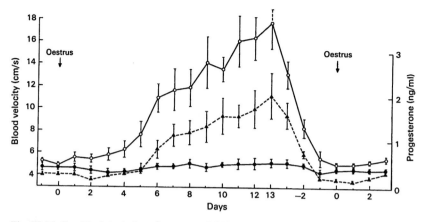

Fig. III.16. Graphical depiction of mean arterial blood velocity to sheep ovaries with corpora lutea (open circles) or without corpora lutea (solid circles) according to day of the oestrous cycle. Mean peripheral plasma progesterone concentrations are plotted in the middle curve as solid triangles. (After Brown *et al.*, 1980.)

be drawn concerning blood flow to the ovulatory follicle. Overall, the Doppler technique used in this study was less than perfect for estimating blood flow because the signal-to-noise ratio of the transducers influenced the accuracy with which the velocity could be measured. In addition, the diameter of the ovarian artery was not known precisely and would of course be changing.

In a study involving gonadotrophin releasing hormone (GnRH)-induced ovulation in anoestrous sheep, Brown *et al.* (1988) showed that a phase of progesterone treatment before oestradiol treatment produced a significant rise in ovarian stromal capillary blood flow whereas treatment with oestradiol alone or progesterone alone had no such influence. Together with the finding that ovarian stromal blood flow normally increases at oestrus in cyclic ewes (Brown, Hales & Mattner, 1974), this would suggest that the rate of ovarian blood flow around the time of ovulation could be a critical factor underlying normal ovarian function. The purpose of increased capillary blood flow just before ovulation is not known but, as suggested above, it may support the substrate requirements of the pre-ovulatory follicle for the final phase of maturation and pave the way for establishment of fully functional luteal cells, with enhanced secretion of steroid hormones.

Capillary blood flow in the ovarian stroma of sheep was significantly greater in the ovulatory ovary than in the non-ovulatory ovary (Brown *et al.*, 1974, 1988). As the pre-ovulatory oestradiol is derived principally from the Graafian follicle destined to ovulate (Bjersing *et al.*, 1972; Baird, 1978), oestradiol may prompt the local vasodilatory effect in the ovulatory ovary by acting directly on the stromal capillary walls or by diffusion into the stromal tissue, although the latter is controversial. Instead, local transfer mechanisms may be involved. In marked contrast to other reproductive organs (e.g. Fallopian tubes or uterus), the ovarian vasculature appears relatively insensitive to peripheral concentrations of oestradiol. This remark is supported by the absence of a simultaneous increase in capillary blood flow in the contralateral non-ovulatory ovary at ovulation. The local vascular response in the ovulatory ovary is enhanced almost two-fold if the ovulatory stimulus is preceded by progesterone treatment, and this highlights the important sensitising rôle of this hormone on the ovarian vasculature.

The preceding paragraphs have been concerned largely with ovarian blood flow in general and partition between the ovulating and non-ovulating ovaries, and the distribution to stroma and follicles in particular. Most of the techniques employed for measurement of blood flow were of an invasive nature, especially in domestic animals, and they were usually performed under general anaesthesia. Interpretation of the findings therefore requires caution. The same remark applies to a study in women in which mean ovary blood flow was calculated from the secretion rate of oestradiol using the Fick principle (Fraser, Baird & Cockburn, 1973).

By contrast, more recent studies have applied the non-invasive (i.e. non-surgical) approach of trans-vaginal colour Doppler ultrasonography to measurements in women. Such methodology enabled detection of an increase in blood flow to the dominant follicle with imminent ovulation (Bourne *et al.*, 1991; Collins *et al.*, 1991). One study using this approach of Doppler imaging has highlighted prominent changes in regional blood flow to the ovulatory follicle (Brännström *et al.*, 1998). In the final stages of follicular maturation after the LH surge, there was a marked increase of flow to the base of the follicle and a concomitant decrease to the apex. During days 10–12 of the menstrual cycle, imaging clearly revealed blood flow around the entire circumference of the dominant follicle. However, during the phase when ovulation was approaching or imminent, blood flow (peak systolic velocity) at the base of the follicle was significantly increased compared with that seen during days 10–12 (Fig. III.17; Table III.3). Whilst there might be a variety of explanations for this change, not least those concerned with a remodelling of the apex of the follicle preparatory to release of the oocyte and accompanying fluid, high blood flow at the base of the follicle might be partly associated with angiogenesis and proliferation of the capillary bed towards the granulosa cells, with incipient luteinisation.

Doppler ultrasonographic studies have also been performed in dogs, with results supporting the concept of cyclic changes in ovarian blood flow.

Table III.3. *Results obtained from Doppler imaging of regional blood flow to the base, side or apex of human Graafian follicles according to stage of the menstrual cycle. Peak systolic velocity at the base of the follicle increased significantly with the approach of ovulation*

Phase of cycle	Time-averaged maximum velocity (cm/s)		
	Base	Lateral	Apex
Days 10–12	6.9 ± 1.3	6.0 ± 1.0	6.5 ± 1.2
Early ovulatory	16.3 ± 3.3	8.8 ± 2.0	1.5 ± 1.1
Late ovulatory	10.1 ± 1.7	6.3 ± 1.0	0

Values represent means ±SEM.
Adapted from Brännström *et al.*, 1998.

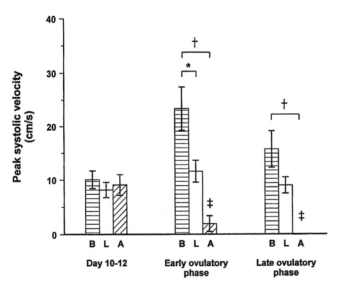

Fig. III.17. Changes in one component of blood flow according to stage of cycle. The figure illustrates peak systolic velocity in the base of a follicle (B), the lateral aspect (L) and at the apex (A). The early ovulatory phase is <20 hours after onset of the LH surge, whereas the late ovulatory phase is >20 hours after onset of the LH surge. (After Brännström *et al.*, 1998.)

Intra-ovarian perfusion increased gradually during pro-oestrus whilst, in the pre-ovulatory period, there was a marked enhancement of blood flow velocity and decline in the indices for pulsatility and resistance. Maximum perfusion was observed at ovulation and during the early luteal phase (Köster, Poulsen Nautrup & Günzel-Apel, 2001).

Temperature gradients within the ovaries

Set against a background of the well-documented mid-cycle rise in human body temperature, measurement of ovarian temperatures *in situ* by several research groups has produced intriguing observations. Studies in rabbits, humans, pigs and cattle have all suggested that pre-ovulatory follicles differ in temperature from neighbouring ovarian stroma and from deep body temperature. Before going into detail, it is worth considering germinal strategies in the two sexes. If, in most mammals, testes need to assume a cooler, scrotal location in order to become fully functional at puberty and produce competent spermatozoa, then questions arise concerning female gonadal function. Might not compartments within an ovary, especially those concerned with maturing and liberating gametes, also require some form of temperature differential at appropriate stages of the oestrous or menstrual cycle? Graafian follicles are largely responsible for the endocrine changes that prompt characteristic mid-cycle responses in primates or oestrous responses in farm and laboratory animals, but they must also liberate a potentially fertile oocyte.

In fact, reports from Copenhagen more than 20 years ago indicated that rabbit pre-ovulatory follicles could be 1.4 ± 0.2 deg.C cooler than ovarian stroma when examined by infra-red scanning or microelectrodes introduced during mid-ventral laparotomy (Grinsted *et al.*, 1980). The antral fluid temperature of human follicles was likewise cooler than ovarian stroma, on occasions by as much as 2.3 deg.C (Grinsted *et al.*, 1985). In domestic farm animals, temperatures were examined with thermistor probes sited in the follicular antrum or adjacent stroma, an invasive approach which nonetheless suggested that pig Graafian follicles of pre-ovulatory diameter (8–10 mm) could be ≥1.0 deg.C cooler than stroma (Hunter *et al.*, 1995). In a subsequent approach using infra-red thermo-imaging of pig ovarian tissues (Plate 2), the mean follicular temperature was 1.7 ± 0.4 deg.C cooler than stroma (Hunter *et al.*, 1997) or 1.3 ± 0.1 deg.C in a much larger and more refined study (Hunter *et al.*, 2000). Graafian follicles of 7–10 mm diameter were cooler than ovarian stroma in all experimental models examined (Table III.4 and Fig. III.18), and both compartments were cooler than jugular and deep rectal temperatures. Comparable studies in cattle revealed a follicle–stroma difference of 1.5 deg.C (Greve *et al.*, 1996).

Introduction into the abdomen of a specially designed endoscope that transmitted infra-red rays to the attached thermo-camera enabled further measurements to be made on pig ovaries. In fourteen sets of measurements in ten different animals, follicles were cooler than adjoining stroma by mean values ranging from 0.6 ± 0.1 deg.C to 1.1 ± 0.1 deg.C (Table III.5 and Fig. III.19; Hunter *et al.*, 2000). When using a completely gas-tight preparation with at least 5 minutes of equilibration for the pre-warmed laparoscope *in situ*, the mean

Table III.4. *The summarised findings concerning deep rectal and ovarian temperatures from 16 sets of observations (ovaries) in 14 animals examined during mid-ventral laparotomy*

	Temperature in °C			
	Deep rectal probe	Ovarian stroma	Graafian follicles of 7–10 mm diameter	Follicles cooler than stroma (differential)
Mean ± SEM	38.0 ± 0.4	37.3* ± 0.2	35.6** ± 0.3	1.7 ± 0.4
Range	37.5–38.6	35.7–38.5	34.5–37.2	0.7–2.6

*and **differ significantly ($P < 0.01$) when tested by one-way ANOVA (analysis of variance).
Adapted from Hunter *et al.*, 1997.

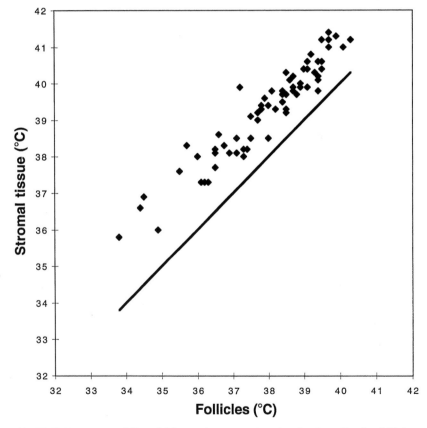

Fig. III.18. Temperature differential for ovarian stroma plotted against large Graafian follicles. The solid line indicates temperature equality of the two tissues. In no instance did follicular temperature exceed that of stroma ($n = 73$). (After Hunter *et al.*, 2000.)

Table III.5. *Summarised measurements of the temperature differentials between ovarian tissues when examined via mid-ventral laparotomy or during endoscopy of mature pigs*

Mean temperature (±SEM) by which Graafian follicles were cooler than stroma (deg.C)	
1.4 ± 0.1	Initial values before measuring cooling rates. Laparotomy
1.3 ± 0.1	Initial values before ligating ovarian blood supply. Laparotomy
1.1 ± 0.1	Ovary viewed deep within incision. Endoscopy
0.7 ± 0.1	Series of values from one animal only. Endoscopy
0.6 ± 0.1	Small (5–7 mm diameter) follicles rather than 9–10 mm diameter follicles. Endoscopy

Adapted from Hunter *et al.*, 2000.

Mean cooling rate with endoscope within the incision site (sow1578)

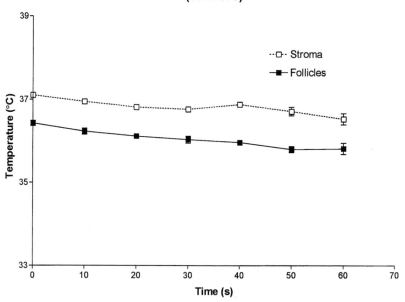

Fig. III.19. Mean cooling rates (±SEM) of three distinct areas of ovarian stroma and in at least three large Graafian follicles in one animal, as determined with an endoscope introduced into the abdomen via a mid-ventral incision. (After Hunter *et al.*, 2000.)

temperature difference between follicles and stroma was 0.6 ± 0.1 deg.C. It is important to emphasise that the animals available for this part of the study had Graafian follicles of only 5–7 mm diameter. Conceivably, pre-ovulatory follicles 9–10 mm in diameter would have shown a greater temperature differential with the stroma. This study suggested that a realistic value for the temperature by which pre-ovulatory porcine follicles were cooler than adjoining stroma might lie between 0.6 and 1.1 deg.C. Even so, the limitations of anaesthesia and a surgical approach were noted, and a technique of non-invasive whole body scanning of a fully conscious animal is needed to obtain a physiological reading (Hunter *et al.*, 2000).

If the assumption is made that pre-ovulatory Graafian follicles are indeed cooler than neighbouring ovarian compartments and extra-ovarian tissues, then questions that arise concern: (1) the means of generating and maintaining such differences in temperature, and (2) the physiological function(s) of such temperature modifications shortly before ovulation. The precise means of generating a temperature gradient remains unknown, over and above the general point that cells generate heat whereas a pre-ovulatory follicle is predominantly a fluid-filled structure with an ever-increasing volume of antral fluid and thus increasing fluid : cell ratio as ovulation approaches. This being so, it appears reasonable to accept that a mature Graafian follicle could be cooler than neighbouring stromal tissue or regressed luteal tissue, although maintenance of such putative temperature gradients would also require explanation.

As to a means of cooling, there is seemingly a need to invoke endothermic reactions within Graafian follicles, reactions that actively remove heat from a system rather than generate heat (exothermic). As discussed in Chapter IV, ovarian follicular fluid contains a wide range of chemical constituents and commenting on those that might be involved in endothermic reactions remains somewhat speculative, but there is physico-chemical evidence derived from studies of bovine follicular fluid upon saline dilution (Luck & Griffiths, 1998; Luck, Nutley & Cooper, 1999) and similarly of human follicular fluid (Luck *et al.*, 2001). There is also provisional evidence that hydration of large molecular weight components such as glycoprotein molecules could act to depress follicular temperature (Luck & Gregson, 2000). Such hydration might occur principally during remodelling of the cell layers composing the follicle wall and especially during mucification and expansion of the cumulus oophorus that accompanies resumption of meiosis (see below). There is a specific interest in the hydration of proteoglycans, a principal intercellular cement substance that is doubtless influenced in a key way by the antral oocyte. Other potential sources of endothermic reactions could be related to the synthesis and secretion of diverse peptides, proteins and prostaglandins at progressively changing

Local transfer of heat between small blood vessels
within the ovary maintains the temperature gradient
between follicle and stroma

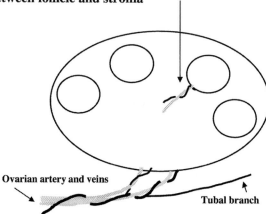

Ovarian artery and veins

Tubal branch

Local transfer of hormones from ovarian veins to artery
induces local high concentrations to ovary and tube

Fig. III.20. Diagrammatic representation of an ovary to indicate potential regions of heat exchange (1) in the blood supply to individual Graafian follicles and (2) perhaps also in the ovarian pedicle where a counter-current transfer of hormones is widely accepted to occur. Arteries, solid fill; veins, stippled. (After Hunter *et al.*, 2000.)

concentrations. Portions of the various biosynthetic pathways in these follicular activities might contribute to cooling in conjunction with the extensive glyco-protein modifications shortly before ovulation (see Hunter *et al.*, 1997).

If it is accepted that endothermic reactions are involved in the establishment of intra-ovarian temperature gradients, then a means must be put forward for the maintenance of gradients deep within the abdomen. Two physiological systems may function to avoid local cooling of follicles being over-ridden by an influence of the systemic circulation. First, a counter-current exchange of heat might be taking place in the ovarian pedicle so that returning venous blood could cool incoming arterial blood, but such arterial blood would presumably flow to both follicles and stroma. Second, there would certainly need to be a counter-current exchange of heat in the blood supply to individual follicles (Fig. III.20). An exchange of heat in the ovarian plexus could take place between the highly coiled and intimate arrangement of the ovarian artery and ovarian vein. This shows close parallels with the male gonadal blood supply, in which a counter-current heat exchange has been clearly demonstrated in species with scrotal testes (Setchell, 1978; Waites & Setchell, 1990). It is already established that the ovarian plexus facilitates an exchange of hormones between vein and artery, especially of relatively small molecules such as prostaglandins and steroids, as well as inert gases (McCracken, 1971; McCracken, Baird & Goding, 1971; Einer-Jensen, 1988), so there would seem to be no problem in principle for other forms of transfer across the walls of the ovarian artery and vein. Heat can be transferred 10–100 times more effectively than hormones (N. Einer-Jensen; personal communication).

As for the blood supply to individual follicles, although experimental demonstrations of a micro counter-current transfer have yet to be made at such a fine level, drawings of follicles in pig ovaries (Andersen, 1926) and more recent scanning electron micrographs of vascular corrosion casts in rabbit ovaries (Macchiarelli *et al.*, 1997) illustrate the intimate network of blood vessels and lymphatics and thus the potential for an exchange of heat – and thereby the local maintenance of temperature gradients.

Concerning the possible physiological functions of reduced temperatures in Graafian follicles approaching ovulation, a principal suggestion would be that they act to reduce the frequency of mutation in the germ cell line: the meiotic vulnerability of maturing germ cells has long been appreciated (Ehrenberg, von Ehrenstein & Hedgram, 1957). Reduced follicular temperature could likewise be beneficial to cytoplasmic maturation of oocytes, enabling a characteristic spectrum of vitelline proteins to be expressed and become functional during resumption of meiosis. An inference here is that conventional deep body temperature during the phase of pre-ovulatory meiotic activity could perturb the vitelline content of macromolecules, especially proteins involved in subsequent development. As an extension, inappropriate culture temperatures for *in vitro* maturation of oocytes might compromise both nuclear and cytoplasmic compartments of the cell, leading to reduced viability at subsequent stages of development. The scope for damage to the oocyte cytoskeleton and microtubules of the meiotic spindle was discussed by Hunter *et al.* (1997).

The above emphasis on lowered follicular temperature during the phase of resumption of meiosis should not detract from a possible involvement in actual shedding of the oocyte – that is, the process of ovulation itself. Considerations here might include the following:

1. The proteolytic enzymes such as collagenase that act to depolymerise the wall of a Graafian follicle shortly before ovulation (see Chapter VIII) may function most effectively at values below those of deep body temperature.
2. The increasing coagulability of follicular fluid as ovulation approaches, especially prominent in species such as the pig, may be a physical reaction requiring a particular temperature below that of deep body temperature to function most incisively. Specific control of changes in

coagulability with impending ovulation would seem essential, otherwise too viscous a follicular fluid could impede shedding of the oocyte from the follicular antrum into the Fallopian tube.

3. The risk of haemorrhage from the vascular bed during its extensive proliferation through the cell layers of the follicle wall shortly before ovulation may be reduced at lower temperature. Conversely, the process of blood clotting may be accelerated after discharge of the oocyte and release of much of the follicular fluid (Hunter *et al.*, 1997).

As a postscript, the above findings on ovarian follicular temperatures could be placed in a context of practical significance; this would be one of *in vitro* fertilisation of mammalian oocytes, perhaps preceded by *in vitro* maturation. Ever since the report of Cheng, Moor & Polge (1986) that incubating maturing oocytes and a sperm suspension at 39 °C rather than 37.5 or 38.0 °C gave a higher incidence of fertilisation, *in vitro* fertilisation laboratories have been attracted to the culture temperature of 39 °C. Although incubation at this temperature clearly facilitates sperm activation and penetration of the oocytes, the overall yield of the technique in terms of viable fetuses seldom exceeds 15–20%, especially in our own species. Accordingly, *in vitro* procedures may need to mimic more closely the physiological temperature shifts in ovarian follicular tissues, perhaps employing a computerised programme of temperature changes in the culture oven. Temperatures incorrect by as little as 1.0 deg.C or less could have devastating influences on cell organelles and especially on the protein chemistry and folding that regulate development (Hunter *et al.*, 1997; Hunter, 2000; Hunter, Einer-Jensen & Greve, 2002).

Why do Graafian follicles grow so large?

When the dimensions of a pre-ovulatory oocyte are considered, commonly a diameter of 100–120 μm or so in humans and the domestic farm species but suspended in an expanded mucified mass of cumulus cells, then the size of a pre-ovulatory follicle appears enormous. By way of example, human Graafian follicles may attain diameters of 20–22 or even 23 mm, individual porcine Graafian follicles 9–10 mm (but there may be 12–20 of them), and equine follicles as much as 45 or even 50 mm shortly before ovulation. So the question inevitably arises as to why follicles grow so large and, as a subsidiary, is such a size essential for normal ovulation? A variety of different responses can be offered.

1. It could simply be a question of proportions, of scaling. That is, larger animals would have larger ovaries, which, in turn, would be expected to

have larger follicles. Such an explanation is not completely adequate (see Parkes, 1932; Brambell, 1956), but most variation between species in the sizes of Graafian follicles can be accounted for by differences in adult body weight (Gosden & Telfer, 1987).

2. Related to the preceding point but possibly more persuasive, size might determine the endocrine functions of a pre-ovulatory follicle. In order to secrete a sufficient quantity of hormones, not least key steroid hormones to initiate appropriate hypothalamic and pituitary responses together with tonic responses in the reproductive tract, follicles might need to attain a certain threshold dimension. However, hormones are synthesised and secreted by cells rather than by follicles, and cell numbers (theca and granulosa cells) do not appear to be closely correlated with follicular diameter.

3. Follicles may need to achieve the approximate size of a mature corpus luteum – the structure that will subsequently develop in its place. In one sense, this would be so that a collapsed follicle can be filled with proliferating luteal cells. Viewed from this angle, the dimensions of a mature Graafian follicle may be dictated in part by those required by the mature corpus luteum.

4. Dimensions may reflect preparations for eventual release of an oocyte at ovulation. As follicles grow larger, they will tend to protrude more from the ovarian surface – which may offer an evolutionary advantage in terms of 'egg capture' at ovulation. A protruding follicle would enable more intimate contact with the fimbriated infundibulum of the Fallopian tube, thereby promoting the chances of successful oocyte transfer from the gonad into the genital duct (see Chapter IX).

5. Mature Graafian follicles may grow to achieve dimensions that facilitate the generation of intra-ovarian temperature gradients – that is, to enable expression of cooling (endothermic) reactions within appropriate follicles (see preceding section). Growth of a Graafian follicle will alter the ratio between the volume of antral fluid and the volume of granulosa cells lining the cavity. Whereas cells generate heat, a mature follicle is largely a fluid reservoir, even if in a dynamic state. Altering the ratio of the two components, fluid and cells, could be related to the expression of intra-ovarian temperature gradients.

As to whether the diameters described for mature Graafian follicles are essential for ovulation, the answer is seemingly 'No'. Two lines of evidence can be mounted. First, anticipating the endogenous gonadotrophin surge by administering a single injection of hCG can provoke ovulation of follicles far smaller than mature diameter. This is true in a wide range of species. Second, in breeds of sheep with significant differences in the number of ovulations (i.e. ovulation rate), the larger the number of ovulations, the smaller the follicular diameter at ovulation (Driancourt *et al.*, 1986).

Concluding remarks

Although discussion of ovarian vasculature, lymphatics and innervation may have a somewhat classical flavour, there can be little doubt that these systems play a critical rôle in the development and regression of Graafian follicles and in selection of ovulatory follicles. The vascular and lymphatic networks in the theca are especially impressive and presumably reflect in part a need to provide substrate to the developing structure and its encompassed oocyte. Conversely, they emphasise the local synthesis of angiogenic factors by a developing follicle and also a local influence of relatively high titres of oestradiol-17β. The extent to which the ovarian microvasculature permits specific communication between different follicles in a wave or cohort remains to be examined with contemporary imaging and analytical techniques. Regulation of blood flow to individual follicles could clearly have an incisive influence on their developmental potential. This remark should be taken to include counter-current systems of exchange within the microcirculation.

As to ovarian innervation, this is seen traditionally in a context of muscular regulation, including that of the vasculature, and less so in terms of modulating steroidogenesis. However, the local release of factors from nerve cells other than catecholamines (e.g. peptides, cytokines, nitric oxide) needs to be anticipated in any modern interpretation of follicular development and selection. Nervous pathways provide vital links between follicular structures and indeed between the two ovaries, and should not be discounted as unimportant simply because of incidental surgical observations.

Finally, the discussion of ovarian temperature gradients is drawn from studies that continue, and a last word on their existence has not yet been written. Nonetheless, mathematical modelling of rates of cooling of ovarian tissues during different forms of experimental intervention has indicated that the reported intra-ovarian temperature gradients cannot be accounted for by artefact alone and certainly rely on an intact blood supply. Further work on endothermic reactions in maturing follicles would therefore seem to hold promise and could have important implications in clinical procedures of *in vitro* maturation and *in vitro* fertilisation.

References

Amsterdam, A., Lindner, H.R. & Gröschel-Stewart, U. (1977). Localisation of actin and myosin in the rat oocyte and follicular wall by immunofluorescence. *Anatomical Record*, **187**, 311–17.

Andersen, D.H. (1926). Lymphatics and blood vessels of the ovary of the sow. *Contributions to Embryology of the Carnegie Institution, Washington*, **17**, 107–23.

Andersen, D.H. (1927). Lymphatics of the Fallopian tube of the sow. *Contributions to Embryology of the Carnegie Institution, Washington*, **19**, 135–48.

Bahr, J., Kao, L. & Nalbandov, A.V. (1974). The role of catecholamines and nerves in ovulation. *Biology of Reproduction*, **10**, 273–90.

Baird, D.T. (1978). Pulsatile secretion of LH and ovarian estradiol during the follicular phase of the sheep estrous cycle. *Biology of Reproduction*, **18**, 359–64.

Baird, D.T., Webb, R., Campbell, B.K., Harkness, L.M. & Gosden, R.G. (1999). Long-term ovarian function in sheep after ovariectomy and transplantation of autografts stored at −196 °C. *Endocrinology*, **140**, 462–71.

Bassett, D.L. (1943). The changes in the vascular pattern of the ovary of the albino rat during the estrous cycle. *American Journal of Anatomy*, **73**, 251–91.

Bendz, A., Hansson, H.A., Svendsen, P. & Wiqvist, N. (1982). On the extensive contact between veins and arteries in the human ovarian pedicle. *Acta Physiologica Scandinavica*, **115**, 179–82.

Bjersing, L., Hay, M.F., Kann, G., *et al.* (1972). Changes in gonadotrophins, ovarian steroids and follicular morphology in sheep at oestrus. *Journal of Endocrinology*, **52**, 465–79.

Bourne, T.H., Jurkovic, D., Waterstone, J., Campbell, S. & Collins, W.P. (1991). Intrafollicular blood flow during human ovulation. *Ultrasound in Obstetrics and Gynaecology*, **1**, 53–9.

Brambell, F.W.R. (1956). Ovarian changes. I. Development of the ovary and oogenesis. In *Marshall's Physiology of Reproduction*, vol. 1, part 1, ed. A.S. Parkes, pp. 397–542. London & New York: Longmans, Green & Co.

Brännström, M., Zackrisson, U., Hagström, H.G., *et al.* (1998). Preovulatory changes of blood flow in different regions of the human follicle. *Fertility and Sterility*, **69**, 435–42.

Brown, B.W., Cognie, Y., Chemineau, P., Poulin, N. & Salama, O.A. (1988). Ovarian capillary blood flow in seasonally anoestrous ewes induced to ovulate with GnRH. *Journal of Reproduction and Fertility*, **84**, 653–8.

Brown, B.W., Emery, M.J. & Mattner, P.E. (1980). Ovarian arterial blood velocity measured with Doppler ultrasonic transducers in conscious ewes. *Journal of Reproduction and Fertility*, **58**, 295–300.

Brown, B.W., Hales, J.R.S. & Mattner, P.E. (1974). Capillary blood flow in sheep ovaries measured by iodoantipyrine and microsphere techniques. *Experientia*, **30**, 914–15.

Bruce, N.W. & Moor, R.M. (1975). Ovarian follicular blood flow in the sheep. *Journal of Reproduction and Fertility*, **43**, 392–393.

Bruce, N.W. & Moor, R.M. (1976). Capillary blood flow to ovarian follicles, stroma and corpora lutea of anaesthetised sheep. *Journal of Reproduction and Fertility*, **46**, 299–304.

Burden, H.W. (1972). Adrenergic innervation in ovaries of the rat and guinea-pig. *American Journal of Anatomy*, **133**, 455–62.

Burden, H.W. (1978). Ovarian innervation. In *The Vertebrate Ovary: Comparative Biology and Evolution*, ed. R.E. Jones, pp. 615–38. New York: Plenum Press.

Burden, H.W., Lawrence, I.E., Louis, T.M. & Hodson, C.A. (1981). Effects of abdominal vagotomy on the estrous cycle of the rat and the induction of pseudopregnancy. *Neuroendocrinology*, **33**, 218–22.

Burr, J.H. & Davies, J.I. (1951). The vascular system of the rabbit ovary and its relationship to ovulation. *Anatomical Record*, **111**, 273–97.

Calza, L. (1807). Über den Mechanismus der Schwangerschaft. *Archiv für die Physiologie*, Reils **7**, 341–401.

Cavender, J.L. & Murdoch, W.J. (1988). Morphological studies of the microcirculatory system of periovulatory ovine follicles. *Biology of Reproduction*, **39**, 989–97.

Chaichareon, D.P., Rankin, J.H. & Ginther, O.J. (1976). Factors which affect the relative contributions of ovarian and uterine arteries to the blood supply of reproductive organs in guinea pigs. *Biology of Reproduction*, **15**, 281–90.

Cheng, W.K.T., Moor, R.M. & Polge, C. (1986). *In vitro* fertilisation of pig and sheep oocytes. *Theriogenology*, **25**, 146 (Abstract).

Clark, J.G. (1900). The origin, development and degeneration of the blood vessels of the human ovary. *Johns Hopkins Hospital Reports*, **9**, 593–676.

Collins, W., Jurkovic, D., Bourne, T., Kurjak, A. & Campbell, S. (1991). Ovarian morphology, endocrine function and intra-follicular blood flow during the periovulatory period. *Human Reproduction*, **6**, 319–24.

Delson, B., Lubin, S. & Reynolds, S.R.M. (1949). Vascular patterns in human ovary. *American Journal of Obstetrics and Gynecology*, **57**, 842–53.

Dissen, G.A. & Ojeda, S.R. (1999). Ovarian innervation. In *Encyclopedia of Reproduction*, vol. 3, ed. E. Knobil & J.D. Neill, pp. 583–9. San Diego & London: Academic Press.

dos Santos-Ferreira, A., Motta, P., Esperanca-Piña, J.A., Didio, L.J.A. & Cherney, D.D. (1974). Arterial microcirculation of the ovary. *International Surgery*, **59**, 100–2.

Driancourt, M.A., Gauld, I.K., Terqui, M. & Webb, R. (1986). Variations in patterns of follicle development in prolific breeds of sheep. *Journal of Reproduction and Fertility*, **78**, 565–75.

Driancourt, M.A., Gougeon, A., Monniaux, D., Royère, D & Thibault, C. (2001). Folliculogenèse et ovulation. In *La Reproduction chez les Mammifères*, 2nd edn, ed. C. Thibault & M.C. Levasseur, pp. 316–47. Paris: Ellipses.

Edwards, R.G. (1980). *Conception in the Human Female*. London & New York: Academic Press.

Ehrenberg, L., von Ehrenstein, G. & Hedgran, A. (1957). Gonadal temperature and spontaneous mutation rate in man. *Nature (London)*, **180**, 1433–4.

Einer-Jensen, N. (1988). Counter-current transfer in the ovarian pedicle and its physiological implications. *Oxford Reviews of Reproductive Biology*, **10**, 348–81.

Einer-Jensen, N., Cicinelli, E. & Galantino, P. (2000). The ovarian artery supplies the tubal part of the uterus in man. *Journal of Reproduction and Fertility*, No. 25, 17, 25, Abstract Series.

Einer-Jensen, N. & McCracken, J.A. (1981). Physiological aspects of corpus-luteum blood flow and of the counter system in the ovarian pedicle of the sheep. *Acta Veterinaria Scandinavica, Supplement*, **77**, 89–101.

Ellinwood, W.E., Nett, T.M. & Niswender, G.D. (1978). Ovarian vasculature: structure and function. In *The Vertebrate Ovary: Comparative Biology and Evolution*, ed. R.E. Jones, pp. 583–614. New York & London: Plenum Press.

Fink, G. & Schofield, G.C. (1971). Experimental studies on the innervation of the ovary in cats. *Journal of Anatomy*, **109**, 115–26.

Fraser, I.S., Baird, D.T. & Cockburn, F. (1973). Ovarian venous blood PO_2, PCO_2 and pH in women. *Journal of Reproduction and Fertility*, **33**, 11–17.

Gillet, J.Y. (1971). La microvascularisation de l'ovaire. *Gynécologie et Obstétrique (Paris)*, **70**, 251–72.

Gillet, J.Y., Koritké, J.G., Muller, P. & Juliens, C. (1968). On the microvascularisation of rabbit ovary. *Comptes Rendus de la Société de Biologie*, **162**, 762–6.

Goding, J.R., McCracken, J.A. & Baird, D.T. (1967). The study of ovarian function in the ewe by means of a vascular autotransplantation technique. *Journal of Endocrinology*, **39**, 37–52.

Gosden, R.G. (2000). Low temperature storage and grafting of human ovarian tissue. *Molecular and Cellular Endocrinology*, **163**, 125–9.

Gosden, R.G., Baird, D.T., Wade, J.C. & Webb, R. (1994). Restoration of fertility to oophorectomised sheep by ovarian autografts stored at −196 °C. *Human Reproduction*, **9**, 596–603.

Gosden, R.G. & Telfer, E. (1987). Scaling of follicular sizes in mammalian ovaries. *Journal of Zoology*, **211**, 157–68.

Greenwald, G.S. & Roy, S.K. (1994). Follicular development and its control. In *The Physiology of Reproduction*, 2nd edn, ed. E. Knobil & J.D. Neill, pp. 629–724. New York: Raven Press.

Greve, T., Grøndahl, C., Schmidt, M., Hunter, R.H.F. & Avery, B. (1996). Bovine preovulatory follicular temperature: implications for *in vitro* production of embryos. *Archiv für Tierzücht, Dummerstorf*, **39** (Special Issue), 7–14.

Grinsted, J., Blendstrup, K., Andreasen, M.P. & Byskov, A.G. (1980). Temperature measurements of rabbit antral follicles. *Journal of Reproduction and Fertility*, **60**, 149–55.

Grinsted, J., Kjer, J.J., Blendstrup, K. & Pedersen, J.F. (1985). Is low temperature of the follicular fluid prior to ovulation necessary for normal oocyte development? *Fertility and Sterility*, **43**, 34–9.

Harrison, R.J. & Weir, B.J. (1977). Structure of the mammalian ovary. In *The Ovary*, 2nd edn, vol. 1, ed. Lord Zuckerman & B.J. Weir, pp. 113–217. New York & London: Academic Press.

Heap, R.B., Fleet, I.R. & Hamon, M. (1985). Prostaglandin $F_{2\alpha}$ is transferred from the uterus to the ovary in the sheep by lymphatic and blood vascular pathways. *Journal of Reproduction and Fertility*, **74**, 645–56.

Hunter, R.H.F. (2000). Temperature considerations for *in vitro* studies with equine gametes. In *FetoMaternal Control of Pregnancy*, Havemeyer Foundation Monograph Series No. 2, ed. T.A.E. Stout & J.F. Wade, pp. 3–4. Newmarket, Suffolk: R & W Publications.

Hunter, R.H.F., Bøgh, I.B., Einer-Jensen, N., Müller, S. & Greve, T. (2000). Pre-ovulatory Graafian follicles are cooler than neighbouring stroma in pig ovaries. *Human Reproduction*, **15**, 273–83.

Hunter, R.H.F., Einer-Jensen, N. & Greve, T. (2002). Presence and significance of temperature gradients among different ovarian tissues. *Microscopy Research and Technique*. In press.

Hunter, R.H.F., Grøndahl, C., Greve, T. & Schmidt, M. (1997). Graafian follicles are cooler than neighbouring ovarian tissues and deep rectal temperatures. *Human Reproduction*, **12**, 95–100.

Hunter, R.H.F., Grøndahl, C., Schmidt, M. & Greve, T. (1995). Graafian follicles are cooler than ovarian stroma. *Journal of Reproduction and Fertility*, **16**, 8–9, Abstract Series.

Janson, P.O. (1975). Effects of the luteinising hormone on blood flow in the follicular rabbit ovary, as measured by radioactive microspheres. *Acta Endocrinologica (Copenhagen)*, **79**, 122–33.

Jdanov, D.A. (1960). Nouvelles données sur la morphologie fonctionnelle du système lymphatique des glandes endocrines. *Acta Anatomica*, **41**, 240–59.

Jdanov, D.A. (1969). Anatomy and function of the lymphatic capillaries. *Lancet*, **2**, 895–9.

Keller, L. (1966). Beobachtungen an Nerven des Eierstockes. *Zeitschrift für Zellforschung*, **69**, 284–7.

Koritké, J.G., Gillet, J.Y. & Pierti, J. (1967). Les artères de la trompe utérine chez la femme. *Archives d'Anatomie, d'Histologie et d'Embryologie*, **40**, 49–70.

Köster, K., Poulsen Nautrup, C. & Günzel-Apel, A.R. (2001). A Doppler ultrasonographic study of cyclic changes of ovarian perfusion in the Beagle bitch. *Reproduction*, **122**, 453–61.

Lee, B.H., Kim, E.A., Cho, K.J., Choi, W.S., Baik, S.H. & Ojeda, S.R. (1996). Central integration of reproductive function via a neural connection between hypothalamic oxytocinergic and ovarian neuraxes. *Society for Neuroscience*, **22**, 1790, Abstract No. 704.14.

Lee, C.S. & O'Shea, J.D. (1976). The extrinsic blood vessels of the ovary of the sheep. *Journal of Morphology*, **148**, 287–304.

Lindner, H.R., Sass, M.B. & Morris, B. (1964). Steroids in the ovarian lymph and blood of conscious ewes. *Journal of Endocrinology*, **30**, 361–76.

Luck, M.R. & Gregson, K. (2000). Mathematical modelling of potential endothermic reactions in cool follicles. *Journal of Reproduction and Fertility*, **26**, 14, Abstract Series.

Luck, M.R. & Griffiths, S.D. (1998). Endothermic response of follicular fluid to aqueous dilution. *Journal of Reproduction and Fertility*, **22**, 12–13, Abstract Series.

Luck, M.R., Griffiths, S., Gregson, K., Watson, E., Nutley, M. & Cooper, A. (2001). Follicular fluid responds endothermically to aqueous dilution. *Human Reproduction*, **16**, 2508–14.

Luck, M.R., Nutley, M. & Cooper, A. (1999). Microcalorimetric detection of the endothermic response of follicular fluid to aqueous dilution. *Journal of Reproduction and Fertility*, **24**, 15, Abstract Series.

Macchiarelli, G., Nottola, S.A. & Motta, P.M. (1997). On the presence of extensive arteriovenous contacts in the rabbit ovarian hilum and medulla. Ultrastructural observations by scanning electron microscopy of vascular corrosion casts. In *Recent Advances in Microscopy of Cells, Tissues and Organs*, ed. P.M. Motta, pp. 519–25. Rome: Antonio Delfino Editore.

Macchiarelli, G., Nottola, S.A., Vizza, E., *et al.* (1993). Microvasculature of growing and atretic follicles in the rabbit ovary: a SEM study of corrosion casts. *Archives of Histology and Cytology*, **56**, 1–12.

Macchiarelli, G., Nottola, S.A., Vizza, E., Kikuta, A., Murakami, T. & Motta, P.M. (1991). Ovarian microvasculature in normal and hCG stimulated rabbits. A study of vascular corrosion casts with particular regard to the interstitium. *Journal of Submicroscopical Cytology and Pathology*, **23**, 391–5.

Mattner, P.E., Brown, B.W. & Hales, J.R.S. (1981). Evidence for functional arterio-venous anastomoses in the ovaries of sheep. *Journal of Reproduction and Fertility*, **63**, 279–84.

Mattner, P.E. & Thorburn, G.D. (1969). Ovarian blood flow in sheep during the oestrous cycle. *Journal of Reproduction and Fertility*, **19**, 547–9.

Mayerhofer, A., Dissen, G.A., Costa, M.E. & Ojeda, S.R. (1997). A role for neurotrans-mitters in early follicular development: induction of functional follicle stimulating hormone receptors in newly formed follicles of the rat ovary. *Endocrinology*, **138**, 3320–9.

McCracken, J.A. (1971). Prostaglandin $F_{2\alpha}$ and corpus luteum regression. *Annals of the New York Academy of Sciences*, **180**, 456–72.

McCracken, J.A., Baird, D.T. & Goding, J.R. (1971). Factors affecting the secretion of steroids from the transplanted ovary of the sheep. *Recent Progress in Hormone Research*, **27**, 537–82.

Mezzogiorno, V., Fusaroli, P. & Cavallucci, G. (1967). Caratteri anatomo-microscopici degli ampi seni venosi del meso e dell'ilo ovarico umano negli ultimi mesi di vita fetale. *Anatomia e Chirurgia*, **12**, 5–17.

Morris, B. & Sass, M.B. (1966). The formation of lymph in the ovary. *Proceedings of the Royal Society, London, Series B*, **164**, 577–91.

Mossman, H.W. & Duke, K.L. (1973). *Comparative Morphology of the Mammalian Ovary*. Madison, WI: University of Wisconsin Press.

Neilson, D., Jones, G.S., Woodruff, J.D. & Goldberg, B. (1970). The innervation of the ovary. *Obstetrics and Gynecological Survey*, **25**, 889–904.

Niswender, G.D., Moore, R.T., Akbar, A.M., Nett, T.M. & Diekman, M.A. (1975). Flow of blood to the ovaries of ewes throughout the estrous cycle. *Biology of Reproduction*, **13**, 381–8.

Niswender, G.D., Reimers, T.J., Diekman, M.A. & Nett, T.M. (1976). Blood flow: a mediator of ovarian function. *Biology of Reproduction*, **14**, 64–81.

Novy, M.J. & Cook, M.J. (1973). Redistribution of blood flow by prostaglandin $F_{2\alpha}$ in the rabbit ovary. *American Journal of Obstetrics and Gynecology*, **117**, 381–5.

Oktay, K., Newton, H. & Gosden, R.G. (2000). Transplantation of cryopreserved human ovarian tissue results in follicle growth initiation in SCID mice. *Fertility and Sterility*, **73**, 599–603.

Owman, C., Kannisto, P., Liedberg, F., *et al.* (1992). Innervation of the ovary. In *Local Regulation of Ovarian Function*, ed. N.-O. Sjöberg, L. Hamberger, P.O. Janson & C. Owman, pp. 149–70. Park Ridge, NJ: Parthenon.

Owman, Ch., Rosengren, E. & Sjöberg, N.-O. (1967). Adrenergic innervation of the human female reproductive organs: a histochemical and chemical investigation. *Obstetrics and Gynecology*, **30**, 763–73.

Owman, Ch. & Sjöberg, N.-O. (1966). Adrenergic nerves in the female genital tract of the rabbit: with remarks on cholinesterase-containing structures. *Zeitschrift für Zellforschung*, **74**, 182–97.

Owman, Ch., Sjöberg, N.-O., Svensson, K.-G. & Walles, B. (1975). Autonomic nerves mediating contractility in the human Graafian follicle. *Journal of Reproduction and Fertility*, **45**, 553–6.

Parkes, A.S. (1932). The reproductive processes of certain mammals. II. The size of the Graafian follicle at ovulation. *Proceedings of the Royal Society, London, Series B*, **109**, 185–96.

Poirier, P. & Charpy, A. (1901). Lymphatiques de l'ovaire. In *Traité de l'Anatomie Humaine*, vol. 2, p. 1199, and vol. 5, p. 370. Paris.

Polano, O. (1903). Beiträge zur Anatomie der Lymphbahnen im menschlichen Eierstock. *Monatsschrift für Geburtshelfen und Gynaekologie*, **17**, 281–95.

Prénant, A. & Bouin, P. (1911). *Traité d'Histologie*. Paris: Masson.

Rathmacher, R.P. & Anderson, L.L. (1968). Blood flow and progesterone levels in the ovary of cycling and pregnant pigs. *American Journal of Physiology*, **214**, 1014–18.

Redmer, D.A. & Reynolds, L.P. (1996). Angiogenesis in the ovary. *Reviews of Reproduction*, **1**, 182–92.

Reiffenstuhl, G. (1964). *The Lymphatics of the Female Genital Organs*. Philadelphia: Lippincott.

Reynolds, S.R.M. (1973). Blood and lymph vascular systems of the ovary. In *American Handbook of Physiology*, section 7, vol. II, *Female Reproductive System*, Part 1, ed. R.O. Greep & E.B. Astwood, pp. 261–316. Washington, DC: American Physiological Society.

Schultea, T.D., Dees, W.L. & Ojeda, S.R. (1992). Postnatal development of sympathetic and sensory innervation of the rhesus monkey ovary. *Biology of Reproduction*, **47**, 760–7.

Setchell, B.P. (1978). *The Mammalian Testis*. London: Paul Elek.

Sjöberg, N.-O., Bodelsson, G. & Stjernquist, M. (1997). Innervation of the female reproductive tract. In *Recent Advances in Microscopy of Cells, Tissues and Organs*, ed. P.M. Motta, pp. 535–46. Rome: Antonio Delfino Editore.

Staples, L.D., Fleet, I.R. & Heap, R.B. (1982). Anatomy of the utero-ovarian lymphatic network and the composition of afferent lymph in relation to the establishment of pregnancy in the sheep and goat. *Journal of Reproduction and Fertility*, **64**, 409–20.

Täher, E.S. (1964). Das innere Lymphgefässsystem des Eierstocks. *Tierärztliche Umsch*, **19**, 194–7.

Waites, G.M.H. & Setchell, B.P. (1990). Physiology of the mammalian testis. In *Marshall's Physiology of Reproduction*, 4th edn, vol. 2, *Reproduction in the Male*, ed. G.E. Lamming, pp. 1–105. Edinburgh: Churchill Livingstone.

Walles, B., Gröschel-Stewart, U., Owman, Ch., Sjöberg, N.-O. & Unsicker, K. (1978). Fluorescence histochemical demonstration of a relationship between adrenergic nerves and cells containing actin and myosin in the rat ovary with special reference to the follicle wall. *Journal of Reproduction and Fertility*, **52**, 175–8.

Wehrenberg, W.B., Chaichareon, D.P., Dierschke, D.J., Rankin, J.H. & Ginther, O.J. (1977). Vascular dynamics of the reproductive tract in the female rhesus monkey: relative contributions of ovarian and uterine arteries. *Biology of Reproduction*, **17**, 148–53.

Wislocki, G.B. & Dempsey, E.W. (1939). Remarks on the lymphatics of the reproductive tract of the female monkey (*Macaca mulatta*). *Anatomical Record*, **75**, 341–63.

Zondek, B., Sulman, F. & Black, R. (1945). The hyperemia effect of gonadotrophins on the ovary and its use as a rapid pregnancy test. *Journal of the American Medical Association*, **128**, 939–44.

IV
Ovarian follicular-antral fluid

Introduction

The extracellular fluid that accumulates in the antral cavity of a Graafian follicle, conventionally referred to as follicular fluid, could be contributing significantly to diverse reproductive events; this would be locally within the ovary and less immediately in the adjacent genital duct system. Although follicular fluid has been examined in considerable detail in terms of its content of steroid and protein hormones, prostaglandins and peptide growth factors, there remains scope for a more systematic characterisation of binding proteins at different stages of follicular maturation, since these will influence the available local concentrations of a hormone. Such hormones have been measured in aspirates of follicular fluid obtained at laparoscopy in attempts to predict the incidence of fertilisation under *in vitro* conditions and, better still, the subsequent viability of an embryo. Interactions of follicular fluid with the oocyte and its granulosa cell investment also deserve further study, especially during resumption of meiosis and mucification of the cumulus oophorus in the final hours before ovulation. Follicular fluid might be contributing to the reprogramming of such cells as granulosa lutein cells, and the oocyte itself would certainly be influencing the composition of follicular fluid. Consideration of constituents of the fluid that might be involved in endothermic reactions would also seem to be a research topic of current significance (see Chapter III). And the nature of follicular fluid has still not been fully resolved in the sense of the relative contributions (quantitative and qualitative) of blood plasma ultrafiltration and specific secretion at various stages of maturation up to the moment of follicular collapse, not least during the phase of enhanced vascular permeability following the gonadotrophin surge.

These comments clearly refer to pre-ovulatory events, but there remain questions of the ultimate fate of follicular fluid after ovulation, the nature of responses in the Fallopian tube upon initial fluid entry, and the extent to which it contributes to a privileged microenvironment surrounding the gametes at fertilisation. Ovarian follicular fluid has been demonstrated to influence the capacitation of spermatozoa *in vitro*, so consideration of its contribution within the Fallopian tube should be informative. At the very least, a modulating influence on the ampullary epithelium and its secretions would be anticipated, not least owing to its potent content of hormones at the time of ovulation (see Chapter IX).

Previous reviews concerned with the formation and/or composition of ovarian follicular fluid include those by Lipner (1973), Edwards (1974, 1980), McNatty (1978, 1982), Guraya (1985), Sinosich (1987), Gosden *et al.* (1988), Lenton *et al.* (1988) and Ménézo & Guérin (2001). Follicular fluid is also examined in the slim volume by Peters & McNatty (1980) and in the chapter by Baker (1982).

Formation of follicular fluid

Extracellular fluid within a growing follicle becomes readily apparent once the antral stage has been reached. In general, pools of follicular fluid begin to form when the granulosa cells have passed through about 11–12 mitotic cycles and a substantial follicle containing some 2000–3000 cells has developed (Gosden *et al.*, 1988). Remodelling of the follicle enables coalescence of the separate pools of fluid in a central antrum, which is limited by a microscopically uniform layer of granulosa cells (Fig. IV.1). This layer is 'interrupted' close to its polar position by protrusion of the mass of cells supporting the oocyte, the cumulus oophorus, a population of granulosa cells that continues to proliferate when the mitotic activity of other granulosa cells has ceased.

Although follicular fluid is considered to play an important rôle in ovulation and displacement of an oocyte into the Fallopian tube, it is instructive to recall that certain mammals do not develop vesicular follicles. Various insectivores such as the Madagascan tenrec (family Tenrecidae) are said to shed their eggs from solid follicles of dimensions not much larger than those at which an antrum would conventionally form (Strauss, 1950; Mossman & Duke, 1973). If this should be consistently so, then it begs a question as to the precise involvement of follicular fluid at ovulation in most mammals so far examined. It also raises the possibility of smooth muscle elements in the theca participating in actual displacement of the oocyte, even though modification of tissues in the ripe follicle is thought to be a primary factor in release of the egg. In fact, the granulosa layer of the mature tenrec follicle becomes spongy and oedematous owing to swelling and fluid infiltration between the individual cells (Strauss, 1950).

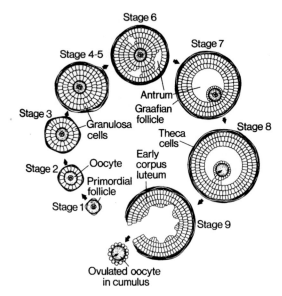

Fig. IV.1. A diagrammatic illustration of the extensive remodelling of follicular components during formation of a Graafian follicle. Individual pools of intercellular fluid gradually coalesce to form a conspicuous antrum. (Courtesy of Professor T.G. Baker.)

Discussions of follicular fluid generally refer to three so-called phases, a convention stemming from the paper of Robinson (1918). Antral fluid that forms until close to the time of ovulation is termed primary fluid or 'liquor folliculi'. When the rate of fluid accumulation increases markedly during final maturation of the follicle as a response to the gonadotrophin surge, this is referred to as secondary fluid. Fluid that accumulates in the collapsed follicle after ovulation as a mixture of blood, transudate and secretion is termed tertiary fluid. Despite such long-standing terminology, actual analyses of follicular fluid components have seldom followed such a classification and tertiary fluid has seemingly received a negligible amount of attention (see Chapter IX). Primary fluid is accepted to be largely of intracellular origin whereas secondary fluid would represent extensive transudation from the thecal capillary bed (Burr & Davies, 1951), presumably involving increased blood flow after the gonadotrophin surge and increased permeability of blood vessels. However, even during the final pre-ovulatory phase, fluid accumulation is slow compared with movement across many other membranes in the body (Gosden *et al.*, 1988). For example, aqueous humour in the anterior chamber of the eye is formed at a rate of approximately 2 µl/min, which stands in contrast to ovine follicular fluid accumulation during the pre-ovulatory phase at 20 µl/day.

Table IV.1. *Electrical properties of the wall of a Graafian follicle as compared with those of other mammalian epithelial cells*

Nature of epithelium	Species	Classification	Resistivity (Ω cm^2)
Ovarian follicle (unstripped)	Mouse	Leaky	49
	Pig	Leaky	60
Gall bladder	Rabbit	Leaky	28

Adapted from Gosden *et al.*, 1988.

Formation of fluid occurs across a membrane or membranes. Although traditionally referred to as the membrana granulosa (Young, 1961), granulosa cells can be regarded as epithelial cells with the outermost layer resting on a delicate basement membrane, the membrana propria. Transudation from thecal capillaries must cross both basement membrane and the granulosa cells to reach the antrum, but the follicular epithelium is highly permeable to water and dissolved substances. Occlusive junctions are not found between cells, although other forms of junctional complex do exist for maintaining structural integrity and for intercellular communication (Albertini & Anderson, 1974). Channels separating granulosa cells permit molecules of up to M_r 500 000 to penetrate and reach the antrum. In terms of molecular access to follicular fluid, there is therefore a concern that harmful substances entering the blood may have ready access to the microenvironment of the oocyte. Arising in this manner, long-term influences on germ cells cannot be overlooked, nor can the potential hazards for a fertilised egg and developing embryo.

Although granulosa cell secretion is taken as the source of most primary follicular fluid, with an important contribution of transudation from the capillary bed to secondary fluid, further specific mechanisms may be involved in the formation of antral fluid, especially as influenced by gonadotrophic hormones. A rôle for osmotic gradients has been examined by Gosden *et al.* (1988), and active transport considered with sodium (Na) ion as a prime candidate, but the results remain equivocal. Evidence in favour of active transport under rigorous experimental conditions is still awaited. A non-invasive technique will be required for making direct measurements of Na^+ and Cl^- fluxes *in situ* in follicles at different stages of maturity. As to the resistivity of the follicle wall, this may be sufficiently low for the epithelium to be regarded as 'electrically leaky' (Table IV.1; Gosden & Hunter, 1988; Gosden *et al.*, 1988).

Assuming that osmotic potential determines the rate and direction of net water transport, it is worth reflecting also that chemical potential will be influenced by differences in temperature. Recalling the work of Grinsted *et al.* (1980, 1985) on both rabbit and human Graafian follicles, referred to in Chapter III, and more recent studies in Copenhagen on temperature gradients within the ovaries of large domestic species (Hunter *et al.*, 1997, 2000), then clearly follicular temperature could be influencing formation of follicular fluid. In addition to this physical aspect, an involvement of macrophages and other white blood cells in modifying the composition of follicular fluid should not be overlooked. There is a major traffic in white cells through the tissues of pre-ovulatory follicles (see Chapter VIII), and secretions from such cells may enter the pool of follicular fluid. Because the passage of white cells within the follicle wall is most prominent with imminent ovulation, the greatest potential contribution of macrophage secretions would presumably be to secondary and tertiary follicular fluid.

Physical condition of follicular fluid

Although analysis of follicular fluid will tend to concentrate on chemical constituents, it needs to be emphasised that the fluid is not physically homogeneous (Zachariae & Jensen, 1958; Luck *et al.*, 2000). As a general rule, the fluid tends to be more viscous close to granulosa cell surfaces as compared with a more watery fluid in the central part of the antrum. Because the fluid is usually collected by aspiration through an appropriate gauge needle into a hypodermic syringe, such differences are frequently overlooked. Differences in the physical state of follicular (antral) fluid become prominent with impending ovulation. As judged from sampling in the large domestic species, a viscous fluid may reflect in part an enhanced synthesis of glycosaminoglycans (see next section), especially after the pre-ovulatory surge of gonadotrophic hormones and mucification of the cumulus mass. According to Edwards (1980), follicular fluid in monkeys also becomes highly viscous as ovulation approaches whereas human follicular fluid is said to be only slightly viscous. It has an osmotic pressure close to that of plasma. The large amounts of colloidal material accumulating in some pre-ovulatory follicles may produce a distinct colloid osmotic pressure (Fig. IV.2). Lipner (1973) related osmotically active particles in follicular fluid to the process of depolymerisation preceding ovulation, thereby leading to an increased colloid osmotic pressure. However, Zachariae & Jensen (1958) pointed out that colloid osmotic pressure constitutes only a small part of the total osmotic pressure of follicular fluid.

A coagulable component may also become apparent in follicular fluid examined *in vitro* if sampled close to the time of ovulation. Although there is no specific evidence to hand, a reasonable interpretation here would be that this late stage change in the condition of follicular fluid is related to modifications

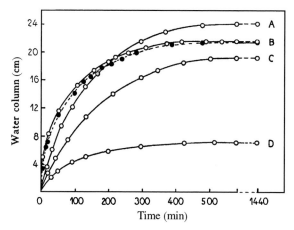

Fig. IV.2. Diagrammatic representation of the colloid osmotic pressure in a Graafian follicle. (After Zachariae & Jensen, 1958.)

in the vascular bed. The thecal capillaries become extremely prominent shortly before ovulation and, at least in farm animals and various laboratory species, show an increased permeability and petechial haemorrhages as the walls of the vessels gradually break down (Betteridge & Raeside, 1962; Hunter, 1967). Under such circumstances, not only would the process of transudation be facilitated but a more direct passage of blood components into the follicular fluid could occur. More specifically, as a consequence of increased filtration pressure and vascular permeability, fluid traverses the endothelium, enters the intercellular spaces, and accumulates within the follicular antrum. Proteins leak out of the vascular bed and there is a deposition of fibrin (Zachariae & Jensen, 1958; Cavender & Murdoch, 1988). The coagulability that can frequently be detected in fluid sampled at the terminal stages of follicular growth, particularly when follicles have commenced to become flaccid (Hunter, 1980, 1984a), would seem to be related to the clotting process of blood components. Extensive clotting may be held in check by the action of heparin, prostacyclin or plasmin (Cavender & Murdoch, 1988) or there may be an active anticoagulant with properties similar to those of heparin (Stangroom & Weevers, 1962). Stangroom & Weevers (1962) demonstrated the presence of an anti-thrombin in equine follicular fluid.

In a key review of follicular fluid, albeit more than a quarter of a century ago, Edwards (1974) suggested that there were differences between species in the tendency of follicular fluid to clot when removed from the antrum, as previously noted by Zachariae & Jensen (1958). He indicated that fluids taken from pre-ovulatory cow and pig follicles clot easily whereas those from human follicles do not, yet fibrinogen is present in the fluid of many species. Human follicular fluid contains approximately 40% of the fibrinogen present in serum (Shalgi *et al.*, 1973), but an anticoagulant has been postulated, as above. However, considerations of time and temperature are also relevant to this discussion on coagulability of follicular fluid. The sampling interval before ovulation will certainly influence the properties of the fluid under consideration. The temperature at which the fluid is removed and examined will also be critical. Most human follicular fluid specimens will have been collected during laparoscopy, with precautions against rapid cooling. By contrast, many of the farm animal specimens, especially in the early studies of Edwards (1962, 1965) on aspiration of oocytes and resumption of meiosis *in vitro*, were based on post mortem collection of ovaries. As of writing, one would need to be persuaded that species differences are more important than conditions of sampling when judging the coagulability and clotting potential of follicular fluid.

Composition of follicular fluid

Accumulating between granulosa cells in growing follicles, this ovarian fluid is a mixture of novel secretion, especially of steroid hormones, peptides and glycosaminoglycans (GAGs), and of plasma exudate, especially its proteins, although usually at a lower concentration. Labelled serum proteins can be demonstrated in follicular fluid (Mancini *et al.*, 1963). The novel secretion is primarily of granulosa cell origin, a statement intended to include products of the highly specialised cumulus oophorus. However, follicular fluid would also contain secretions of the theca interna that have crossed the basement membrane and, all too often overlooked, molecular products of the oocyte itself. There would be the further influence of dead and dying cells within or close to the antrum.

A major part of this chapter could be devoted to the composition of follicular fluid, describing in detail all the constituents that have been measured in different reproductive states. This would anaesthetise or exhaust the reader and is not the approach to be followed here. Instead, similarities and differences between follicular fluid and blood plasma will be commented upon briefly, other constituents will be highlighted, and the paragraphs that follow will be concerned with the diverse range of hormones and peptides identified in follicular fluid. At the very outset, it is worth emphasising that the concentrations of a number of classical hormones found in follicular fluid, for example oestradiol, progesterone and various prostaglandins, may far exceed their concentrations in systemic blood. One is therefore prompted to consider the local influence of such elevated hormone concentrations on those cells and membranes immediately exposed to the fluid, such as the granulosa cells surrounding the antrum, the cumulus oophorus, and the oocyte itself. In addition to influencing the availability of cell surface receptors, there could be local programming influences on the synthetic activity and differentiation of exposed somatic cells.

Most attention has been paid to the antral fluid of large follicles; that is, primary or secondary fluid (see above). Although separated from blood vessels by the inner layers of the follicle, including a relatively permeable basement membrane (Szöllösi *et al.*, 1978), passage of molecules across the follicle wall changes as the time of ovulation approaches. This is not least due to gradual dissolution of the basement membrane after the gonadotrophin surge and an altered permeability of the capillary bed, i.e. of the blood–follicular fluid barrier (see Zachariae, 1958). Size and shape will influence the movement of protein molecules into follicular fluid. Small molecules in blood plasma are able to equilibrate rapidly with follicular fluid, and even larger proteins may enter within a few minutes (Edwards, 1980). As judged from human samples, most plasma proteins and steroid-binding proteins are found in follicular fluid (Beier, Beier-Hellwig & Delbos, 1983), including an androgen-binding protein in fluid from polycystic ovaries. Overall, about 80% of the protein content in plasma is found in follicular fluid, i.e. qualitatively similar but quantitatively less (Lipner, 1973). However, the total protein content may sometimes be only about half the serum protein level (Sinosich, 1987), perhaps due to fluid sampling from small and immature follicles. A disparity in total quantity of protein had earlier been reported by Perloff *et al.* (1955) for human follicular fluid and by Shivers, Metz & Lutwak-Mann (1964) for domestic animals. Secretion of proteins by the granulosa cells and oocyte would offer one explanation for the presence of non-serum proteins.

The proteins include enzymes, complement factors, immunoglobulins (Ig), transport proteins (e.g. albumin, α_2-macroglobulin, pregnancy-associated plasma protein A (PAPP-A), protease inhibitors, and proteins with uncertain function such as placental protein 5 (PP5) and α-follicular fluid antigen (α-FF antigen). Follicular fluid enzyme activity includes glycolytic and proteolytic enzymes. Some such as galactosyl transferase demonstrate marked changes in isoenzyme activity in pre-ovulatory follicular fluid. Others are listed in Table IV.2. Protease inhibitors constitute a significant proportion of the intra-follicular protein environment. Many of the protease inhibitors may be derived from the peripheral circulation, having entered the antrum by passive diffusion across the basement membrane. By contrast, PAPP-A and PP5 are not detected in the peripheral circulation (Sinosich, 1987). Other antigens found in greater concentration within the pre-ovulatory follicle than in the peripheral circulation

Table IV.2. *A comparison of enzymes detected (+) or seemingly absent (−) in human, bovine and porcine follicular fluid*

	Presence in different animals		
Enzyme	Human	Cow	Pig
Endopeptidase	+	−	+
Proteinase	−	+	+
Plasmin	−	+	−
Aminopeptidase (cytosol)	+	−	+
Alkaline phosphatase	+	+	+
Acid phosphatase	+	+	−
Fructose bisphosphate aldolase	+	−	+
Lactate dehydrogenase	+	+	+
Aspartate aminotransferase	+	+	+
Alanine aminotransferase	−	+	+
Collagenase	−	−	+
Hyaluronoglucosidase	+	+	−
Pyrophosphatase	+	−	−

Adapted from Baker, 1982.

include α-FF antigen and plasminogen. Some physico-chemical characteristics of these antigens are given in Table IV.3.

As to physiological functions, α-FF antigen has many similarities to inhibin (see Chapter V). PAPP-A may be part of the intra-follicular protease inhibitor pool, functioning to maintain proteolytic homeostasis during the pre- and peri-ovulatory phases. In conjunction with other protease inhibitors, PAPP-A may provide local protection against the maternal phagocytic–proteolytic defence system, i.e. as part of a general immunosurveillance system in the peri-ovulatory period (Sinosich, 1987). Especially concerned with the events of ovulation is collagenase in the follicle wall (see Chapter VIII). Plasminogen and its activator are also closely involved in the process of ovulation.

Various forms of angiotensin (e.g. angiotensins II and III) have been documented in follicular fluid during growth and maturation of the surrounding structure (Jarry *et al.*, 1988). Precisely why these molecules should accumulate in the antrum and what their physiological significance might be for the germ and somatic cells are unknown. A regulatory influence on the vascular bed of the follicle comes to mind and likewise pressure within the antral cavity, but a specific involvement has yet to be demonstrated.

GAGs are long, straight-chain carbohydrates associated with protein, and are a prominent component of follicular fluid (for a review, see Luck, 1994). Their properties of flow, filtration and adhesion are seemingly essential to the

Table IV.3. *Some physico-chemical characteristics of pregnancy-associated plasma protein A (PAPP-A), placental protein 5 (PP5) and α-follicular fluid (α-FF) antigen*

	PAPP-A	PP5	α-FF antigen
Relative molecular mass (M_r)			
Gel filtration	820 000	140 000, 36 000, 18 000	55 000
SDS-PAGE (− reduction)	410 000	NA	38 000, 14 000
SDS-PAGE (+ reduction)	210 000	NA	38 000, 14 000
Electrophoretic mobility	α 2	β 1	α 1
Isoelectric point	4.5	4.6	4.3
Affinity interactions			
Heparin	+	+	−
Matrex red A	±	±	+
Thrombin	d	+	NA
Zinc chelate	+	−	NA
Concanavalin A	+	+	−

SDS-PAGE, sodium dodecyl sulphate–polyacrylamide gel electrophoresis; NA, not assayed; +, reversible interaction; ±, minor proportion interacting with ligand; − no interaction.
After Sinosich, 1987.

developing follicle and may play a key rôle at the time of ovulation. At least two major forms of GAG exist in follicular fluid. One is rich in chondroitin sulphuric acid and present in the fluid of larger growing follicles. The other is viscous, rich in hyaluronic acid, and located in the cement substance between granulosa cells. Hyaluronidase is essential for depolymerising this cement. Changes in overall GAG composition are found as a response to gonadotrophic hormones during the development of Graafian follicles, and have a profound influence on the viscosity of follicular fluid.

The elemental composition of fluid from large follicles is, in general, similar to that of ovarian venous plasma (Table IV.4), but there is clearly scope for further studies here with fresh and precisely timed samples. In addition to values presented in Table IV.4, divalent zinc ions are found in equimolar distribution across the basement membrane (Sinosich, 1987). Findings in the large farm species agree with those in rabbits, and may indicate the existence of active ion transport by the sodium pump. Bicarbonate concentrations were slightly lower in follicular fluid, which could suggest that the follicular wall is a leaky epithelium (Gosden & Hunter, 1988). Any charge resulting from net ion transport would therefore be shunted along low resistance pathways.

Electrophysiological evidence that active transport can occur across the follicle wall was obtained by McCaig (1985). He described a small, variable,

Table IV.4. *Values for the concentration of electrolytes (mmol/l) in the antral fluid of large Graafian follicles (FF) as compared with those in plasma (P) or serum (S)*

Species	Na^+ FF	Na^+ P/S	K^+ FF	K^+ P/S	Cl^- FF	Cl^- P/S	Ca^{2+} FF	Ca^{2+} P/S	Mg^{2+} FF	Mg^{2+} P/S
Human	124	145	4.4	4.6	109	104	0.94	1.04	0.76	0.68
	143	154	5.4	5.4	140	146				
Rabbit	133	125	7.0	4.3	136	127	3.56	3.85	1.40	1.67
	140	136	6.2	5.7	144	139				
Sheep	149	149	4.7	4.9	107	106	2.29	2.28	0.89	0.87
Pig	128	143	15.9	5.2						
	142	147	7.6	7.1						
	145	140	4.9	4.8			10.3	10.8		
	141	138	3.8	3.8	97.3	95.7	2.3	2.27	0.75	0.77
Cow	132		9.2		149.5		3.1			

Adapted from Gosden *et al.*, 1988.

transmural potential difference in mouse follicles, and showed that this was metabolically coupled and changed with physiological status during the oestrous cycle. This stands in contrast to the findings of P. Mathews & H. Lipner (unpublished, cited by Lipner, 1973), who were unable to detect a potential difference in either rabbit or rat follicles. Across species, the observations might be interpreted as negation of the active transport hypothesis for formation of follicular fluid, at least during the more rapid phase of pre-ovulatory expansion. The major pathway for entry of water into the antrum is probably paracellular, although the forces involved require clarification.

As to small organic molecules, excluding steroid hormones (see below), the technique of proton nuclear magnetic resonance spectroscopy has been applied to the fluid aspirated from pre-ovulatory follicles. In specimens obtained from cows, sheep and pigs, the concentrations of low molecular mass, non-protein-bound metabolites (e.g. acetate, alanine, creatinine/creatine, glycine, lactate and valine) were similar to those of blood plasma (Gosden *et al.*, 1990), supporting the view that the wall of Graafian follicles is highly permeable to small organic molecules (Szöllösi *et al.*, 1978).

Turning now to endocrine aspects, there are numerous reports on the concentration of various steroid hormones and peptide molecules in follicular fluid during spontaneous cycles and following stimulation with gonadotrophic preparations or their analogues (Gore-Langton & Armstrong, 1988). The contents of cystic follicles have also been examined in detail. Such analytical work effectively commenced during the late 1950s, prompted by the application of paper chromatography to the separation of individual steroids (e.g. Short, 1957, 1958, 1960, 1962). The improved assays of the late 1960s and early 1970s, involving either radioimmunoassay or protein binding assays, offered a marked increase in sensitivity and a means of measuring hormones in peripheral blood. Studies on follicular fluid blossomed in many experimental situations, not least those of laparoscopic sampling during procedures of oocyte retrieval in *in vitro* fertilisation clinics. Despite the wealth of results, there remains one major shortcoming. It is not possible to make sequential measurements on fluids from the same Graafian follicle in a physiological manner, i.e. meaningful longitudinal studies are not possible. Aspiration of antral fluid punctures the basement membrane of the follicle, compromises its physiological integrity, and permits invasion of the granulosa cell layers by actively proliferating thecal capillaries. Accordingly, comparisons of hormone concentrations at different stages of follicular development must depend on sampling a series of different follicles.

An extensive review of steroid hormones in human follicular fluid was presented by Lenton *et al.* (1988). Such hormones were being used to monitor growth of Graafian follicles during induced or stimulated cycles, the concentration of hormone varying according to the stage of follicular development (Table IV.5). Amongst the points emphasised was that the large oestrogenic follicle is the healthy pre-ovulatory follicle, i.e. the one destined to ovulate. A ratio of high oestradiol:low androgen concentrations was considered to reflect a healthy follicle (McNatty, 1982). The concentration of progesterone in follicular fluid rises rapidly soon after the pre-ovulatory surge of LH, usually for about 25 hours, following which the rate of increase slows down and reaches a plateau of approximately 10 000–12 000 ng/ml follicular fluid. However, if a follicle is left undisturbed beyond the normal aspiration time for human *in vitro* fertilisation, then progesterone concentrations may rise to 17 500 ng/ml follicular fluid (Lenton *et al.*, 1988). As can be noted from Table IV.5, the follicular fluid concentrations of androstenedione and testosterone are much lower than those of oestradiol in healthy specimens. These androgens can therefore be viewed as intermediaries in the biosynthetic pathway to oestradiol. By contrast, high androgen concentrations and a high ratio of androstenedione : oestradiol in follicular fluid are usually indicative of impending atresia, at least for the oocyte (Liu, Sandow & Rosenwaks, 1986).

The relative abundance of steroid hormones in follicular fluid changes markedly as ovulation approaches, although the concentration of oestradiol always exceeded that of oestrone (Baird & Fraser, 1975). Concentrations may rise to such an extent that some mechanism must exist to maintain steroids in

Table IV.5. *The mean concentrations (±SEM) of key steroid hormones in human follicular fluid before and after the spontaneous luteinising hormone (LH) surge during an unstimulated cycle*

Steroid hormone (ng/ml)	Before LH surge (n = 8)	Hours from LH surge					Average mid-cycle peripheral plasma concentration
		0–10 (n = 8)	10–20 (n = 8)	20–30 (n = 8)	30–40 (n = 16)	>40[a] (n = 4)	
Androstenedione	1280 ± 16	190 ± 40	180 ± 30	200 ± 50	170 ± 22	150 ± 40	6.0
Testosterone	110 ± 10	45 ± 4	52 ± 12	39 ± 4	28 ± 2	37 ± 8	0.6
Oestradiol	3730 ± 360	3580 ± 270	2790 ± 230	1450 ± 80	810 ± 45	450 ± 60	0.3
Progesterone	2900 ± 380	5410 ± 710	8650 ± 260	9880 ± 580	12 200 ± 530	17 400 ± 350	1.0

[a]Ovulation (follicular rupture) was artificially delayed (by giving indomethacin) in three out of the four women whose follicular fluids were not recovered until >40 hours from the start of the spontaneous LH surge.
Adapted from Lenton *et al.*, 1988.

solution. Association of steroids with protein molecules offers one means of maintaining homogeneity. Oestradiol in bovine follicular fluid was largely bound to albumin (Takikawa, 1966), and high affinity oestrogen-binding proteins were subsequently demonstrated in human follicular fluid (Takikawa & Yoshinaga, 1968). Indeed, there is a specific binding protein in follicular fluid with a higher affinity for oestradiol than for oestrone (Giorgi, Addis & Colombo, 1969; Baird & Fraser, 1975). Various questions arise. For example, do the oocyte and/or its somatic cell investment need some form of follicular oestrogen–protein complex to facilitate maturation or is the local steroid–protein association of such low affinity that the steroid can be liberated to interact freely with the oocyte and somatic cells within the follicle? Although oocytes appear to need high concentrations of oestradiol in their bathing follicular fluid before they can resume nuclear and cytoplasmic maturation (Hunter, Cook & Baker, 1976; Table IV.6), no unusual steroid-binding proteins that might react with the oocyte or its investments were detected in the follicular fluid of domestic animals (Cook, Hunter & Kelly, 1977). The bulk of the steroid in follicular fluid is bound to albumin with low affinity, indicating that steroid molecules can be readily released.

There remains the question of the extent to which steroid-binding globulins from blood (sex hormone-binding globulin (SHBG) and corticosteroid-binding globulin) are generally present in follicular fluid. And, if so, are they of any significance in modulating the availability of steroids within the follicular compartments? ^{131}I-labelled whole serum albumin and γ-globulins entered follicular fluid rapidly and reached peak concentrations within 3 hours (Mancini *et al.*, 1963). Oestrogen-binding β-globulin was also referred to by Takikawa (1966) and Giorgi *et al.* (1969).

Pituitary hormones are, of course, present in follicular fluid and their concentrations usually vary with the stage of follicular development. Both FSH and LH concentrations increase in pre-ovulatory follicles, although remaining below values in systemic blood. Prolactin is also present in follicular fluid, and may generally decline in concentration with increasing follicular growth or close to ovulation, but there is considerable variability (McNatty, 1978). Values for peptide hormones in follicular fluid and for cAMP are also given in Table IV.7. Of particular note, peak concentrations of what was termed inhibin by the assays available in the late 1970s were found at a time when oestradiol was maximal, suggesting a granulosa cell source. Follicular cAMP concentrations were high before the start of the LH surge, falling to low or undetectable values in a majority of follicles more than 20 hours after the LH surge. The cAMP was thought to originate in theca cells, being transmitted across the basement membrane to the granulosa cells, the oocyte and its cumulus

Table IV.6. *Ovarian responses to systemic injection of human chorionic gonadotrophin (hCG), indicating increasing maturity of oocytes shed in the late follicular phase of the oestrous cycle. Values for oestradiol concentrations in the follicular fluid are reported in Hunter et al. (1976)*

Figures show results obtained from 88 post-puberal pigs illustrating five aspects of the response to a single intramuscular injection of hCG given at various stages of the follicular phase of the oestrous cycle. These results can be considered in terms of a series of biological systems which undergo maturation during the latter part of the oestrous cycle.

Day of oestrous cycle[a] when hCG injected	System under consideration: No. of animals injected	Behavioural oestrus: Animals showing oestrus		Ovarian response: Recent ovulations		Maturity of oocyte nucleus: Primary oocytes (%)	Block to polyspermy: Polyspermic eggs (%)	Availability of vitelline factors (MPGF): Eggs containing transformed sperm heads (%)
		No.	(%)	Total no.	Mean per animal			
17	7	0	0	130	18.6	81.2	91.7	<10
18	9	1	11.1	112	12.4	17.4	38.0	
19	45	39	86.7	581	12.9	2.1	2.4	
20	27	25	92.6	308	11.4	1.4	2.8	>90

MPGF, male pronucleus growth factor.
[a] Only animals with oestrous cycles lasting 20 or 21 days were assigned to this experiment.

Table IV.7. *The mean concentration (±SEM) of various peptides and cAMP in ovarian follicular fluid and peripheral plasma before and after the spontaneous luteinising hormone (LH) surge during an unstimulated cycle*

	Before LH surge (n = 8)	Hours from LH surge					Average mid-cycle peripheral plasma concentration
		0–10 (n = 8)	10–20 (n = 8)	20–30 (n = 8)	30–40 (n = 16)	>40[a] (n = 4)	
Peptides							
LH (U/l)	2.4 ± 0.4	2.4 ± 0.4	7.6 ± 13	15.2 ± 1.5	13.6 ± 1.1	7.4 ± 0.5	38.0
FSH (U/l)	1.8 ± 0.3	2.0 ± 0.3	3.3 ± 0.5	4.9 ± 0.3	6.4 ± 0.6	8.1 ± 2.4	12.5
Inhibin (U/mg)	9.9 ± 1.3	13.5 ± 2.5	11.3 ± 1.3	9.1 ± 0.8	9.3 ± 1.0	6.9 ± 2.3	–
Prolactin (U/l)	390 ± 30	280 ± 13	360 ± 30	330 ± 20	390 ± 32	400 ± 6.0	180
cAMP							
Mean conc. (nmol/l)	61 ± 12	50 ± 8	33 ± 15	14.5 ± 5.0	10.6 ± 4.0	5.8 ± 0.7	
Follicles with detectable levels (%)	100	100	50	38	19	12	

FSH, follicle stimulating hormone; cAMP, cyclic AMP; conc., concentration.
[a] Ovulation (follicular rupture) was artificially delayed (by giving indomethacin) in three out of the four women whose follicular fluids were not recovered until >40 hours from the start of the spontaneous LH surge.
Adapted from Lenton *et al.*, 1988.

investment. As to FSH, it begins to enter the follicle in increasing amounts about 25 hours after the LH surge. It is thought to suppress intercellular coupling and to promote secretion of GAGs from granulosa cells. In follicles of the rabbit, an induced ovulator, cAMP rises initially and then declines abruptly as the time of ovulation approaches (Marsh, Mills & LeMaire, 1972).

There is much evidence from diverse mammalian species, including primates, that ovulation is dependent upon a pre-ovulatory rise in the concentration of prostaglandins within the Graafian follicle (Behrman, 1979; Poyser, 1981). Early observations in support of this statement came from rats (LeMaire, Leidner & Marsh, 1975; Brown & Poyser, 1984), rabbits (Yang, Marsh & LeMaire, 1974), gonadotrophin-treated immature pigs (Ainsworth, Baker & Armstrong, 1975; Tsang *et al.*, 1979) and women (Edwards, 1980). Despite the wealth of evidence, including that from prostaglandin inhibitor studies (see Chapter VIII), the precise manner in which these low molecular weight compounds contribute to terminal events in a pre-ovulatory follicle is uncertain. Their rôle is probably multi-compartmental: prostaglandins have an ability to influence diverse physiological and biochemical systems.

The interval between the gonadotrophin surge and ovulation of approximately 40 hours in pigs is one of the longest reported for domestic species (Hunter, 1980) and is closely similar to the interval in women (Edwards, 1980). There is thus a special interest in noting the concentration of prostaglandins in pig follicular fluid during the pre-ovulatory interval in animals with spontaneous oestrous cycles (Hunter & Poyser, 1985). Values for PGE_2, $PGF_{2\alpha}$ and 6-oxo-$PGF_{1\alpha}$ (the chemically hydrated product of PGI_2) are given in Table IV.8. The concentrations of PGE_2 showed the most dramatic change, rising gradually at first during the pre-ovulatory interval and then steeply to 166 ng/ml shortly before ovulation. Concentrations of $PGF_{2\alpha}$ and 6-oxo-$PGF_{1\alpha}$ in follicular fluid were also significantly elevated shortly before ovulation. However, once ovulation had commenced or if it was overdue, as perhaps in follicles destined to become cystic, then the concentration of prostaglandins did not rise (Table IV.8). Hence, the findings in this, as in other species, are consistent with follicular prostaglandins contributing critically to the process of ovulation and possibly to maturation of the oocyte.

Intra-follicular pressure, PO_2, PCO_2, pH

Measurements on physico-chemical characteristics of follicular fluid *in situ* within the antrum are limited, especially so for intra-follicular pressures at precise times before ovulation (see McNatty, 1978; Guraya, 1985). Nonetheless, a consistent finding in the large domestic species is that intra-follicular pressure

Table IV.8. *The mean concentration of three prostaglandins in follicular fluid aspirated at laparotomy or autopsy from six different groups of pigs according to the time interval elapsing from the onset of standing oestrus or a pro-oestrous injection of human chorionic gonadotrophin (hCG)*

		Concentration in ng/ml					
	No. of animals sampled	PGE_2		$PGF_{2\alpha}$		6-oxo-$PGF_{1\alpha}$	
		Mean	(Range)	Mean	(Range)	Mean	(Range)
Interval from onset of oestrus or hCG (h)							
8–12	4	1.4	(0.7–1.9)	2.3	(1.6–2.8)	2.1	(0.9–3.1)
20–30	8	4.0	(1.1–9.5)	1.6	(0.3–3.9)	1.4	(0.3–3.5)
31–39	10	6.2	(2.6–16.7)	1.8	(0.3–4.9)	1.8	(0.3–5.1)
40–42	8	166.1	(8.9–529.2)	24.6	(2.3–94.6)	11.2	(0.8–54.4)
Ovulation commenced	3	1.1	(0.5–1.9)	0.8	(0.6–1.1)	1.5	(1.0–1.8)
Ovulation overdue	3	2.0	(1.4–2.8)	0.2	(0.1–0.2)	0.2	(0.1–0.4)

PG, prostaglandin.
After Hunter & Poyser, 1985.

does not increase as the time of ovulation approaches but rather is conspicuously reduced. Instead of the follicle wall becoming tense and thereby seeming to rupture at ovulation, it gradually becomes flaccid and is better described as collapsing, notably during ovulation in pigs, sheep, cows and horses (see Hunter, 1980; and Plate 5). This major modification in the condition of a pre-ovulatory follicle may be principally a consequence of specific enzymatic activity in the wall of the follicle (see Chapter VIII). Collagenase in particular is acting to depolymerise that region of the follicular apex through which the oocyte will eventually be liberated at ovulation. Softening of the follicle wall is extensive in the final hour before ovulation, and it is this modification that seems to underlie reduced intra-follicular pressure, perhaps associated with seepage of follicular fluid.

Again, measurements of PO_2, PCO_2 and pH in follicular fluid are limited (see Lipner, 1973; Edwards, 1980), seldom precise with respect to the time of ovulation, and usually subject to some degree of error due to sampling in anaesthetised patients. Epidural anaesthesia is considered to minimise such error (Moir & Mone, 1964). Even so, there is reasonable consistency in the reported values, and an indication that these may be close to the values in ovarian venous blood (Table IV.9).

Human follicular fluid has a pH of 7.2–7.3 (Edwards, 1974), a figure of 7.27 being recorded by Shalgi, Kraicer & Soferman (1972) and 7.34 by Fraser, Baird & Cockburn (1973). Values for pH in a comparable large domestic species are given in Table IV.10. Although a single value is generally recorded for a sample of follicular fluid, it should be recognised that there will be microenvironments within the antrum and not least at cell surfaces, perhaps with a significantly modified pH towards the oocyte itself.

At the theoretical level, a mathematical model was developed to examine the gradient in oxygen concentration across the follicular epithelium (Gosden & Byatt-Smith, 1986). Most dissolved oxygen entering the follicle by diffusion was consumed in the outer layer of cells. Little reached the oocyte. The model predicted that a large pre-antral follicle with a radius of 0.15 mm would consume oxygen at the rate of 0.22 nmol/min. However, major differences would be anticipated in a fully vascularised Graafian follicle and especially so after the pre-ovulatory surge of LH. Dissolution of the basement membrane would enable capillaries to colonise the granulosa cells.

Contribution to Fallopian tube and peritoneal fluids

Bearing in mind the relatively large volumes of follicular fluid referred to in the opening sections of this chapter, and the fact that the bulk of the follicular fluid

Table IV.9. *Values for PO₂, PCO₂ and pH in the follicular fluids of*
sixteen women

Patient no.	Follicle size[a]	Follicle side	PO_2(mmHg)	PCO_2(mmHg)	pH
1	Medium	L	121.0	48.5	7.3
2	Medium	R	53.5	46.0	7.3
3	Medium	R	54.0	39.0	7.3
4	Large	L	65.0	57.5	7.3
5	Medium	R	53.0	40.5	7.4
	Medium	L	97.0	33.5	7.4
6	Medium	R	54.2	40.0	7.3
	Medium	R	90.0	39.0	7.3
7	Large	R	51.5	54.5	7.2
	Large	R	49.5	55.0	7.3
	Medium	L	87.5	37.0	7.4
8	Medium	R	123.0	33.0	7.3
9	Medium	L	141.0	–	7.4
10	Medium	L	138.0	–	7.3
	Small	R	138.0	–	7.4
11	Medium	L	98.0	38.5	7.3
	Small	L	146.0	–	–
12	Medium	L	95.0	56.0	7.4
	Small	R	130.0	47.5	7.5
13	Large	L	128.5	29.0	7.4
	Medium	R	121.5	34.5	7.3
	Small	L	151.5	–	–
14	Small	R	129.0	31.5	7.4
	Small	L	121.0	–	–
15	Small	L	140.5	48.5	–
16	Medium	L	114.0	54.0	7.2

L, left; R, right.
[a]Large, >20 mm diameter; medium, 10–20 mm diameter; small, <10 mm diameter.
Adapted from Fraser *et al.*, 1973.

Table IV.10. *Values for pH in the follicular fluid of*
spontaneously cycling pigs within 24 hours of the
anticipated time of ovulation

pH	Reference
7.41	Knudsen *et al.*, 1978
7.34[a]	R.G. Gosden & R.H.F. Hunter, unpublished
7.32[a]	R.H.F. Hunter & P.C. Léglise, unpublished
7.37[a]	R.H.F. Hunter & R. Nichol, unpublished

[a]Graafian follicles of 8–10 mm diameter.

is displaced from the collapsing Graafian follicle at the time of ovulation, then the fate of such fluid needs to be considered. In species in which the ovary is not completely enclosed within a bursa, there are two major routes of disposal for follicular fluid – passage into the lumen of the Fallopian tube, entry into the peritoneal cavity, or a combination of both. The first of these would seem to be the principal destination for released follicular fluid, at least initially so. This is because the fimbriated extremity of the Fallopian tube embraces the ovary at ovulation, and contractile movements of the fimbria together with the enhanced beat of dense tracts of cilia on its inner surface displace the oocyte, its granulosa cells and follicular fluid into the tubal ostium and so to the upper reaches of the ampulla (see Hafez & Blandau, 1969; Hunter, 1988). Indeed, in a number of polytocous (polyovular) laboratory species such as the golden hamster and mouse, distension of the ampulla soon after ovulation indicates the location of the eggs, their aggregated cumulus investments and much of the follicular fluid. These products of ovulation can be released into a culture dish simply by puncturing the wall of that portion of the Fallopian tube. Despite these remarks, Lenton *et al.* (1988) suggested the sudden release of human follicular fluid directly into the peritoneal cavity, but without citing specific observations or references.

Although the newly ovulated eggs continue to be displaced to the site of fertilisation at the ampullary–isthmic junction by both cilial beat and myosalpingeal contraction, a process requiring 9–13 min in rabbits (Harper, 1961), this would seem not to be the destination for most of the follicular fluid. Evidence from large domestic species such as sheep, in which it is straightforward to cannulate the two extremities of the Fallopian tube, indicates that bulk flow of luminal fluid is ad-ovarian at the time of ovulation (Fig. IV.3) and thus out of the tubal ostium into the peritoneal cavity (Bellve & McDonald, 1970). Such fluid will include follicular fluid, although already diluted by mixing with luminal fluid of the Fallopian tubes. One reason for this ad-ovarian direction of bulk fluid displacement is that the muscle layers of the isthmus are extremely tightly constricted at this time, offering resistance to ab-ovarian passage of luminal fluid. Viscous secretions that accumulate within the isthmus – especially its distal (caudal) portion – also act to impede a free flow of fluid. These considerations are emphasised in the case of mares, with their surprisingly small and tight Fallopian tube isthmus and yet a pre-ovulatory Graafian follicle that may achieve 5 cm in diameter.

It is not intended to suggest that all follicular fluid is refluxed from the Fallopian tube in this manner, for a minor quantity will remain in contact with the eggs and their surrounding somatic cells. However, most follicular fluid will be displaced into the peritoneal cavity within a short time of

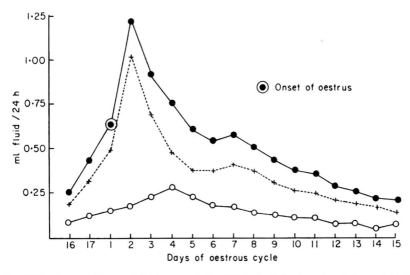

Fig. IV.3. The overall rates of fluid accumulation (●) in the Fallopian tube of sheep plotted according to the day of the oestrous cycle. The direction of flow is indicated by the lower two curves, as estimated by cannulation of the ampullary (+) and isthmic (○) extremities. (After Bellve & McDonald, 1970.)

ovulation. Together with any fluid that has already leaked from the ovarian surface (pre-ovulatory seepage) and escaped under the embracing fimbria, such follicular fluid will contribute to the accumulated peritoneal fluid, and its content of hormones – e.g. follicular fluid prostaglandins – may exert potent influences on the peritoneal membrane itself. In women examined soon after the time of ovulation, aspiration of straw-coloured fluid from the pouch of Douglas will, in part, be sampling diluted follicular fluid. The volume of fluid in the human peritoneal cavity may vary from 5 to 20 ml, the volume changing during the menstrual cycle to reach a maximum value after ovulation (Oral, Olive & Arici, 1996).

Involvement of follicular fluid in events of fertilisation

As noted in the preceding section, a minor quantity of follicular fluid undoubtedly reaches the site of fertilisation associated with the oocyte(s) and its expanding investment of loosened and mucified cumulus cells. This statement may underestimate the passage of follicular fluid immediately after ovulation, but reflux flow of 'bulk fluid' from the ampulla into the peritoneal cavity referred to above will be taken as a working model here. In addition to the presence of antral fluid from the Graafian follicle(s) at the site of fertilisation, many of the granulosa cells (both cumulus oophorus and corona radiata) will remain viable at the site of fertilisation. Hence, their synthetic and secretory activities will also contribute to the fluid microenvironment at the initial meeting place of eggs and spermatozoa (Hunter, 1988). Whilst the volume of fluid so generated may appear minor, this will not be so relative to the size of a sperm cell and, in particular, to the surface area of membranes overlying the sperm head.

Before examining cellular interactions at the site of fertilisation, there is one further important consideration. Following preliminary observations in rabbits and bats on pre-ovulatory sperm distribution in the Fallopian tubes (Harper, 1973; Racey, Suzuki & Medway, 1975), it has become widely appreciated that the caudal (distal) region of the Fallopian tube isthmus serves as the so-called *functional sperm reservoir* in diverse mammalian species (for a review, see Hunter, 1995a), including perhaps in our own species (Hunter, 1995b, 1998). Because the fertilising spermatozoon is activated from the functional sperm reservoir at or close to the time of ovulation, there is a general impression that it is follicular fluid that prompts a primary release of spermatozoa from the isthmus reservoir. This seems improbable for a number of reasons.

1. Initial activation and release of a discrete number of spermatozoa from the functional reservoir commence shortly before ovulation rather than as a consequence of this process (Hunter, 1981, 1984b; Hunter & Nichol, 1983). Although some leakage of follicular fluid from the follicular surface may occur before actual collapse or rupture of the follicle at ovulation, it is doubtful whether differing degrees of leakage of follicular fluid could generate the rather precise initial release of spermatozoa from arrest in the caudal isthmus.

2. As reasoned above, follicular fluid is unlikely to progress down the Fallopian tube(s) towards the uterus, and especially so beyond the ampullary–isthmic junction. Not only is the patency of the isthmus scarcely detectable shortly before or at ovulation, but cilial beat and waves of myosalpingeal contraction at this time proceed principally towards the site of fertilisation (Blandau & Gaddum-Rosse, 1974; Hunter, 1977; Battalia & Yanagimachi, 1979). In addition, there is usually an accumulation of viscous glycoproteins as a form of mucus in the caudal isthmus, and this would further inhibit any putative passage of follicular fluid soon after ovulation.

3. Direct surgical introduction of microdroplets of follicular fluid into the sperm reservoir in the caudal isthmus prompts a major release of spermatozoa leading to extensive polyspermic fertilisation (Hunter, Hagbard-Petersen & Greve, 1999). Such specific evidence would once

again strongly suggest that follicular fluid does not reach the functional sperm reservoir in the caudal isthmus to cause the initial highly controlled activation and release of fertilising spermatozoa.

4. Using the concentration of progesterone in follicular fluid as a tracer, Hansen, Srikandakumar & Downey (1991) estimated that only 0.5% of the anticipated volume of follicular fluid was present in the Fallopian tubes shortly after ovulation, and that this decreased 10–12-fold within 4 hours of ovulation.

Despite these cautionary remarks, follicular fluid almost certainly does play a rôle in the process of capacitation and the ensuing acrosome reaction. Spermatozoa mature towards the fully capacitated state as they (1) ascend through different compartments of the female tract, (2) are liberated from the influence of seminal plasma, and (3) come increasingly into contact with specialised female tract fluids of differing regional composition. However, completion of the capacitation process seemingly does not occur in an arbitrary or fixed number of hours after a pre-ovulatory mating or coitus. Rather, completion of capacitation is an ovulation-related event (Hunter, 1987; Smith & Yanagimachi, 1989). In other words, the relatively unstable and short-lived sperm cell representative of full capacitation achieves this vulnerable state only at the time of ovulation, although it clearly requires a minimum time or threshold even after post-ovulatory insemination (Austin, 1951; Chang, 1951).

There is the further consideration that the process of capacitation may only be completed for the fertilising spermatozoon in the region of the ampullary–isthmic junction and in the presence of the egg investments and associated follicular fluid (Hunter, 1995a, 1997). Follicular fluid is known to promote capacitation under conditions of *in vitro* fertilisation (Yanagimachi, 1969), and it would therefore seem highly probable that it participates under physiological circumstances. As to the ensuing vesiculation of the destabilised membranes around the anterior portion of the sperm head termed the acrosome reaction, there is extensive evidence of an involvement of ovarian follicular fluid. Indeed, the steroid hormone progesterone present in high titres in follicular fluid at the time of ovulation has long been invoked as a potential stimulus for the acrosome reaction (Austin, Bavister & Edwards, 1973). The remodelling influence of progesterone on the sperm surface membranes has more recently been referred to as a non-genomic influence, since it is acting on the plasma and outer acrosomal membranes rather than on the highly condensed sperm nucleus.

It is worth considering that other components of ovarian follicular fluid, for example prostaglandins present in high titres and GAGs and indeed catecholamines, could all be acting on the sperm cell to facilitate the process of fertilisation. Not only may constituents of follicular fluid be involved directly in the events of fertilisation, but they will also be modifying the composition of tubal luminal fluid, especially the regional environment at the ampullary–isthmic junction. This would be initially by dilution and mixing but it is highly probable that components of follicular fluid also act on the endosalpinx in this region to alter the nature of transudation and specific secretion, thereby modifying indirectly the microenvironment of fertilisation. Not only would this be influential in the completion of capacitation, but it might also assist hyperactivation of the sperm flagellum required for penetration of the head through the zona pellucida. Prostaglandins may have a crucial rôle in this regard.

Concluding remarks

This chapter has ranged widely over topics as diverse as the origins of follicular fluid, the nature of its constituents, and the potential contribution of the fluid to physiological events in the Fallopian tubes. Much more could have been written under each heading but the objective was to give a flavour rather than an all-embracing review. One cannot stress too strongly that most considerations of follicular fluid are still at a 'macro' level, whereas fluid composition and fluid influences at the surface of cells – both somatic and the oocyte – should be the current physiological focus. This remark concerns not only microenvironments within a Graafian follicle but also those in different portions of the neighbouring Fallopian tube. As suggested above, the release of follicular fluid, cumulus cells and oocyte at the time of ovulation could have a major programming influence on the duct system, preparing it locally for the events of fertilisation – that is, for final maturation of spermatozoa, penetration of the egg, and development of the zygote into a competent embryo. After arguing the case strongly for such a post-ovulatory contribution of the cumulus (e.g. Hunter, 1988, 1995a), it is pleasing that the follicular cells and fluid released with an oocyte at ovulation – and perhaps also polymorphonuclear leucocytes – are now being appreciated as an important paracrine tissue, exerting a potent influence on sperm–egg interactions at the site of fertilisation. Such has been judged to be the case on the basis of elegant scanning and transmission electron microscopic observations, coupled with persuasive immunofluorescence detail (Motta, Nottola & Familiari, 1997; Motta *et al.*, 1999). A local influence of the cumulus mass on the wall of the duct, especially myosalpinx, has also been proposed (Sato, 1997).

Prior to ovulation and clearly at a micro-level, granulosa cells in the early stages of degeneration together with those that are moribund or already dead will release breakdown products – specific molecules – that in turn will influence the local composition of follicular fluid. Such modification could provide a subtle system of messages both to the oocyte ànd to neighbouring Graafian follicles via the microcirculation in the vascular and lymphatic beds (see Chapter III). Patterns of follicular development, in reality regulation of the follicular hierarchy, might be modulated with sensitivity in this manner, although as part of a much larger signalling programme. As has long been suspected, the chemical and hormonal environment within a follicle will be a key factor in determining the developmental potential of its oocyte. And in this specific context, it is essential to appreciate that different sub-populations of granulosa cells within an individual follicle can secrete a different spectrum of molecules.

Readers with an evolutionary turn of mind will be left dwelling on the advantages of formation of a Graafian follicle containing a significant quantity of antral fluid, and wondering why some mammals have not adopted this feature of ovarian development. There are diverse forms of communication between a gonad and the neighbouring genital duct, and follicular fluid as a medium for well-modulated endocrine conversations would of course be part of a much more extensive vascular, lymphatic and nervous scheme. Due to its steroid hormone and peptide content, follicular fluid may have evolved to play a crucial rôle in follicle selection and dominance (in contrast to a massive unselected wave of ovulation in aquatic ancestral species), and in regulating the growth, meiotic arrest and then resumption of meiosis prior to ovulation.

Over and above these vital aspects, follicular fluid may also have evolved to regulate the developmental potential and fate of granulosa cells, particularly to discourage their tendency to luteinise until an appropriate stage of follicular maturity has been achieved and the pre-ovulatory surge of gonadotrophic hormones provoked by positive feedback of oestradiol has occurred. In this interpretation, formation of follicular fluid might have as much to do with preventing the formation of an incipient corpus luteum as in supporting the cells of its precursor structure.

References

Ainsworth, L., Baker, R.D. & Armstrong, D.T. (1975). Pre-ovulatory changes in follicular fluid prostaglandin F levels in swine. *Prostaglandins*, **9**, 915–25.

Albertini, D.F. & Anderson, E.J. (1974). The appearance of intercellular connections during the ontogeny of the rabbit ovarian follicle with particular reference to gap junctions. *Journal of Cell Biology*, **63**, 234–50.

Austin, C.R. (1951). Observations on the penetration of the sperm into the mammalian egg. *Australian Journal of Scientific Research, B*, **4**, 581–96.

Austin, C.R., Bavister, B.D. & Edwards, R.G. (1973). Components of capacitation. In *The Regulation of Mammalian Reproduction*, pp. 247–54. NIH Bethesda Conference, Bethesda, Maryland.

Baird, D.T. & Fraser, I.S. (1975). Concentration of oestrone and oestradiol in follicular fluid and ovarian venous blood of women. *Clinical Endocrinology*, **4**, 259–66.

Baker, T.G. (1982). Oogenesis and ovulation. In *Reproduction in Mammals*, 2nd edn, vol. 1, ed. C.R. Austin & R.V. Short, pp. 17–45. Cambridge: Cambridge University Press.

Battalia, D.E. & Yanagimachi, R. (1979). Enhanced and coordinated movement of the hamster oviduct during the periovulatory period. *Journal of Reproduction and Fertility*, **56**, 515–20.

Behrman, H.R. (1979). Prostaglandins in hypothalamo-pituitary and ovarian function. *Annual Review of Physiology*, **41**, 685–700.

Beier, H.M., Beier-Hellwig, K. & Delbos, R. (1983). Hormones and proteins involved in uterine preparation for implantation. In *Fertilisation of the Human Egg In Vitro*, ed. H.M. Beier & H.R. Lindner, pp. 307–27. Berlin & Heidelberg: Springer-Verlag.

Bellve, A.R. & McDonald, M.F. (1970). Directional flow of Fallopian tube secretion in the ewe at the onset of the breeding season. *Journal of Reproduction and Fertility*, **22**, 147–9.

Betteridge, K.J. & Raeside, J.I. (1962). Observation of the ovary by peritoneal cannulation in pigs. *Research in Veterinary Science*, **3**, 390–8.

Blandau, R.J. & Gaddum-Rosse, P. (1974). Mechanism of sperm transport in pig oviducts. *Fertility and Sterility*, **25**, 61–7.

Brown, C.G. & Poyser, N.L. (1984). Studies on ovarian prostaglandin production in relation to ovulation in the rat. *Journal of Reproduction and Fertility*, **72**, 407–14.

Burr, J.H. & Davies, J.I. (1951). The vascular system of the rabbit ovary and its relationship to ovulation. *Anatomical Record*, **111**, 273–97.

Cavender, J.L. & Murdoch, W.J. (1988). Morphological studies of the microcirculatory system of periovulatory ovine follicles. *Biology of Reproduction*, **39**, 987–97.

Chang, M.C. (1951). Fertilising capacity of spermatozoa deposited into the Fallopian tubes. *Nature (London)*, **168**, 697–8.

Cook, B., Hunter, R.H.F. & Kelly, A.S.L. (1977). Steroid-binding proteins in follicular fluid and peripheral plasma from pigs, cows and sheep. *Journal of Reproduction and Fertility*, **51**, 65–71.

Edwards, R.G. (1962). Meiosis in ovarian oocytes of adult mammals. *Nature (London)*, **196**, 446–50.

Edwards, R.G. (1965). Maturation *in vitro* of mouse, sheep, cow, pig, rhesus monkey and human ovarian oocytes. *Nature (London)*, **208**, 349–51.

Edwards, R.G. (1974). Follicular fluid. *Journal of Reproduction and Fertility*, **37**, 189–219.

Edwards, R.G. (1980). *Conception in the Human Female*. London & New York: Academic Press.

Fraser, I.S., Baird, D.T. & Cockburn, F. (1973). Ovarian venous blood PO_2, PCO_2 and pH in women. *Journal of Reproduction and Fertility*, **33**, 11–17.

Giorgi, E.P., Addis, M. & Colombo, G. (1969). The fate of free and conjugated oestrogens injected into the Graafian follicle of equines. *Journal of Endocrinology*, **45**, 37–50.

Gore-Langton, R.E. & Armstrong, D.T. (1988). Follicular steroidogenesis and its control. In *The Physiology of Reproduction*, ed. E. Knobil & J.D. Neill, pp. 331–85. New York: Raven Press.

Gosden, R.G. & Byatt-Smith, J.G. (1986). Oxygen concentration gradient across the ovarian follicular epithelium: model, predictions and implications. *Human Reproduction*, **1**, 65–8.

Gosden, R.G. & Hunter, R.H.F. (1988). Electrophysiological properties of the follicle wall in the pig ovary. *Experientia*, **44**, 212–14.

Gosden, R.G., Hunter, R.H.F., Telfer, E., Torrance, C. & Brown, N. (1988). Physiological factors underlying the formation of ovarian follicular fluid. *Journal of Reproduction and Fertility*, **82**, 813–25.

Gosden, R.G., Sadler, I.H., Reed, D. & Hunter, R.H.F. (1990). Characterisation of ovarian follicular fluids of sheep, pigs and cows using proton nuclear magnetic resonance spectroscopy. *Experientia*, **46**, 1012–15.

Grinsted, J., Blendstrup, K., Andreasen, M.P. & Byskov, A.G. (1980). Temperature measurements of rabbit antral follicles. *Journal of Reproduction and Fertility*, **60**, 149–55.

Grinsted, J., Kjer, J.J., Blendstrup, K. & Pedersen, J.F. (1985). Is low temperature of the follicular fluid prior to ovulation necessary for normal oocyte development? *Fertility and Sterility*, **43**, 34–9.

Guraya, S.S. (1985). *Biology of Ovarian Follicles in Mammals*. Berlin, Heidelberg & New York: Springer-Verlag.

Hafez, E.S.E. & Blandau, R.J. (eds.) (1969). *The Mammalian Oviduct*. Chicago: University of Chicago Press.

Hansen, C., Srikandakumar, A. & Downey, B.R. (1991). Presence of follicular fluid in the porcine oviduct and its contribution to the acrosome reaction. *Molecular Reproduction and Development*, **30**, 148–53.

Harper, M.J.K. (1961). Egg movement through the ampullar region of the Fallopian tube of the rabbit. In *Proceedings of the 4th International Congress on Animal Reproduction*, The Hague, pp. 375–80.

Harper, M.J.K. (1973). Stimulation of sperm movement from the isthmus to the site of fertilisation in the rabbit oviduct. *Biology of Reproduction*, **8**, 369–77.

Hunter, R.H.F. (1967). Porcine ovulation after injection of human chorionic gonadotrophin. *Veterinary Record*, **81**, 21–3.

Hunter, R.H.F. (1977). Function and malfunction of the Fallopian tubes in relation to gametes, embryos and hormones. *European Journal of Obstetrics, Gynaecology and Reproductive Biology*, **7**, 267–83.

Hunter, R.H.F. (1980). *Physiology and Technology of Reproduction in Female Domestic Animals*. London & New York: Academic Press.

Hunter, R.H.F. (1981). Sperm transport and reservoirs in the pig oviduct in relation to the time of ovulation. *Journal of Reproduction and Fertility*, **63**, 109–17.

Hunter, R.H.F. (1984a). Ovarian follicular development, maturation and atresia in pigs. In *La Période Péri-Ovulatoire*, Colloque de la Société Française pour l'Étude de la Fertilité, Séminaire Panthéon, pp. 47–54. Paris: Masson et Cie.

Hunter, R.H.F. (1984b). Pre-ovulatory arrest and peri-ovulatory redistribution of competent spermatozoa in the isthmus of the pig oviduct. *Journal of Reproduction and Fertility*, **72**, 203–11.

Hunter, R.H.F. (1987). The timing of capacitation in mammalian spermatozoa – a reinterpretation. *Research in Reproduction*, **19**, 3–4.

Hunter, R.H.F. (1988). *The Fallopian Tubes: Their Rôle in Fertility and Infertility*. Berlin, Heidelberg & New York: Springer-Verlag.

Hunter, R.H.F. (1995a). Ovarian endocrine control of sperm progression in the Fallopian tubes. *Oxford Reviews of Reproductive Biology*, **17**, 85–125.

Hunter, R.H.F. (1995b). Human sperm reservoirs and Fallopian tube function: a rôle for the intra-mural portion? *Acta Obstetricia et Gynecologica Scandinavica*, **74**, 677–81.

Hunter, R.H.F. (1997). Sperm dynamics in the female genital tract: interactions with Fallopian tube microenvironments. In *Microscopy of Reproduction and Development: A Dynamic Approach*, ed. P.M. Motta, pp. 35–45. Rome: Antonio Delfino Editore.

Hunter, R.H.F. (1998). Have the Fallopian tubes a vital rôle in promoting fertility? *Acta Obstetricia et Gynecologica Scandinavica*, **77**, 475–86.

Hunter, R.H.F., Bøgh, I.B., Einer-Jensen, N., Müller, S. & Greve, T. (2000). Pre-ovulatory Graafian follicles are cooler than neighbouring stroma in pig ovaries. *Human Reproduction*, **15**, 273–83.

Hunter, R.H.F., Cook, B. & Baker, T.G. (1976). Dissociation of response to injected gonadotrophin between the Graafian follicle and oocyte in pigs. *Nature (London)*, **260**, 156–8.

Hunter, R.H.F., Grøndahl, C., Greve, T. & Schmidt, M. (1997). Graafian follicles are cooler than neighbouring ovarian tissues and deep rectal temperatures. *Human Reproduction*, **12**, 95–100.

Hunter, R.H.F., Hagbard-Petersen, H. & Greve, T. (1999). Ovarian follicular fluid, progesterone and Ca^{2+} ion influences on sperm release from the Fallopian tube reservoir. *Molecular Reproduction and Development*, **54**, 283–91.

Hunter, R.H.F. & Nichol, R. (1983). Transport of spermatozoa in the sheep oviduct: preovulatory sequestering of cells in the caudal isthmus. *Journal of Experimental Zoology*, **228**, 121–8.

Hunter, R.H.F. & Poyser, N.L. (1985). Ovarian follicular fluid concentrations of prostaglandins E_2, $F_{2\alpha}$ and I_2 during the pre-ovulatory period in pigs. *Reproduction, Nutrition et Développement*, **25**, 909–17.

Jarry, H., Meyer, B., Holzzapfel, G., Hinney, B., Kuhn, W. & Wuttke, W. (1988). Angiotensin II/III and substance P in human follicular fluid during IVF: relation of the peptide content with follicular size. *Acta Endocrinologica, Copenhagen*, **119**, 277–82.

Knudsen, J.F., Litkowski, L.J., Wilson, T.L., Guthrie, H.D. & Batta, S.K. (1978). Concentration of hydrogen ions, oxygen, carbon dioxide and bicarbonate in porcine follicular fluid. *Journal of Endocrinology*, **79**, 249–50.

LeMaire, W.J., Leidner, R. & Marsh, J.M. (1975). Pre- and post-ovulatory changes in the concentration of prostaglandins in rat Graafian follicles. *Prostaglandins*, **9**, 221–9.

Lenton, E.A., King, H., Thomas, E.J., *et al.* (1988). The endocrine environment of the human oocyte. *Journal of Reproduction and Fertility*, **82**, 827–41.

Lipner, H. (1973). Mechanism of mammalian ovulation. In *Handbook of Physiology*, Section 7: *Endocrinology*, vol. II, *Female Reproductive System*, Part I, pp. 409–37. Washington, DC: American Physiological Society.

Liu, H.-C., Sandow, B.A. & Rosenwaks, Z. (1986). Follicular fluid and oocyte maturation. In *In Vitro Fertilisation*, ed. H.W. Jones, G.S. Jones, G.D. Hodgen & Z. Rosenwaks, pp. 106–25. Baltimore, MD: Williams & Wilkins.

Luck, M.R. (1994). The gonadal extracellular matrix. *Oxford Reviews of Reproductive Biology*, **16**, 33–85.

Luck, M.R., Ye, J., Almislimani, H. & Hibberd, S. (2000). Follicular fluid rheology and the duration of the ovulatory process. *Journal of Reproduction and Fertility*, **120**, 411–21.

Mancini, R.E., Vilar, O., Heinrich, J.J., Davidson, O.W. & Alvarez, B. (1963). Transference of circulating labelled serum proteins to the follicle of the rat ovary. *Journal of Histochemistry and Cytochemistry*, **11**, 80–8.

Marsh, J.M., Mills, T.M. & LeMaire, W.J. (1972). Cyclic AMP synthesis in rabbit Graafian follicles and the effect of luteinising hormone. *Biochimica et Biophysica Acta*, **273**, 389–94.

McCaig, C.D. (1985). A potential difference across mouse ovarian follicle. *Experientia*, **41**, 609–11.

McNatty, K.P. (1978). Follicular fluid. In *The Vertebrate Ovary. Comparative Biology and Evolution*, ed. R.E. Jones, pp. 215–59. New York: Plenum Press.

McNatty, K.P. (1982). Ovarian follicular development from the onset of luteal regression in humans and sheep. In *Follicular Maturation and Ovulation*, ed. R. Rolland, E.V. Van Hall, S.G. Hillier, K.P. McNatty & J. Shoemaker, pp. 1–8. Amsterdam: Excerpta Medica.

Ménézo, Y. & Guérin, P. (2001). Biochimie de l'environnement *in vivo* et *in vitro* de l'ovocyte et du jeune l'embryon. In *La Reproduction chez les Mammifères et l'Homme*, ed. C. Thibault & M.C. Levasseur, pp. 410–24. Paris: Ellipses.

Moir, D.D. & Mone, J.G. (1964). Acid–base balance during epidural anaesthesia. *British Journal of Anaesthesia*, **36**, 480–5.

Mossman, H.W. & Duke, K.L. (1973). *Comparative Morphology of the Mammalian Ovary*. Madison, WI: University of Wisconsin.

Motta, P.M., Nottola, S.A. & Familiari, G. (1997). Cumulus corona cells at fertilisation and segmentation: a paracrine organ. In *Microscopy of Reproduction and Development: A Dynamic Approach*, ed. P.M. Motta, pp. 173–82. Rome: Antonio Delfino Editore.

Motta, P.M., Nottola, S.A., Familiari, G., Macchiarelli, G., Correr, S. & Makabe, S. (1999). Structure and function of the human oocyte–cumulus–corona cell complex before and after ovulation. *Protoplasma*, **206**, 270–7.

Oral, E., Olive, D.L. & Arici, A. (1996). The peritoneal environment in endometriosis. *Human Reproduction Update*, **2**, 385–98.

Perloff, W.H., Schultz, J., Farris, E.J. & Balin, H. (1955). Some aspects of the chemical nature of human ovarian follicular fluid. *Fertility and Sterility*, **6**, 11–16.

Peters, H. & McNatty, K.P. (1980). *The Ovary*. London & Toronto: Paul Elek, Granada Publishing.

Poyser, N.L. (1981). *Prostaglandins in Reproduction*. Chichester: John Wiley & Sons Ltd.

Racey, P.A., Suzuki, F. & Medway, L. (1975). The relationship between stored spermatozoa and the oviducal epithelium in bats of the genus *Tylonycteris*. In *The Biology of Spermatozoa*, ed. E.S.E. Hafez & C.G. Thibault, pp. 123–33. Basel: Karger.

Robinson, A. (1918). The formation, rupture and closure of ovarian follicles in ferrets and ferret–polecat hybrids, and some associated phenomena. *Transactions of the Royal Society of Edinburgh*, **52**, 303–62.

Sato, E. (1997). Cumulus differentiation and its possible physiological roles. In *Microscopy of Reproduction and Development: A Dynamic Approach*, ed. P.M. Motta, pp. 129–35. Rome: Antonio Delfino Editore.

Shalgi, R., Kraicer, P., Rimon, A., Pinto, M. & Soferman, N. (1973). Proteins of human follicular fluid: the blood–follicle barrier. *Fertility and Sterility*, **24**, 429–34.

Shalgi, R., Kraicer, P.F. & Soferman, N. (1972). Gases and electrolytes of human follicular fluid. *Journal of Reproduction and Fertility*, **28**, 335–40.

Shivers, S.A., Metz, C.B. & Lutwak-Mann, C. (1964). Some properties of pig follicular fluid. *Journal of Reproduction and Fertility*, **8**, 115–20.

Short, R.V. (1957). Steroids related to progesterone in the human and mare placenta. *Journal of Endocrinology*, **15**: i–ii.

Short, R.V. (1958). Progesterone in blood. I. The chemical determination of progesterone in peripheral blood. *Journal of Endocrinology*, **16**, 415–25.

Short, R.V. (1960). Steroids present in the follicular fluid of the mare. *Journal of Endocrinology*, **20**, 147–56.

Short, R.V. (1962). Steroids present in the follicular fluid of the cow. *Journal of Endocrinology*, **23**, 401–11.

Sinosich, M.J. (1987). Human ovarian follicular antigens. In *Future Aspects in Human In Vitro Fertilisation*, ed. W. Feichtinger & P. Kemeter, pp. 64–76. Berlin, Heidelberg & New York: Springer-Verlag.

Smith, T.T. & Yanagimachi, R. (1989). Capacitation status of hamster spermatozoa in the oviduct at various times after mating. *Journal of Reproduction and Fertility*, **86**, 255–61.

Stangroom, J.E. & Weevers, R.G. (1962). Anticoagulant activity of equine follicular fluid. *Journal of Reproduction and Fertility*, **3**, 269–82.

Strauss, F. (1950). Ripe follicles without antra and fertilisation within the follicle: a normal situation in a mammal. *Anatomical Record*, **106**, 251–252.

Szöllösi, D., Gérard, M., Ménézo, Y. & Thibault, C. (1978). Permeability of ovarian follicle: corona cell–oocyte relationship in mammals. *Annales de Biologie Animale, Biochimie, Biophysique*, **18**, 511–21.

Takikawa, H. (1966). Binding of steroids to follicular fluid proteins. In *Steroid Dynamics*, ed. G. Pincus, T. Nakao & J.F. Tait, pp. 217–35. New York: Academic Press.

Takikawa, H. & Yoshinaga, S. (1968). Isolation of oestrogen binding proteins from follicular fluid of the cow ovary. In *Proceedings of the 3rd International Congress of Endocrinology, Mexico*, International Congress Series No. 157, Abstract No. 21, p. 9.

Tsang, B.K., Ainsworth, L., Downey, B.R. & Armstrong, D.T. (1979). Pre-ovulatory changes in cyclic AMP and prostaglandin concentrations in follicular fluid of gilts. *Prostaglandins*, **17**, 141–8.

Yanagimachi, R. (1969). In vitro capacitation of hamster spermatozoa by follicular fluid. *Journal of Reproduction and Fertility*, **18**, 275–86.

Yang, N.S.T., Marsh, J.M. & LeMaire, W.J. (1974). Post-ovulatory changes in the concentration of prostaglandins in rabbit Graafian follicles. *Prostaglandins*, **6**, 37–44.

Young, W.C. (1961). The mammalian ovary. In *Sex and Internal Secretions*, 3rd edn, ed. W.C. Young, pp. 449–96. Baltimore, MD: Williams & Wilkins.

Zachariae, F. (1958). Studies on the mechanism of ovulation: permeability of the blood–liquor barrier. *Acta Endocrinologica, Copenhagen*, **27**, 339–42.

Zachariae, F. & Jensen, C.E. (1958). Studies on the mechanism of ovulation: histo-chemical and physico-chemical investigations on genuine follicular fluids. *Acta Endocrinologica, Copenhagen*, **27**, 343–55.

V

Endocrine potential and function
of a Graafian follicle

Introduction

A principal concern of this chapter lies in the ability of theca interna and granulosa cells to synthesise steroid hormones, and the manner in which these two cell types are programmed by gonadotrophic hormones. Not least, there is an interest in the way in which the biosynthetic ability of these two cell types changes according to the maturity of a Graafian follicle. Because Chapter IV was concerned with follicular fluid, the extent to which steroid hormones accumulate in the follicular antrum versus the extent to which they are liberated into the vascular and lymphatic beds also comes under review. Two further aspects require close examination: first, the ability of locally produced peptide factors and cytokines to mediate and alter the responses of theca and granulosa cells to trophic hormones from the anterior pituitary gland; and second, the ability of the oocyte itself to influence the secretory potential of follicular somatic cells and indeed the differentiation and destiny of such cells (see Westman, 1934).

In recent years, the cell layers on either side of the basement membrane of a Graafian follicle have more and more come to be seen as a functional unit, working in concert to synthesise specific steroid hormones. In fact, the idea of such synergism dates from at least 1959 when Falck proposed – and demonstrated in an experimental model – that a theca interna-derived steroid precursor molecule passed to the granulosa cells for enzymatic conversion down the biosynthetic pathway. In one sense, however, the notion of synergism between follicular cell types had existed for very much longer, for Corner (1919) appreciated that theca and granulosa cells together gave rise to lutein cells and differentiated into a functional corpus luteum. This realisation pre-dated discovery of the gonadotrophic hormones by a number of years (Aschheim & Zondek, 1927; Smith & Engle, 1927).

Anticipating and responding to the question as to why hormones of the Graafian follicle are so important: they serve to integrate reproductive events at several different levels. Viewed classically, gonadal hormones programme the hypothalamus and anterior pituitary gland, especially in terms of oestrous behaviour (sexual receptivity) and the pre-ovulatory surge of gonadotrophic hormones, which itself determines the time of ovulation (see Chapter VIII). To this extent, and by means of a 'long' feedback loop, follicular hormones act to programme the destiny of the cells from which they originate. They also influence the tissues of the reproductive tract, more immediately by local mechanisms as well as via the systemic circulation, and impose tissue changes that are characteristic of oestrous or menstrual cycles, most conspicuously in the uterus. Secretion of hormones by a maturing follicle will influence its oocyte and somatic cell investments, and there is doubtless also local modulation of the follicular hierarchy in terms of recruitment, selection and dominance (see Chapter VI). And finally in this introductory context, follicular hormones will have a significant local influence on the process of angiogenesis (for reviews, see Findlay, 1986; Reynolds, Killilea & Redmer, 1992; Redmer & Reynolds, 1996).

Steroids and gonadotrophins: 'two-cell' theory of oestradiol synthesis

Graafian follicles are known as a major site of oestrogen synthesis and secretion. Depending on species, the corpus luteum and placenta may also be important sources of oestrogens in the systemic circulation, even though progesterone is considered to be their primary steroid product. Oestradiol is the principal hormone secreted by a maturing Graafian follicle, although, in the hours following the pre-ovulatory surge of gonadotrophic hormones, synthesis of progesterone gradually assumes prominence. This change from ovarian oestradiol to progesterone secretion reflects the process of luteinisation and is a consequence of loss of aromatisation potential. Whereas precursor molecules in the circulating blood and ovarian cells themselves provide the substrate for specific stepwise enzymatic conversion down the biosynthetic pathway for steroid hormones (Fig. V.1), synthesis of oestradiol is frequently considered simply in terms of its derivation from androstenedione. This is because most follicular oestradiol synthesis is by the granulosa cells, and androstenedione substrate in particular furnished by the theca interna diffuses across the basement membrane for aromatisation by the granulosa cells. Indeed, this represents the essence of the 'two-cell' theory of oestrogen biosynthesis in the mammalian ovary (Falck, 1959). Before reviewing this theory, consideration must be given to two important questions.

cholesterol   20α-hydroxycholesterol   20,22-dihydroxycholesterol

20α-hydroxy-4-pregnen-3-one   progesterone   pregnenolone

17α-hydroxyprogesterone   17α-hydroxypregnenolone

testosterone   androstenedione   dehydroepiandrosterone

17β-oestradiol   oestrone

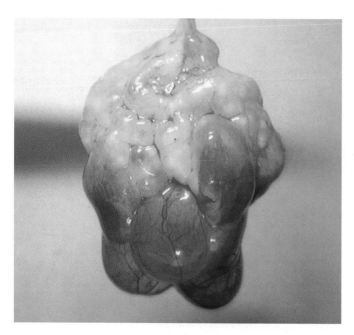

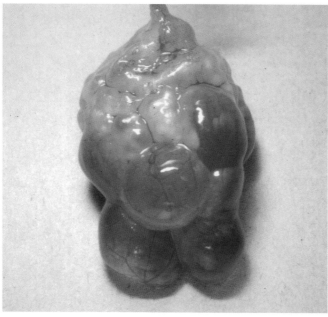

PLATE 01

Pig ovaries viewed at the start of the process of ovulation to show deformation of previously spherical follicles of 9–10mm diameter. These highly haemorrhagic and now flaccid structures appear pendulous when the ovary is raised. Note the absence of other Graafian follicles larger than 2–3mm diameter.

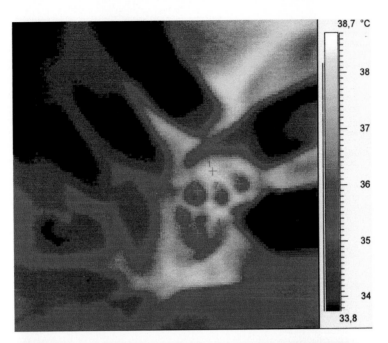
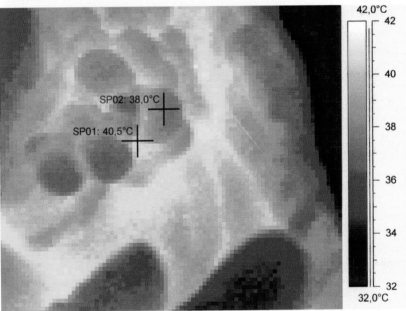

PLATE 02

Thermo-imaging of pig ovarian tissues examined during mid-ventral laparotomy with the fimbria displaced. Note the temperature differential between pre-ovulatory Graafian follicles of 9–10mm diameter (cooler) and adjoining ovarian stroma (warmer).

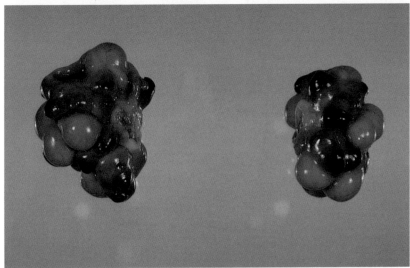

PLATE 03
Upper An example of superovulation (total of 74 recent ovulations) obtained in response to a single injection of 1500 i.u. PMSG given at the onset of the follicular phase of the oestrous cycle.
Lower Recent ovulations induced during the luteal phase of the oestrous cycle by means of 1500 i.u. PMSG on day 5 followed four days later by 500 i.u. hCG.

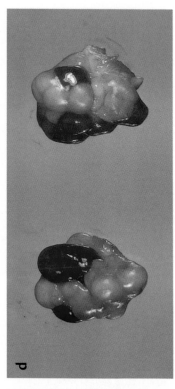

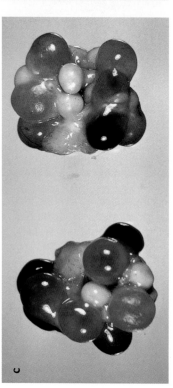

PLATE 04 To portray the sequence of changes in the wall of mature Graafian follicles during the last hour before ovulation. Note the progressively increasing degree of haemorrhage in (a), (b) and (c), and the collapsed follicles in (d). The spherical yellow bodies are corpora albicantia from the preceding oestrous cycle.

First, bearing in mind the prominence of the theca interna and its extensive vascular bed, in marked contrast to the avascular granulosa cell layers (Fig. V.2), is it reasonable and logical to assume that cells of the theca interna do not themselves contribute to oestradiol synthesis? A proliferative influence of oestradiol on the capillary bed is well documented (see Sturgis, 1961; Greenwald & Roy, 1994). Second, if a major strategy underlying oestradiol secretion is for circulating steroid hormone to reach target tissues in the reproductive tract and brain, then are the tactics implicit in the 'two-cell' theory the most effective means of achieving this end? In other words, to shunt androgen precursor out of theca cells and across the basement membrane of the follicle into granulosa cells for conversion to oestradiol, and then for this steroid product to have to leave the granulosa cells, once more traverse the basement membrane in order to reach the theca capillary and lymphatic beds and so to the systemic circulation, represents a *tour de force*! This is especially so when regarding the architecture of a Graafian follicle and appreciating that only one layer of granulosa cells actually lies on the basement membrane, so further diffusion of precursor molecules would be required if the inner layers of granulosa cells were to participate actively in the process of aromatisation.

A related question in this discussion is precisely why should androgens – notably androstenedione – be mobilised inwards from the theca interna to find their way to appropriate granulosa cells instead of passing immediately into blood and lymphatic capillaries in the original state or as a conversion product? Taken together, these various lines of thought should alert the reader to a possible direct contribution of theca cells themselves in the secretion of circulating oestradiol. A critical point would be whether aromatase activity or potential (i.e. its mRNA) can be demonstrated in cells of the theca interna. Current molecular techniques may not describe all the relevant species (isomeric forms) of mRNA.

Reflections along the above lines prompted the author in a general review of the topic to suggest that 'cells of the theca interna predominate in the biosynthesis of oestrogens' (Hunter, 1980), a view that failed to be supported by findings in cultured preparations of dispersed theca or granulosa cells. Such *in vitro* studies on the separated components of dissected follicles supported the 'two-cell' theory of Falck (1959). In essence, and with considerable technical skill, Falck transplanted different types of follicle cell into the anterior chamber of the eye. By observing morphological changes in strips of vaginal epithelial tissue introduced alongside the transplanted follicle cells, he noted

Fig. V.1. The biosynthetic pathway for gonadal steroid hormones illustrating the derivation of progestins, androgens and oestrogens by progressive enzymatic modification of the cholesterol molecule. (Courtesy of Dr B. Cook.)

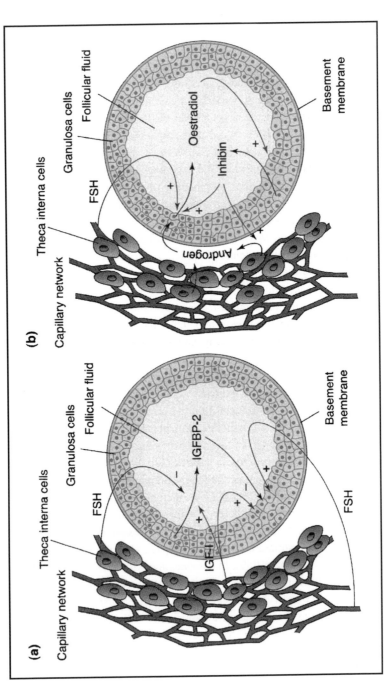

Fig. V.2. (a and b) Diagrammatic representation of a Graafian follicle to highlight the prominent vascular bed in the theca interna. A basement membrane separates it from the avascular granulosa cells. Shortly before ovulation (b), as a response to the LH surge, the basement membrane undergoes dissolution, enabling capillaries to colonise the granulosa cells. FSH, follicle stimulating hormone; IGF-1, insulin-like growth factor 1; IGFBP-2, IGF-binding protein 2. (After Armstrong & Webb, 1997.)

that maximum synthesis of oestrogen(s) took place when granulosa and theca cells were sited next to each other. Accordingly, he deduced that some form of synergism was occurring between the two cell types during follicular oestrogen synthesis *in situ*. Following these impressive studies, and in effect their endorsement by others (e.g. Short, 1962; Baird, 1977; Fortune, 1986; Hillier, Whitelaw & Smyth, 1994), the 'two-cell' theory has assumed the status of dogma: androstenedione, and to a lesser extent testosterone, from the theca pass as substrate to granulosa cells for conversion into oestradiol-17β owing to the action of cytochrome P450 aromatase. However, in recent considerations of steroidogenesis in Graafian follicles, the concept that cells of the theca interna might themselves synthesise oestradiol has been given a new lease of life, and demonstrated as fact in porcine follicular tissue (Hunter *et al.*, 1994; Shores & Hunter, 1999). Even so, it is accepted that granulosa cells of fully developed Graafian follicles have high aromatase activity and will convert most androgens diffusing from the theca into oestrogens. Such conversion finds expression in the concentration of oestradiol in follicular fluid (see Hunter, Cook & Baker, 1976).

Remaining within the title of this chapter, there could now be an extensive discussion of steroid hormones and the specific enzymes required for programming molecular changes and thereby progression down the biosynthetic pathway. But this would be familiar territory to many readers and somewhat repetitive of classical reviews (Savard, Marsh & Rice, 1965; Baird, 1984; Gore-Langton & Armstrong, 1988). As a concise reminder, it could be noted that the biosynthetic pathway for steroid hormones is described conventionally with cholesterol (a 27-carbon compound) as the starting point. Subsequent enzymatic conversions enable the molecule to progress to the stage of progestins (21-carbon compounds; pregnane family), androgens (19-carbon compounds; androstane family), or oestrogens (18-carbon compounds; oestrane family). Both low and high density lipoprotein are sources of precursor cholesterol in the ovary. Bound to such lipoproteins, cholesterol is available in the systemic circulation, so assuming no limitation in its movement from the theca capillary bed into theca cells and across the basement membrane to the granulosa cells, then the activity of enzymes will regulate the actual steroid hormone generated. The first of these is pregnenolone, a precursor to all ovarian steroid hormones. The steroidogenic enzymes are located principally in the mitochondria or endoplasmic reticulum.

One of the functions of gonadotrophic stimulation is to programme enzymes in the biosynthetic pathway. The availability of binding sites for FSH and LH on theca and granulosa cells at different stages of maturity is therefore of immediate relevance. At the very least, the availability of specific receptors, i.e. the number and affinity, on respective follicular cell types gives an indication of potential biosynthetic activities. Binding of gonadotrophic hormones to the inner cell layers of a pre-ovulatory follicle was clearly demonstrated with radio-labelled tracer preparations (Rajaniemi & Vanha-Pertulla, 1972). However, responses of the ovary to gonadotrophic hormones are modulated locally by autocrine factors within a Graafian follicle such as steroids, peptide factors and proteins. That the gonadotrophins themselves are essential for follicular development and maturation is demonstrable by both hypophysectomy and immunisation against GnRH: there are no ovulatory follicles in the absence of FSH and LH (Driancourt *et al.*, 1993).

Theca cells develop LH receptors and granulosa cells develop FSH receptors (Fig. V.3). LH receptors are also acquired in due course by stimulated granulosa cells of mature Graafian follicles. In reality, FSH acts in concert with oestradiol to stimulate the development of LH receptors on the granulosa cells of a maturing follicle. Both theca and granulosa cells have the potential to synthesise progesterone from cholesterol precursor, the conversion being stimulated by LH for cells of the theca interna and FSH/LH for granulosa cells. Once LH receptors have developed on granulosa cells, such cells secrete progesterone more prominently. The potential for further stepwise conversion to androgens is seemingly limited to theca cells under the influence of LH (Fig. V.4), LH acting at the same time to stimulate ovarian blood flow, thereby enhancing the availability of substrate (Janson, 1975). As outlined above, the final step in the pathway, the conversion of androstenedione to oestradiol, occurs principally in the granulosa cells due to the availability of aromatase under the influence of FSH stimulation. In general, theca cells are considered to have a limited ability to synthesise oestrogens. To the extent that this last statement is based on *in vitro* observations, with cultured cells dissected free from follicles and thus deprived of their intrinsic innervation, it should be treated with caution. As suggested at the outset, synthesis and secretion of oestradiol by cells of the theca interna under *in vivo* conditions may be of physiological significance in a fully developed follicle. What is not in dispute is that theca cells have only a limited aromatase activity, in contrast to the relatively abundant aromatase of granulosa cells.

Overall, the increasing frequency of LH pulses during the late follicular phase of an oestrous or menstrual cycle stimulates an increasing secretion of androgens from the theca cells. These pass principally as androstenedione to the granulosa cells and thereby indirectly increase production of oestradiol. As to FSH, it stimulates aromatisation and its influence is enhanced by an autocrine action of IGF-1 and inhibin produced by the granulosa cells themselves, but reduced by a paracrine action of EGF from the theca cells (Driancourt *et al.*, 1993). The influence of FSH is also amplified by activin but reduced by IGF-binding

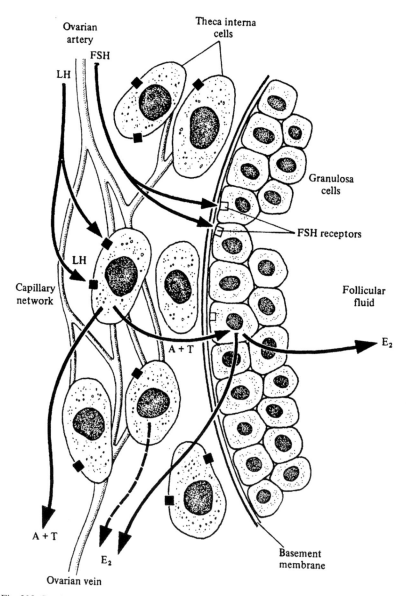

Fig. V.3. Semi-diagrammatic portrayal of the presence of receptors (binding sites) for LH on cells of the theca interna and FSH on cells of the granulosa in a maturing Graafian follicle. E_2, oestradiol; A, androstenedione; T, testosterone. (After Baird, 1984.)

CONTROL OF STEROIDOGENESIS IN GRANULOSA AND THECA CELLS

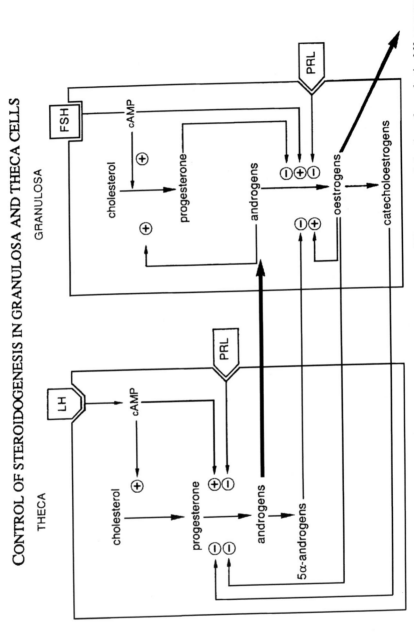

Fig. V.4. Diagrammatic portrayal of theca and granulosa cells with their respective LH and FSH receptors, and also those for prolactin. LH promotes the formation of androgens in theca cells of a Graafian follicle for subsequent aromatisation in a granulosa cell under the influence of FSH. cAMP, cyclic

proteins (IGFBPs) reducing the availability of IGF (see below). With increasing follicular growth, inhibin production also increases steadily and, as the Graafian follicle becomes fully mature, oestradiol and inhibin feed back negatively on the secretion of FSH. The resultant fall in the concentration of circulating FSH may assist selection of the dominant follicle (see Chapter VI).

Seemingly of relevance here, insulin significantly stimulated aromatase activity in the absence of FSH in an *in vitro* system culturing granulosa cells from small bovine follicles (2–4 mm diameter). However, FSH decreased aromatase activity in the presence of high doses of insulin (Bhatia & Price, 2001). Nonetheless, the interactions of FSH and insulin upon this key steroidogenic enzyme will doubtless change according to the stage of follicular maturity and in the presence of a viable oocyte.

The cellular action of gonadotrophic hormones has been examined at some length in terms of biochemical detail. Upon binding to their specific receptors, both LH and FSH trigger a cAMP-mediated mechanism for a rate-limiting oestrogen production through cAMP-activated protein kinases.

To conclude this section, reference will be made to values for steroid hormones in the circulation of women. Peripheral concentrations of testosterone vary between 0.2 and 0.7 ng/ml according to the stage of the menstrual cycle. Approximately half of the testosterone is synthesised in the ovary, the remainder in the adrenal cortex. The concentration of oestrogens in the serum of cyclic individuals ranges from undetectable to some 700 pg/ml, generally reaching a peak 1 day before ovulation. About 60% of oestrogens secreted into the circulation is transported bound to sex hormone-binding globulin (SHBG), a further 20% is bound to albumin, and 20% remains in the free form. The four oestrogens found in women (oestradiol-17α, oestradiol-17β, oestriol and oestrone) have identical biological activities but different potencies (Ying & Zhang, 1999). The dominant follicle is the major source of oestradiol secreted by the human ovary, contributing more than 90% of this steroid in the systemic circulation. Concentrations of oestradiol in the follicular fluid are also elevated and may exceed 2000 ng/ml. Progesterone is the major progestin and a precursor for androgens and oestrogens. Peripheral concentrations of progesterone in women range from undetectable to 10 ng/ml, with a peak at day 8 after ovulation. A set of values for steroid hormones in human follicular fluid is given in Table V.1.

Steroid acute regulatory protein

A critical rate-limiting step in the acute upregulation of steroid hormone biosynthesis in the gonads and adrenal glands is the presentation of cholesterol

Table V.1. *Changes in the mean (±SEM) concentrations (ng/ml) of steroids in follicular fluid from the selectable to the pre-ovulatory stage of ovarian development in women*

Follicle type	E_2	A + T + DHT	17α-OHP	P
Atretic (1–5 mm)	20 ± 5	794 ± 90		73 ± 13
Selectable	15 ± 5	638 ± 113		130 ± 45
Newly selected (EF)	658 ± 38	487 ± 128	713 ± 318	417 ± 120
Pre-ovulatory (MF)	1270 ± 161	542 ± 176	460 ± 112	440 ± 74
Pre-ovulatory (LF: E_2 peak)	2396 ± 348	203 ± 37	1002 ± 212	1228 ± 228
Pre-ovulatory (LF: $E_2 \ll$ LH surge)	2583 ± 228	287 ± 44	1812 ± 142	2464 ± 226
Pre-ovulatory (after the LH surge)	1109 ± 142	79 ± 21	2034 ± 326	7773 ± 643

EF, early follicular phase; MF, mid-follicular phase; LF, late follicular phase; E_2, oestradiol; A, androstenedione; T, testosterone; DHT, dihyrotestosterone; P, progesterone; OHP, hydroxyprogesterone.
After Gougeon, 1997.

substrate to the inner mitochondrial membrane. It is there converted to pregnenolone by a step requiring cytochrome P450 cholesterol side chain cleavage enzyme (Fig. V.5). A novel protein for transporting cholesterol into the mitochondria is termed steroid or steroidogenic acute regulatory protein (StAR), being especially involved in the acute regulation of steroid synthesis in response to a gonadotrophin surge. Deficiencies of the protein, as in patients with congenital lipoid adrenal hyperplasia (an autosomal recessive disorder associated with mutations in the StAR gene) lead to deficient steroid production by both adrenals and gonads (Miller, 1997). And, of direct relevance, targeted inactivation of the StAR gene in mice causes severe impairment of steroidogenesis (Caron *et al.*, 1997). It is now appreciated that StAR acts on the outer mitochondrial membrane to direct sterol flux to the cholesterol-poor inner membranes (Arakane *et al.*, 1997). In the case of mutations, the absence of StAR allows only limited conversion of cholesterol into pregnenolone, resulting in accumulation of lipid droplets in gonadal cells and those of the adrenal cortex.

StAR was first identified as a 30 kDa phosphoprotein, noted to accumulate rapidly in mitochondria if appropriate cells were stimulated by trophic hormones or analogues of cAMP. The precursor form of StAR has a half-life in minutes, whereas the 30 kDa molecule is longer lived, sufficient for its localised cholesterol transporting function. Analysis of spontaneously occurring mutations and site-directed mutants indicated that the C terminus of the protein is

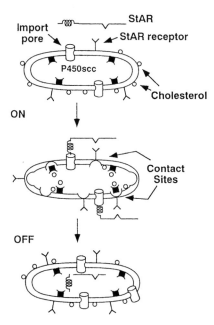

Fig. V.5. A simplified model depicting the action of steroidogenic acute regulatory protein (StAR), indicating initial binding to the outer mitochondrial membrane. Formation of contacts between outer and inner membranes then enables cholesterol to flow down a chemical gradient to reach its side chain cleavage enzyme. It is then processed, removed from its receptor, and the membrane contact sites are dismantled. (After Arakane *et al.*, 1997.)

critical for functional activity. Expression of the StAR gene is dependent upon the orphan nuclear receptor, steroidogenic factor 1. This is also essential for the expression of steroidogenic enzymes.

In human ovaries, granulosa cells contain low amounts of StAR mRNA before ovulation but this is increased by superovulation treatment (Sugawara *et al.*, 1995), which results in large amounts of StAR being revealed during luteinisation. There is thus the suggestion that StAR expression is induced by gonadotrophin treatment. Analogues of cAMP will increase transcription of the StAR gene in granulosa cells (Kiriakidou *et al.*, 1996). In bovine ovaries obtained at ovariectomy, StAR mRNA was present in all follicles examined, and detected in the theca interna but not in granulosa cell layers by northern blotting; this was confirmed by *in situ* hybridisation studies (Bao *et al.*, 1998). Treatment of cows with equine chorionic gonadotrophin increased StAR mRNA in such follicles in comparison with FSH treatment (Soumano & Price, 1997). These findings stand in contrast to those of an earlier bovine study from the same group that reported StAR in both theca and granulosa cells (Pescador *et al.*, 1996), perhaps a reflection of proximity to ovulation and the extent of gonadotrophin treatment or of heterogeneity in the cell populations being examined. Worth bearing in mind here is that StAR mRNA has been revealed in granulosa or luteinised granulosa cells of atretic follicles (Bao *et al.*, 1998).

As to equine follicles, StAR mRNA could not be detected in granulosa cells before treatment with hCG. Such treatment altered StAR expression in both granulosa cells and those of the theca interna. Whereas granulosa expression increased conspicuously 30 hours after hCG treatment, StAR mRNA decreased progressively in the theca cells during this interval (Kerban, Boerboom & Sirois, 1999). This pattern was essentially confirmed by Watson, Thomson & Howie (2000) using the technique of immunostaining for StAR.

The rôle of StAR in facilitating entry of cholesterol to the inner mitochondrial membrane provides a probable intracellular target of the LH, cAMP and protein kinase A (PKA) pathway for the acute regulation of progesterone production by luteinised cells (Zeleznik, 1999). Intracellular cAMP, in response to FSH or LH binding to their receptors, results in the activation of PKA by dissociation of the regulatory subunit of PKA from its catalytic subunit. The active catalytic subunit is then able to phosphorylate cytoplasmic proteins that cause a rapid stimulation of steroidogenesis in ovarian cells by facilitating the entry of cholesterol to the inner mitochondrial membrane in the manner described.

The StAR protein sequence appears to be highly conserved, with greater than 90% similarity among species so far examined (Stocco & Clark, 1996). Consisting of seven exons, the StAR gene has been cloned and the kinetics of responses to LH in cultured theca cells examined. In cells harvested from bovine follicles in the mid-to-late follicular phase of the oestrous cycle, detectable upregulation of StAR gene transcription occurred within 1–2 hours and reached a maximum at 4 hours. mRNA levels had returned to baseline within 12 hours (Ivell *et al.*, 2000). Despite its activity in gonadal tissues and in cells of the adrenal cortex, the StAR gene has been reported not to be expressed in the human placenta (Arakane *et al.*, 1997), a major site of steroid hormone synthesis. Several explanations come to mind. One would be an inappropriate sampling of placental tissue at an inappropriate stage of gestation. Another might be that acute responses are more likely to be critical during relatively short-term events of the menstrual cycle.

Ovarian proteins – inhibin, activin and follistatin

Usually present in large amounts in the fluid of maturing Graafian follicles, these peptides not only influence growth, differentiation and atresia of the follicle but may also act to regulate maturation of the oocyte. They can be viewed as supplementing or, better still, as modulating the key rôle of gonadotrophic hormones in programming follicular development, granulosa cell differentiation and steroid hormone synthesis. Inhibin and activin may function both systemically in a feedback loop and locally within the ovary by autocrine and paracrine routes. There is evidence too for local expression of inhibin/activin molecules within the anterior pituitary gland. The field has received numerous reviews (e.g. Baird & Smith, 1993; Findlay, 1993; Hillier & Miro, 1993; Woodruff & Mather, 1995; Knight, 1996; Matzuk *et al.*, 1996; Mather, Moore & Li, 1997; Knight & Glister, 2001).

Inhibin

Long suspected as a non-steroidal regulator of gonadotrophin secretion from the anterior pituitary gland, inhibin was eventually isolated from follicular fluid and accepted to be produced mainly by granulosa cells. An *in vitro* assay was developed on the basis of its ability to suppress FSH secretion by cultured rat pituitary cells. Inhibin was found selectively to suppress both basal and GnRH-stimulated FSH synthesis and secretion without influencing LH secretion (for a review, see Ying & Zhang, 1999). Often referred to in the singular, inhibins in fact represent a complex family of molecules (Fig. V.6), structurally related to the TGF-β superfamily.

Secreted predominantly by the ovaries, inhibins are dimeric proteins – more specifically, heterodimeric glycoproteins linked by disulphide bonds – consisting of a common α subunit and one of two highly homologous β subunits (A and B). They thereby constitute either α–β_A subunits (inhibin A) or α–β_B subunits (inhibin B). Granulosa cells synthesise an excess of α subunit over β subunit and therefore secrete substantial amounts of free α subunit in addition to related peptides. The concentration of the different forms of inhibin in follicular fluid – the free α subunits and the α–β dimers – changes during development of the follicle, and may influence maturational processes in the oocyte (Silva, Groome & Knight, 1999). Inhibin A is a product of the dominant follicle whereas inhibin B comes from the cohort of growing follicles. All dimeric forms of inhibin appear to suppress FSH secretion by the anterior pituitary gland. In contrast, free α subunit forms do not demonstrate inhibin-like bioactivity, although the full-length α subunit precursor will inhibit the binding of FSH to its receptor. Because there is evidence that inhibin subunits are expressed within the anterior pituitary gland, a paracrine regulation of FSH certainly needs to be considered.

Diverse experimental studies have shown that gonadotrophins and steroid hormones can regulate secretion of immuno-reactive inhibin in various

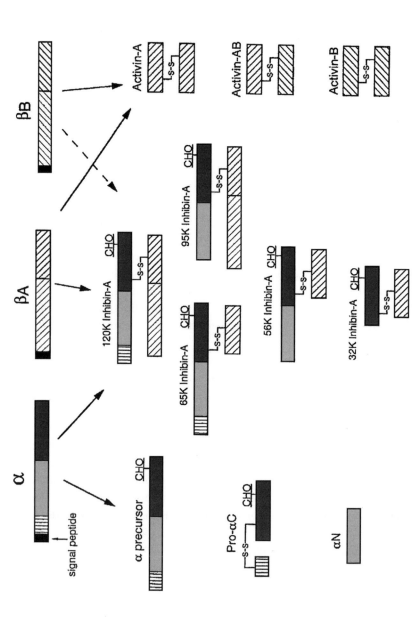

Fig. V.6. Depiction of the synthesis and structure of inhibin, a molecule related to TGF-β. The subunits are encoded by mRNAs that generate far larger precursor molecules, hence the necessary cleavage steps with divergent processing to produce various active molecules. (After Schwall, 1999b.)

mammalian species: human (Hillier *et al.*, 1991a), non-human primate (Hillier *et al.*, 1989), rat (Erickson & Hsueh, 1978; Suzuki *et al.*, 1987) and cow (Henderson & Franchimont, 1983). Gonadotrophic hormones and sex steroids can also influence secretion of bioactive dimeric inhibins by human granulosa-luteal cells *in vitro* (Muttukrishna, Groome & Ledger, 1997). Secretion of both inhibins (A and B) *in vivo* may be regulated by FSH or hCG in our own species (Lockwood, Muttukrishna & Groome, 1996). In cattle, FSH will increase secretion of inhibin A by non-luteinised granulosa cells *in vitro* (Glister *et al.*, 2001).

Evidence that inhibin A is derived from a mature dominant follicle comes from its concentrations in human serum, which peak shortly before the pre-ovulatory gonadotrophin surge. Such peripheral concentrations of inhibin also show a peak in the mid-luteal phase, suggesting the corpus luteum as a further source of inhibin A (Groome, Illingworth & O'Brien, 1994; Muttukrishna *et al.*, 1994). The serum concentration of inhibin B peaks in the early follicular phase (days 4–5), thereafter remaining low for most of the menstrual cycle (Groome *et al.*, 1996). Examination of follicular fluid taken from individual human follicles yielded a linear relationship between concentrations of inhibin A, follicle size and degree of maturation (Magoffin & Jakimiuk, 1997). In a further study on human follicular fluid, the concentration of inhibin A increased whereas that of inhibin B decreased with increasing size of follicle, almost all follicles in the study being > 14 mm diameter (Lau *et al.*, 1999). This needs to be set against the reported increase in concentrations of inhibin B in follicular fluid during human follicle growth from 4–14 mm diameter prior to its decrease in follicles > 14 mm diameter (Magoffin & Jakimiuk, 1997). Hence, the suggestion exists that the concentration of inhibin A in follicular fluid may be a useful marker of the physiological status of a follicle whereas inhibin B could be a useful marker for the reserves of follicles in infertile or older women.

Bovine follicular fluid contains several forms of α subunit that are generated by proteolytic modification of a full-length α subunit precursor. Bovine follicular fluid also contains diverse forms of dimeric inhibin, extending from the largest unprocessed forms (110–120 kDa) down to the fully processed form (32 kDa) (Silva *et al.*, 1999). As a working hypothesis, pro-αC and perhaps other forms of free inhibin α subunit secreted in large amounts by cumulus cells inhibit the developmental potential of cumulus–oocyte complexes. Free inhibin α subunit, together with activin A and follistatin (see below), could play an intra-follicular paracrine rôle in influencing bovine oocyte maturation *in vivo*. However, there is a strong suggestion that inhibins and free α subunits have different paracrine rôles in regulating follicle and oocyte maturation, possibly achieved in combination with the effects of activin and follistatin (Silva *et al.*, 1999). Addition of recombinant inhibin A and activin A to serum-free culture medium had a beneficial influence on *in vitro* maturation of bovine oocytes (Stock, Woodruff & Smith, 1997). As to sheep, inhibin enhances FSH-stimulated oestradiol production by cultured granulosa cells and, in a reciprocal sense, oestradiol and aromatisable androgens will stimulate granulosa cell production of inhibin (Campbell & Webb, 1995).

One further experimental observation concerning inhibin will be mentioned here. It can stimulate synthesis of androgens and will synergise with LH and IGF-1 to increase the secretion of androgens by theca cells in culture. There is good evidence from *in vitro* studies that both inhibin and activin may have a direct effect in modulating androgen synthesis in cells of the theca interna (Hsueh *et al.*, 1987). Theca cells treated with picomolar quantities of recombinant inhibin show enhanced LH/IGF-1 stimulated production of androgens whereas activin reduces androgen synthesis by theca cells.

Activins

Also isolated from ovarian follicular fluid on the basis of their ability to modulate the synthesis and secretion of pituitary FSH, activins are similarly dimeric proteins but, in contrast to inhibins, consist of only two β subunits (Fig. V.7), again linked by disulphide bridges. Activins and inhibins are thus structurally related but whereas inhibins suppress pituitary FSH production, activins oppose such action by stimulating FSH (Ethier & Findlay, 2001). Of the various forms of activin so far isolated, activin A ($\beta_A-\beta_A$), activin AB ($\beta_A-\beta_B$) and activin B ($\beta_B-\beta_B$) are thought to be the most widespread and to have similar biological functions. As is the case for inhibin, granulosa cells are a principal site of activin synthesis, but activin may function to regulate the differentiation and proliferation of granulosa cells. Although inhibin suppresses FSH-mediated oestrogen production by granulosa cells, activin enhances such FSH-stimulated oestrogen production by boosting FSH-induced aromatase activity (Miro, Smyth & Hillier, 1991; Miro & Hillier, 1992) and boosting FSH receptor expression in the early stages of follicular development. Maintenance of FSH receptor levels is probably not dependent on activin. Activin A can itself be regulated by FSH and hCG (Lockwood *et al.*, 1996). Activins may act primarily as intra-ovarian regulators of follicle and oocyte development but, because β_A and β_B subunits are expressed within the pituitary gland, activin may have an autocrine or paracrine influence on FSH secretion.

As to changes during the menstrual cycle, highest mean concentrations of serum activin A are found close to mid cycle and during the late luteal – early follicular phases. Nadirs in the mean serum concentrations of activin A occur in both mid-follicular and mid-luteal phases (Muttukrishna *et al.*, 1996).

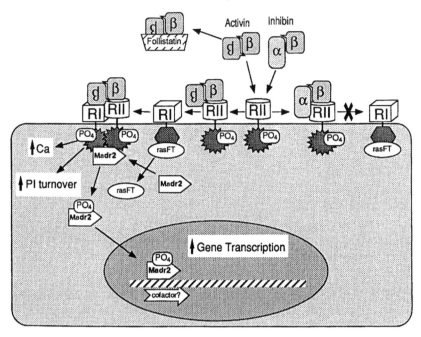

Fig. V.7. Model for activin signal transduction (see text for details). RI, activin type I receptor; RII, activin type II receptor; rasFT, *ras* farnesyl transferase; Mad, mothers against dpp (decapentaplegic); PI, phosphoinositide; Ca, calcium. (After Schwall, 1999a.)

The concentration of activin A in human follicular fluid did not reveal useful correlations with follicle potential, but activin A produced by cumulus–oocyte complexes might be a potential marker of oocyte quality (Lau *et al.*, 1999). Activin A has been reported to stimulate meiotic maturation of human oocytes *in vitro* (Alak *et al.*, 1998).

As indicated above, *in vitro* studies reveal that activins can influence synthesis of androgens by cells of the theca interna (Hsueh *et al.*, 1987). Even when stimulated by LH and IGF-1, human theca cells in culture respond to treatment with recombinant activin A by a reduced production of androgens (Hillier *et al.*, 1991b).

Follistatin

This antagonist of activin, a monomeric protein (single-chain glycoprotein) was also first identified and isolated from follicular fluid (Ueno *et al.*, 1987). It occurs in at least six forms, of molecular mass ranging from 31 to 39 kDa, but has no structural similarities with activin or inhibin. However, a functional link is that follistatin suppresses FSH secretion by the anterior pituitary gland, acting as a high affinity binding protein for activin but less so for inhibin. Follistatin binds to both through their common β subunit, neutralising some properties of activin such as stimulation of FSH secretion but not those of inhibin. Such neutralising potential of follistatin has been observed in various biological systems, most notably in cultures of granulosa cells which represent a major site of production, particularly cumulus cells. Immunostaining in cattle localised follistatin only in granulosa cells, not in the theca interna (Shukovski *et al.*, 1992; Singh & Adams, 1998). The amino acid sequences of follistatin in primates, ruminants, pigs and rats have 97% homology.

The concentration of follistatin in peripheral blood varied little during the oestrous cycle in sheep, whereas its concentration in follicular fluid decreased during the follicular phase of the oestrous cycle in farm animals, especially during the final stages of follicular development (Li, DePaolo & Ford, 1997). In marked contrast, the concentration of follistatin in human follicular fluid did not vary between healthy, atretic and pre-ovulatory follicles in patients selected for *in vitro* fertilisation (IVF) treatment (Erickson *et al.*, 1995). Caution is necessary in interpreting such assay data, since follistatin exists in multiple forms in follicular fluid, not least due to the actions of (1) proteolytic cleavage, (2) various forms of glycosylation, and (3) truncated forms of follistatin binding to cell surface proteoglycans, in particular to heparan sulphate on granulosa cells.

All three inhibin/activin subunits and follistatin are encoded by separate mRNAs. Their expression in the ovary is regulated by gonadotrophic hormones. Inhibin/activin subunit mRNA levels increase in maturing rat follicles as they progress through the oestrous cycle, whereas follicles that have become atretic have little or no inhibin/activin subunit mRNA, suggesting a rôle for these peptides in selection or support of the dominant follicle(s) (Izadyar *et al.*, 1998). Expression of activin receptor genes in mouse cumulus–oocyte complexes and oocytes indicates that activin, synthesised in granulosa cells of maturing follicles, may act locally with its binding protein follistatin to regulate follicular development and oocyte maturation (Izadyar *et al.*, 1998; Silva & Knight, 1998). Activin A increased the developmental potential of both cumulus–oocyte complexes and cumulus-free oocytes, an influence seemingly regulated by the availability of follistatin (Silva & Knight, 1998). mRNA for follistatin is expressed in both healthy and atretic follicles, but follistatin protein is detected only in tertiary follicles and newly formed corpora lutea. The strongest signal for follistatin mRNA expression in cattle was in pre-ovulatory follicles (Shukovski *et al.*, 1992). Two forms of follistatin mRNA are generated by alternative splicing. The follistatin gene is a single copy of approximately

6.0 kilobase-pairs, and consists of five exons plus an initial exon that encodes the signal peptide (Shimasaki *et al.*, 1988; Li *et al.*, 1997). Five different activin subunit genes have been identified, with independent but overlapping patterns of gene expression. Upregulation of subunit gene expression can be stimulated by gonadotrophins through a cAMP-dependent mechanism (Schwall, 1999a).

Bovine cumulus cells of the cumulus–oocyte complex taken from small and medium-sized follicles express inhibin, activin A, follistatin and activin receptor type II mRNA, and synthesise these proteins whereas, under *in vitro* conditions, the oocyte itself can synthesise all except inhibin (Izadyar *et al.*, 1998). As noted above, activin A has been proposed to play a paracrine and/or autocrine rôle in the maturation of bovine oocytes, and this received further support in the study of Silva & Knight (1998). Activin has similarly been noted to enhance the maturation of primate oocytes *in vitro* (Alak *et al.*, 1998), and activin A also stimulates meiotic maturation in rat oocytes.

In porcine follicles, mRNA for the α and β_A subunits of inhibin/activin was detected by dot–blot analysis (Guthrie *et al.*, 1992), expression increasing to the mid stage of the follicular phase of the cycle. However, by the late stage of follicular development, expression of all three inhibin/activin subunits showed a marked reduction. Expression of the follistatin gene and translation of its mRNA decreased during the later stages of the follicular phase, especially as follicles approached 'ovulatory status' (Li *et al.*, 1997).

Targeted deletion of the follistatin gene in embryonic stem cells enabled production of follistatin-deficient homozygous transgenic mice. Such follistatin-deficient animals died soon after birth due to multiple defects: small, growth-retarded fetuses with inadequate respiration, atrophic intercostal muscles and diaphragm, and skeletal abnormalities (see Matzuk *et al.*, 1995). As to deletion of the activin genes, mouse fetuses bearing β_A knockouts develop to term but die shortly after birth owing – in part at least – to failure to suckle. Cranio-facial defects are found in the hard and soft palate. Fetuses bearing a β_B knockout usually survive to term and most of those born appear to develop normally but problems arise in the next generation.

Deletion of the gene for the α subunit of inhibin leads to the development of granulosa cell tumours at a very high incidence in mice, suggesting that inhibin might function as a tumour suppressor gene product with gonadal specificity (Matzuk *et al.*, 1992). The gonadal tumours appear shortly after puberty; thereafter the mice develop cachexia, with lesions in the liver, and soon die. Seemingly of relevance, many ovarian tumours in women secrete large amounts of inhibin and free α subunit, enabling serum inhibin concentrations to be used as a clinical marker.

Peptide growth factors

The ovary produces diverse families of growth factors, seemingly to regulate and coordinate physiological activities within different ovarian compartments. Many of these growth factors, such as IGFs, TGFs and EGFs, can be produced by both germ cells and somatic cells. In terms of promoting follicular development and generating a functional Graafian follicle containing a fully viable oocyte, this complex system of intra-ovarian regulation should be viewed as no less important than the extra-ovarian regulation by hormones of the anterior pituitary gland. Indeed, peptide growth factors modulate responses to gonadotrophic hormones.

Growth factors provide a sensitive and detailed web of communication within the ovary (see Jones & Clemmons, 1995; Ferrara & Davis-Smyth, 1997). In the absence of their synthesis and secretion at appropriate concentrations in the right place and at the correct time, ovarian cells would not be able to interact sufficiently with each other and normal function would be compromised. On the basis of *in vitro* experimentation, growth factors can be shown to act individually. In most circumstances, however, one or more factors would be acting in concert to integrate the cellular events of different ovarian compartments, including an influence on cell death (see Chapter VI). Despite these specific remarks, an absolute physiological requirement for any of the growth factors in reproductive tissues was not convincingly demonstrated until the availability of suitable gene deletion techniques. Knockout models have afforded some clarity (see Nishimori & Matzuk, 1996).

Insulin-like growth factors

Amongst the various families of growth factors that underlie coordinated ovarian activity, the IGF family has received extensive attention. First revealed as factors that would mediate the effects of growth hormone on somatic tissue growth, this finding prompted the name somatomedins. Members of this family have a specific involvement in growth, development and maturation of Graafian follicles, stimulating cell proliferation and steroidogenesis in human granulosa and theca interna cells (Adashi *et al.*, 1985; Giudice, 1992; Armstrong & Webb, 1997). Individual members of the family include IGF-1 and IGF-2 and their receptors, and a corresponding family of binding proteins (IGFBPs) found in the circulation and extracellular compartments. As transport proteins, their action is to control the availability and thus the biological actions of the growth factors. A vital aspect of the regulation of growth factor activity is an association with the extracellular matrix.

IGFs are single-chain polypeptides that show close similarities to insulin. IGF-1 and IGF-2 share a 62% amino acid sequence homology, and have a molecular weight of 7 kDa. IGF-1 is synthesised by various human tissues but principally in the liver. Acting by autocrine and paracrine routes, it would seem to have an important rôle in regulating development of follicles and is present in follicular fluid. mRNAs for IGFs and their receptors are found in human follicles during successive stages of development: mRNA and IGF-1 protein can be detected in granulosa (and theca?) cells. IGF-2 protein has been demonstrated in both theca and granulosa cells, and its receptors are also found in both cell types. Expression of the IGF-1 receptor is upregulated by oestradiol in a variety of tissues. Moreover, oestradiol upregulates mRNA expression for both IGF-1 and IGF-2 receptors in non-steroidogenic tissues.

Human IGF genes are located on different chromosomes. The IGF-1 gene is on chromosome 12, spanning a region of more than 95 kilobase-pairs of chromosomal DNA. There are five exons interrupted by four introns. The human IGF-2 gene is on chromosome 11 and spans 30 kilobase-pairs of chromosomal DNA; it contains nine exons. Immunoreactivity for the human IGF type 1 receptor (IGF-1R) was detected in oocytes of primordial, pre-antral and antral follicles, as well as in granulosa cells of such follicle types, but only in theca cells of pre-antral follicles (Qu *et al.*, 2000).

In rodents, IGF-1 gene expression and protein have been detected predominantly in granulosa cells of developing follicles, suggesting that IGF-1 may be a marker for follicular selection. Both granulosa and theca cells have receptors for IGF-1, and this growth factor may contribute to local amplification of gonadotrophin-stimulated follicular development. IGF-1 will enhance proliferation of follicular cells and their steroid synthetic activity: it will increase FSH-dependent secretion of oestradiol-17β by granulosa cells in rats, and can be viewed as an obligatory mediator of FSH-dependent follicle development.

As demonstrated by means of a knockout model in mice, IGF-1 and FSHR are selectively expressed in growing follicles, and IGF-1 enhances expression of FSHR on granulosa cells (Zhou *et al.*, 1997). Thus, ovarian IGF-1 expression may be acting to promote granulosa cell FSH responsiveness by augmenting FSHR expression. IGF-1 expression is thought to create a positive feedback loop within the follicle in which IGF-1 enhances FSH action and FSH enhances IGF-1 action through complementary receptor upregulation. The IGF-1 influence on FSHR expression in rat granulosa cells is related to cAMP production induced by FSH (Minegishi *et al.*, 2000). In mice, knockout experiments show that adults of both sexes homozygous for a targeted mutation in the IGF-1 gene are infantile dwarfs, unable to ovulate even in response to injections of gonadotrophic hormones. The full extent of IGF-2 involvement in the ovarian physiology of rodents requires clarification.

Expression of mRNA encoding IGF-2 has been detected in the thecal layer of bovine follicles (Armstrong & Webb, 1997) and similar observations have been made for sheep (Perks *et al.*, 1995). Immunohistochemical and cytochemical observations in bovine follicles have revealed IGFBP-2 and -4 on both theca and granulosa cell plasma membranes, in the basement membrane, and within the extracellular matrix surrounding theca cells (Armstrong & Webb, 1997). Both IGF-1 and IGF-2, together with IGF receptors and IGFBPs, are expressed in the porcine ovary (Hammond *et al.*, 1993) and IGF-1 is secreted by granulosa cells in culture. IGF-1 mRNA is expressed in porcine granulosa cells (Gadsby *et al.*, 1996). Altering systemic glucocorticoid concentrations by means of adrenocorticotrophic hormone (ACTH) treatment in pigs will disrupt ovarian IGF-1 synthesis and IGF action *in vivo*, and can impair maturation of follicles (Viveiros & Liptrap, 2000). *In vitro* studies had previously shown that porcine granulosa cells cultured with high levels of cortisol exhibit lower IGF-1 synthesis and reduced IGF-1 stimulated production of progesterone (Viveiros & Liptrap, 1999). In combination with gonadotrophins, IGF-1 will enhance *in vitro* maturation of bovine oocytes (Harper & Brackett, 1992).

As to potential modulatory influences, nutrition and insulin appear to be major regulators of IGF-1 and its binding protein (see Gong *et al.*, 1994, 1997). Decreased nutrient intake leads to growth hormone resistance, decreased IGF-1 and an increase in the plasma level of its binding protein (Lee, Conover & Powell, 1993; Thissen, Ketelslegers & Underwood, 1994).

Transforming growth factors

Polypeptides in this family are classified as either α or β. TGF-α is a single-chain polypeptide composed of 50 amino acid residues. It is structurally similar to EGF (see below) and can bind to the EGF receptor (EGF-R) as avidly as EGF, and thus can activate the receptor (Derynck, 1988). It is synthesised as a large precursor molecule, then enzymatically cleaved to a mature protein of 5.6 kDa. Its gene is located on chromosome 2. TGF-α may be involved in the proliferation of granulosa and theca cells, as is true also for EGF. However, in a surgical preparation (ovarian autotransplant in sheep), TGF-α can act to inhibit ovarian function in terms of steroid hormone and inhibin secretion (Campbell, Gordon & Scaramuzzi, 1994). Evidence is mounting that TGF-α from theca cells can influence proliferation of granulosa cells by a paracrine mechanism (Armstrong & Webb, 1997).

Concerning TGF-β, a dimeric protein of 25 kDa, there are at least five iso-forms (TGF-β1, TGF-β2, TGF-β3, etc.); these are 70–80% identical and have virtually identical biological activities (Hu, Datto & Wang, 1998). TGF-β1, -β2, -β3 and -β5 interact with the same receptor system. TGF-β is produced in both theca and granulosa cells, at least in rodents, and is detectable from an early antral stage in human follicles. Both types of follicular cell respond to TGF-β, which usually enhances cell proliferation and may promote a specific differentiation and maturation of granulosa cells. Such maturation involves production of cAMP, synthesis of steroid hormones, and formation of LH receptors. TGF-β seemingly enhances the stimulatory actions of low levels of FSH whereas it selectively inhibits development of granulosa cells and induction of LH receptors at higher levels of FSH (Kim & Fazleabas, 1999). To add to the complexity of interpretation, TGF-β inhibits granulosa and theca cell proliferation in bovine follicles but enhances gonadotrophin-prompted steroidogenesis (Roberts & Skinner, 1991). TGF-β may also promote the induction of EGFR by FSH and regulate the inhibitory action of EGF during differentiation of granulosa cells.

Overall, TGF-β is involved in the regulation of growth and in the synthesis of steroid hormones during follicular development (Adashi & Resnick, 1986; Dorrington, Chuma & Bendell, 1988); it enhances FSH-dependent secretion by having an upregulatory effect on FSH-stimulated P450 aromatase mRNA expression and P450 aromatase activity (Zachow, Weitsman & Magoffin, 1999). In rat granulosa cells, TGF-β may act as a potent stimulator of *in vitro* maturation of oocytes (Feng, Catt & Knecht, 1988), a finding not confirmed by Tsafriri *et al.* (1991).

Growth differentiation factor 9

GDF-9 is a protein belonging to the TGF-β family, and can be secreted by oocytes. Expression of GDF-9 has been noted in oocytes from diverse groups of species, such as rodents (McGrath, Esquela & Lee, 1995; Hayashi *et al.*, 1999), ruminants (Bodensteiner *et al.*, 1999) and primates (Aaltonen *et al.*, 1999), especially during early stages of follicular development. Its importance in the regulation of folliculogenesis is demonstrated by arrested development beyond the primary follicle stage in animals deficient in GDF-9 (Dong *et al.*, 1996). Conversely, treatment with GDF-9 recombinant protein using an *in vitro* culture preparation of rat tissues stimulated growth of pre-antral follicles (Hayashi *et al.*, 1999). Furthermore, treatment with the recombinant GDF-9 promotes expansion of the cumulus cell mass and reduces mRNA for the LH receptor in cultured mouse granulosa cells (Elvin *et al.*, 1999). In an *in vitro* preparation of rat granulosa cells, GDF-9 treatment stimulates cell proliferation but reduces FSH-induced steroidogenesis and LH receptor formation (Vitt *et al.*, 2000). Moreover, GDF-9 treatment in this latter study increased basal levels of oestradiol and progesterone secretion in granulosa cells from pre-ovulatory follicles. Overall, there is the suggestion that GDF-9, alone or in combination with GDF-9B, might have major responsibility for regulating follicular growth induced by the oocyte (Vitt *et al.*, 2000).

Epidermal growth factor

EGF is part of a large family that includes (TGF-α) and heparin-binding EGF (HB-EGF). The biological actions of these factors are many and diverse. As judged from *in vitro* experiments, EGF stimulates proliferation of granulosa cells and DNA synthesis, prompts changes in cell morphology, and enhances protein phosphorylation. EGF is found in follicular fluid, but the concentration decreases with development of the follicle. EGF and its receptor (EGFR) are expressed in both follicular and stromal cells of the human ovary, and EGFR has been localised in granulosa cells of various species. In rats, EGF will attenuate or inhibit FSH-mediated differentiation of granulosa cells and, in rabbits, EGF can inhibit hCG-induced oestradiol and progesterone secretion *in vitro*. In conjunction with TGF-β, EGF decreases the ability of granulosa cells to express cAMP-induced LH receptors *in vitro*.

EGF is a 6 kDa single-chain polypeptide of 53 amino acid residues, first isolated in 1961 from submandibular gland extracts of newborn mice. EGF is synthesised by proteolytic cleavage of a large precursor molecule of 128 kDa. Its precursor gene in humans is of approximately 110 kilobase-pairs, contains 24 exons, and is located on chromosome 4. The EGF receptor is a 170 kDa monomeric protein of 1186 amino acid residues. Following activation, the EGF receptor–ligand complex is internalised and degraded in lysosomes (Kim & Fazleabas, 1999).

Basic fibroblast growth factor

bFGF represents another family of peptide growth factors of which nine members have so far been revealed. Ranging in molecular weight from 17 to 38 kDa, there is only 14% nucleotide sequence homology in the coding region among the eight human FGF genes (Kim & Fazleabas, 1999). First identified in 1974 as a 146 amino acid residue protein that would cause proliferation of a line of mouse fibroblast cells, bFGF (or referred to as FGF-2) is found in diverse tissues and acts as a mitogen to cells derived from mesoderm. bFGF can be demonstrated in bovine granulosa cells (Neufield *et al.*, 1987), stimulates granulosa cell proliferation, may stimulate theca cell proliferation, seemingly inhibits steroid synthesis by theca cells, and may have an involvement in angiogenesis.

Immunostaining reveals FGF-2 in theca interna cells and in granulosa cells closest to the basement membrane, and also in oocytes from primordial and primary (bovine) follicles. mRNA for FGF-2 and its receptor are present in fetal ovaries and in luteinised granulosa cells after hyperstimulation (Kim & Fazleabas, 1999). Despite such observations, the specific functions of FGF-2 in ovarian physiology remain uncertain. Not least, its interactions with the extracellular matrix during ovarian tissue remodelling require clarification.

Other growth factors such as VEGF, hepatocyte growth factor (HGF) and platelet-derived growth factor (PDGF) have been localised in the gonads and considered to contribute to follicular function. This may well be so at a sensitive and subtle autocrine or paracrine level, but evidence for an essential rôle has yet to be produced. As is the case for all the growth factors discussed above, a contribution to local modulation of follicular development is the best interpretation so far, although potential interactions between the influence of diverse growth factors render such a conclusion less than meaningful. Nonetheless, involvement of growth factors in a system of fine-tuning downstream from gonadotrophic stimulation would seem an appropriate interpretation, not least in the processes of follicular selection and dominance: in other words, as intra-ovarian regulators of folliculogenesis.

Cytokines and eicosanoids

Although various lines of evidence became available during the 1960s (Levey & Medawar, 1966; Kirby, 1967; Jooste, 1968; Harper *et al.*, 1970), it is especially during the last 10–15 years that a contribution of the immune system to the regulation of cyclic ovarian activity has been increasingly recognised. Indeed, this field of research has developed to the extent that there are now many substantial reviews (e.g. Adashi, 1992, 1996; Brännström, 1997; Terranova & Rice, 1997; Bukulmez & Arici, 2000). At the ovarian level, abundant evidence exists for interactions between the immune and endocrine systems, and it is certainly reasonable to propose that cells of the immune system contribute to the processes of follicle selection, development and ovulation. Classical experiments demonstrated a requirement for an intact thymus gland during fetal and postnatal life as a preparation for normal ovarian function after puberty in mice (Nishizuka & Sakakura, 1969; Bukovsky *et al.*, 1977). One manner of

Table V.2. *The nature of leucocytes and their location in ovarian tissues before ovulation, with potential products and possible actions*

Cells	Location	Products and possible actions
Macrophages/ Monocytes	Medulla, theca, follicular fluid	IL-1, IL-6, IL-8, TNF-α, GMCSF, PA, collagenase, prostaglandins, PAF, NO, ROS; promote steroidogenesis, release cytokines and NO
Neutrophils	Medulla, theca, follicular fluid	Cathepsins, collagenase, PA, myeloperoxidase, lysozymes, elastase, TNF-α, IL-1, prostaglandins, PAF, ROS, plasminogen activators, NO; release cytokines, collagenolytic peptides and NO
Lymphocytes	Medulla, theca, follicular fluid	IL-2, IL-6, GMCSF, interferons; secrete cytokines
Eosinophils	Theca	Cationic protein, major basic protein; promote plasminogen activator, disrupt membranes, degrade collagen
Mast cells	Medulla, theca	Histamine, serotonin, TNF-α, kinogenase; release histamine and TNF-α during ovulation

IL, interleukin; TNF, tumour necrosis factor; GMCSF, granulocyte–macrophage colony stimulating factor; PA, plasminogen activator; NO, nitric oxide; PAF, platelet activating factor; ROS, reactive oxygen species.
Adapted from several sources but principally Norman & Brännström, 1996.

viewing the humoral – as distinct from nervous – control of ovarian physiology is that the systemic influences of pituitary gonadotrophins are supplemented by local influences of immune cells within ovarian tissues, their products acting as a form of fine-tuning or modulation. The close interaction between immune and reproductive systems is expressed through the action of cytokines, molecular products of immune cells and other cell types within the ovary. The cyclic changes and prominence of ovarian lymphatic vessels undoubtedly have relevance in this regard (see Chapter III), as may have subtle variations in lymph accumulation and flow. Cells to be focused upon are of white blood cell lineage, not least subsets of macrophages, T lymphocytes, eosinophils and mast cells (Table V.2).

Cytokines are soluble proteins or glycoproteins that act non-enzymatically through specific cell surface receptors to regulate the growth, differentiation and function of a target cell. White blood cells passing through the ovary or becoming resident in ovarian tissues are a potent source of cytokines that can act to modify ovarian physiology (see Chapter VIII). However, cytokines are released not only by immune cells, as initially believed, but also by many epithelial cell types and can act in an autocrine or paracrine manner. They are specifically involved in the events of inflammation. Receptors are present on most cells.

First revealed in the ovary in human follicular fluid, the concentration of at least one family of cytokines, the chemotactic cytokines (chemokines) may be much higher in follicular fluid than in systemic blood. For example, the neutrophil chemotactic and activating chemokines interleukin-8 (IL-8) and growth-regulated oncogene *GRO*α may reach concentrations 14–20-fold higher in follicular fluid (Brännström *et al.*, 1996; Runesson *et al.*, 1996; Bukulmez & Arici, 2000). Monocyte chemoattractant protein 1 (MCP-1) was also detected at higher concentrations in follicular fluid. These findings led to the proposition that specific chemokines are produced within Graafian follicles to attract and activate leucocytes, a suggestion in accord with the previously observed influx of these cells into pre-ovulatory follicles (Brännström, Mayerhofer & Robertson, 1993a; Brännström *et al.*, 1994a).

Detailed studies have been conducted on the distribution of leucocytes in rat ovaries. During much of the oestrous cycle, most leucocytes are seen in theca cells or in the ovarian medulla, but migration of leucocytes into the thecal layers becomes prominent after the pre-ovulatory LH surge (Brännström *et al.*, 1993a). Leucocytes are not detected in the granulosa cell layer until after the LH surge. Comparable findings have been reported for the human ovary. Macrophages are noted mainly in the stroma, being scarce in the walls of follicles during most of the follicular phase. Once again, there is a marked increase in the density of macrophages and of neutrophils in the cell layers of the theca at the time of the pre-ovulatory LH surge. These infiltrating cells produce cytokines (Brännström *et al.*, 1994b).

Interleukins

Activated macrophages produce interleukin 1 (IL-1), a cytokine of importance as an immune mediator and one that may have a local influence on follicular development. The IL-1 system is composed of two ligands (IL-1α and IL-1β), two receptors (types 1 and 2), receptor antagonist, and extracellular binding proteins. IL-1 is produced primarily by mononuclear cells in the peripheral circulation to act as an inflammatory mediator. A complete IL-1 system exists in the human ovary (Hurwitz *et al.*, 1992). Aspirates of human pre-ovulatory follicles, made available at the time of oocyte retrieval for *in vitro* fertilisation, contain IL-1β and IL-1α mRNA but the relative contributions from white cells and from granulosa cells remain uncertain; the latter represent an important source of IL-1 production. As to a local influence on the endocrine system, IL-1 will inhibit both FSH- and hCG-stimulated production of oestradiol in granulosa cells.

Turning to rodents, the ovary secretes IL-1 at ovulation and all the components of the IL-1 system are demonstrable immunologically in the mouse ovary during ovulation (Simón *et al.*, 1994a). As more persuasive evidence, mRNA for the type 1 receptor was also located in large pre-ovulatory rat follicles (Wang *et al.*, 1995), whilst the naturally occurring IL-1 receptor antagonist compromises gonadotrophin-induced ovulation in rats (Peterson *et al.*, 1993). Despite ever-increasing information as to its involvement, the precise manner in which IL-1 facilitates ovulation is still not clear. Induction of mediators such as prostaglandins, progesterone and nitric oxide have all been suggested (Brännström, Mikuni & Hedin, 1997; Davis *et al.*, 1999).

The interleukins IL-1 and IL-8 had previously been associated with softening of reproductive tissues, particularly those of the cervix, so they could conceivably be acting in a comparable fashion in the wall of Graafian follicles (see Chapter VIII). In combination with LH, IL-1 will induce ovulation in perfused preparations of rat and rabbit ovaries (Brännström *et al.*, 1993a; Takehara *et al.*, 1994). IL-1 will also stimulate prostaglandin production in isolated follicles (Brännström, Wang & Norman, 1993b) and accumulation of collagenolytic proteins in rat ovaries (Hurwitz *et al.*, 1993). *In vivo* observations on IL-8 have been referred to above for human Graafian follicles, and granulosa cells can be shown to secrete large quantities of IL-8. A focus has therefore been placed on neutrophils, their activation being modulated by IL-8 (Chang, Gougeon & Erickson, 1998).

Various other interleukins are implicated in ovarian physiology. As examples, IL-2 is a T-cell-derived cytokine that does not influence basal secretion of progesterone from human granulosa–lutein cells in culture, but suppresses hCG-stimulated progesterone production from such cells (Wang *et al.*, 1991). IL-6 reduces proliferation of bovine granulosa cells in culture, and is involved in the regulation of steroid synthesis and secretion (Alpizar & Spicer, 1994). As judged from *in vitro* studies, whether a specific interleukin or macrophage product acts to stimulate or inhibit secretion of steroid hormones, depends on the population of follicle cells in question, their state of maturity, the presence or absence of a corresponding oocyte, and the conditions of culture. Species differences need to be considered. There is also the aspect that IL-1 and IL-2 production by follicular leucocytes may be inhibited by granulosa cell secretions, suggesting local feedback mechanisms (Maccio *et al.*, 1993). And not to be overlooked, oocyte–cumulus complexes produce IL-1α and IL-6 (Tarlatzis & Bili, 1997), with IL-10 being detectable in human oocytes.

Tumour necrosis factor α

TNF-α is another prominent cytokine able to influence ovarian physiology and secreted primarily by activated macrophages and lymphocytes (for a review, see Terranova & Rice, 1997). It is one of the main cytokines involved in final maturation and ovulation of a Graafian follicle (see Chapter VIII). TNF-α is a non-glycosylated protein with a molecular weight of 17 kDa, first described as a tumoricidal factor produced by activated macrophages (Carswell *et al.*, 1975). TNF-α can exert cytostatic effects on many cells and has been suggested to be involved in follicular atresia (Van Voorhis, 1999). It certainly contributes to regression of the corpus luteum, that is the actual process of luteolysis in a variety of species (Benyo & Pate, 1992; Pitzel, Jarry & Wuttke, 1993). Conversely, and under appropriate conditions, TNF-α can promote proliferation of different ovarian cell types (Wang *et al.*, 1992).

Serum concentrations of TNF-α in women fluctuate significantly during the menstrual cycle (Norman & Brännström, 1996; Brännström *et al.*, 1999). TNF-α is present in human follicular fluid (Wang *et al.*, 1992), and TNF-α immunoreactivity and mRNA have been localised in granulosa cells and oocytes in both healthy and atretic Graafian follicles of women and rats (Roby *et al.*, 1990; Sancho-Tello *et al.*, 1992; Marcinkiewicz *et al.*, 1994); they have also been revealed in human theca cells. Human granulosa cells obtained from aspirated pre-ovulatory follicles show reduced FSH-induced oestradiol production and reduced hCG-induced progesterone production in the presence of TNF-α. And, as further evidence of TNF-α involvement in ovarian steroidogenesis, it will inhibit LH-promoted androgen production by reducing LH receptor number, cAMP production and PKA activity (Andreani *et al.*, 1991; Zachow, Tash & Terranova, 1993; Zachow & Terranova, 1994). TNF-α administration in rats inhibits hCG-induced StAR (Chen, Feng & Liu, 1999). Under physiological conditions, however, the relative contributions of macrophage cytokine and granulosa cell cytokine to a particular stage of follicular development are uncertain.

TNF-α has been proposed as an organising molecule in follicular development, since it causes characteristic clustering of theca cells in culture – which may have certain parallels with the formation of the theca interna – an action probably mediated via the protein kinase C pathway (Zachow *et al.*, 1992; Zachow & Terranova, 1993). In cultured preparations of rat theca-interstitial cells, TNF-α stimulated mitotic activity and preferentially increased the number of steroidogenically active cells, responses seemingly independent of those induced by insulin and IGF-1 (Spaczynski, Arici & Duleba, 1999). Stemming from the observed production of TNF-α by oocytes comes the proposal that it organises pre-granulosa and theca cells, not least since TNF-α receptors can be demonstrated on such cell types. Radiating outwards as a molecular gradient from an oocyte source, TNF-α may thus strongly influence the development of follicles. In the final stages of follicular maturation, TNF-α will act in combination with LH to promote ovulation in perfused rat and rabbit ovaries (Brännström *et al.*, 1993b, 1995; Brännström, 1997).

Specific binding sites for TNF-α are found on bovine theca and granulosa cells (Spicer, 1998), and cultured bovine theca and granulosa cells respond to TNF-α with enhanced gonadotrophin-induced steroid synthesis and protein synthesis (Okuda *et al.*, 1999). Porcine theca and granulosa cells likewise respond to TNF-α *in vitro* (Tekpetey, Engelhardt & Armstrong, 1993). TNF-α receptors are membrane bound but may also be found in soluble form achieved by proteolytic cleavage. The TNF-α receptor has been characterised in bovine luteal cells at different stages of the oestrous cycle (Sakumoto *et al.*, 2000).

Leukaemia inhibitory factor

Whilst demonstrable in many biological systems, LIF, a glycoprotein of 43 kDa, is thought to be specifically involved in follicular physiology and may be produced by both macrophages and theca cells. It is demonstrable in follicular fluid and could have an influence on oestradiol synthesis: correlations exist between the concentrations of LIF and those of oestradiol in follicular fluid. The concentration of LIF in human follicular fluid increases in response to treatment with hCG (Arici *et al.*, 1997), such values being markedly higher than those in corresponding samples of serum. Whereas LIF mRNA and protein are expressed in only low amounts in cultures of ovarian stroma, their values can be enhanced by treatment with IL-1 and TNF-α (Arici *et al.*, 1997). Human granulosa cells also express only low amounts of LIF mRNA and protein, such granulosa cell values not responding to IL-1 or TNF-α treatment.

Evidence is accumulating implicating various other cytokines in ovarian physiology. The list will doubtless continue to grow. For example, macrophage colony stimulating factor (MCSF) is another cytokine that appears to contribute to the regulation of ovarian function. Granulocyte–MCSF has been noted to fluctuate during the menstrual cycle in women (Jasper *et al.*, 1996). In mice, MCSF deficient animals have a markedly reduced number of antral and mature follicles at ovulation, a condition corrected by MCSF administration with a corresponding increase in the number of 'local' macrophages (Araki *et al.*, 1996).

Eicosanoids

In marked contrast to cytokines, eicosanoids are oxygenated metabolites of long-chain, polyunsaturated fatty acids. Arachidonic acid is the most common substrate for their synthesis. The eicosanoid family of molecules includes prostaglandins, thromboxanes and leukotrienes. They influence target cells by binding to surface receptors, usually acting in a paracrine or autocrine manner. A prominent involvement of prostaglandins in the process of ovulation has long been appreciated (see Chapter VIII). Concentrations of $PGF_{2\alpha}$ and PGE in the follicular tissues, including follicular fluid, increase within a few hours of the pre-ovulatory gonadotrophin surge or hCG injection (see Table IV.8). Adenylyl cyclase activity becomes markedly enhanced in granulosa cells, emphasising the overall steroidogenic effect of pituitary gonadotrophins. Follicular prostaglandins may stimulate the local release of proteolytic enzymes such as collagenase involved in progressive breakdown of the apex of the follicle. Treatment with inhibitors of prostaglandin synthesis (e.g. indomethacin) will block ovulation in diverse animal species.

Because thromboxanes and leukotrienes have also been noted to accumulate in follicular tissues after the gonadotrophin surge, they are presumed to contribute to the process of ovulation. Circumstantial evidence comes from the inhibition of ovulation when formation of leukotrienes is pharmacologically blocked. Even so, a pre-ovulatory increase in ovarian thromboxane is not an essential component of the ovulatory mechanism, at least not in gonadotrophin-primed immature rats (Wilken *et al.*, 1990).

Endorphins and enkephalins

Although generally viewed as regulatory molecules in the central nervous system, this opioid peptide family is also involved locally in ovarian physiology. A principal member of the group, β-endorphin, has been demonstrated in human ovaries (Aleem, Omar & El Tabbakh, 1986) and in granulosa and interstitial cells of rats (Lolait *et al.*, 1985, 1986). Human ovaries also contain met-enkephalin (Petraglia *et al.*, 1985), as do rabbit tissues (Li, Wu & Kumar, 1991). Peptides of the dynorphin family are found in the granulosa and interstitial cells of rats and in their corpora lutea (Lolait *et al.*, 1986), and α-neoendorphins exist in porcine follicular fluid (Slomczynska *et al.*, 1997). The concentration of β-endorphins in porcine follicular fluid varies cyclically and with size of follicle (Kaminski *et al.*, 2000). The presence of mRNA for pro-enkephalin (Jin *et al.*, 1988), pro-dynorphin (Douglass *et al.*, 1987) and pro-opiomelanocortin (Melner *et al.*, 1986; Sanders, Melner & Curry, 1990) confirms potential ovarian secretion of these molecules.

As to their regulation in ovarian tissues, an involvement of pituitary gonadotrophins was suggested by studies using placental gonadotrophins: both pregnant mare serum gonadotrophin (PMSG) and hCG would enhance β-endorphin secretion in rodent ovaries (Lovegren, Ziminski & Puett, 1991; Kato, Kumai & Okamoto, 1993), and FSH has subsequently been shown to exert sensitive control over porcine granulosa cell secretion of β-endorphin (Kaminski *et al.*, 2000). In addition, there is evidence for local modulation of ovarian opioid influence by ovarian androgens (Melner *et al.*, 1986). Over and above autocrine or paracrine functions of ovarian opioid secretion on theca and granulosa cell activity, they may have a critical influence on secretion of LH by the pituitary gland.

Involvement of nitric oxide

Nitric oxide (NO) is a highly reactive free radical gas synthesised from L-arginine by the catalytic action of NO synthase (NOS). It has demonstrable influences within the reproductive system (for a review, see Rosselli, Keller & Dubey, 1998), and more generally in physiology such as vasodilation, relaxation of smooth muscle, neurotransmission and apoptosis. NO is generated by a family of enzymes collectively termed NOS. There are three distinct isoforms of synthase, and all have been identified in the female reproductive system. Two constitutively expressed NOS enzymes were first identified in neural tissue (neuronal NOS or nNOS) and in vascular endothelium (endothelial NOS or eNOS), and are responsible for the continuous basal release of NO. An inducible form (iNOS) has been revealed in cytokine-stimulated immune cells, and the expression of this form of enzyme generates significant amounts of NO over a long period. The calcium-dependent eNOS and calcium-independent iNOS are hormonally regulated. The NOS/NO system can in turn influence steroidogenesis in human and rat granulosa cells (Van Voorhis *et al.*, 1994; Olson *et al.*, 1996), meiotic maturation of mouse oocytes (Jablonka-Shariff & Olson, 2000), apoptosis in rat follicular cells (Chun *et al.*, 1995), and the process of ovulation in rats and mice (Shukovski & Tsafriri, 1994; Jablonka-Shariff & Olson, 1998). In the latter connection, NOS inhibitors injected intraperitoneally or directly into the periovarian sac (bursa) of rats reduced the number of hCG-induced ovulations (Shukovski & Tsafriri, 1994). Ovulation could also be inhibited by a systemic approach to NO blockade (Bonello *et al.*, 1996).

Working with mature rats, Van Voorhis *et al.* (1995) localised iNOS in the granulosa cells of primary, secondary and small antral follicles, whereas eNOS was detected in ovarian blood vessels. High concentrations of eNOS have been found in the stroma and theca close to the time of ovulation (Zackrisson *et al.*, 1996), and NOS has also been located in granulosa cells by immunocytochemistry. There was a cell-specific pattern of expression that was differentially

FOLLICULAR DEVELOPMENT

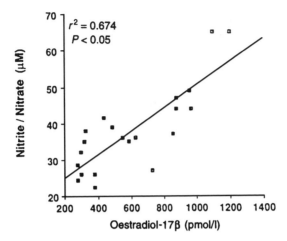

Fig. V.8. Linear regression analysis demonstrating a correlation between increases in serum nitrite/nitrate concentrations and those of oestradiol during the follicular phase of a spontaneous menstrual cycle. (Adapted from Rosselli *et al.*, 1994.)

regulated, but eNOS and iNOS are both thought to participate in the regulation of ovulation and eNOS is expressed on the surface of rat and mouse oocytes (Jablonka-Shariff & Olson, 1998). Moreover, the concentration of NO metabolites in human follicular fluid can be correlated with the size of the follicle (Anteby *et al.*, 1996), and increases in NO can be correlated with increases in oestradiol-17β (Fig. V.8; Rosselli *et al.*, 1994, 1998). NO is generated in response to IL-1 and this cytokine may be involved in ovarian production of NO. Indeed, NO, which is a principal mediator of macrophage tumoricidal and bactericidal activities, is known to participate in inflammatory reactions and facilitate tissue remodelling events within the ovary. The eNOS is the isoform responsible for maintaining systemic blood pressure, and for vascular remodelling and angiogenesis. NO may increase ovarian blood flow close to the time of ovulation.

The cytokine IL-1 will induce NO synthesis in cultured rat ovarian cells (Ben-Shlomo *et al.*, 1994). In addition, IL-1 will increase the LH-induced ovulation rate in an *in vitro* perfused rat ovarian model, whereas such an IL-1-induced increase can be suppressed by NOS inhibitors (Bonello *et al.*, 1996). The latter involved conspicuous influences on oestradiol production by the perfused ovary and on the flow of perfusate.

Although there is now a substantial body of evidence implicating both 'ovarian cell-derived' and 'vascular endothelial cell-derived' NO in ovarian physiology, the precise sequence of the involvement of NO at each stage of follicular development and during ovulation itself has yet to be determined. This will not be a straightforward task, since the contribution of resident and migrating white blood cells within differing ovarian tissues would also need to be taken into account.

Concluding remarks

Putting to one side consideration of inputs from the autonomic nervous system, a means of setting much of the detailed information presented above in perspective would be to view systemic control of ovarian function as residing in gonadotrophic and other hormones of the pituitary gland. Complementary regulation would be provided by gonadal secretions in the form of steroid hormones and diverse peptides acting in endocrine, paracrine or autocrine fashion. An ever-increasing number of such peptides has been revealed and shown to interact with gonadotrophic hormones and with the follicular cells themselves. In striving to understand the involvement of ovarian steroid and peptide hormones in the development of Graafian follicles, it is perhaps not sufficient to suggest that such molecules offer a form of local fine-tuning of the pituitary trophic inputs, for this does not satisfactorily answer exactly how the level of local peptide input is generated nor the manner whereby the interactions of local peptide secretions are coordinated. A way forward here would be to recognise a vital contribution of the oocyte and its surrounding mass of cumulus cells in influencing the level of peptide secretions and – acting rather as a centrally located computer – programming with sensitivity the rôle of neighbouring granulosa and theca cells. Not only would these conversations be heard within the Graafian follicles in question, but they might also be monitored by other less mature follicles in the same ovary that could then adjust their developmental strategies accordingly.

References

Aaltonen, J., Laitinen, M.P., Vuojolainen, K., *et al.* (1999). Human growth factor 9 (GDF-9) and its novel homolog GDF-9B are expressed in oocytes during early folliculogenesis. *Journal of Clinical Endocrinology and Metabolism*, **84**, 2744–50.

Adashi, E.Y. (1992). The potential relevance of cytokines to ovarian physiology. *Journal of Steroid Biochemistry and Molecular Biology*, **5**, 439–44.

Adashi, E.Y. (1996). Immune modulators in the context of the ovulatory process: a role for interleukin-1. *American Journal of Reproductive Immunology*, **35**, 190–4.

Adashi, E.Y. & Resnick, C.E. (1986). Antagonistic interactions of transforming growth factors in the regulation of granulosa cell differentiation. *Endocrinology*, **119**, 1879–81.

Adashi, E., Resnick, C., D'Ercole, J. *et al.* (1985). Insulin-like growth factors as intraovarian regulators of granulosa cell growth and function. *Endocrine Reviews*, **6**, 400–20.

Alak, B.M., Coskun, S., Friedman, C.I., Kennard, A., Kim, M.H. & Seifer, D.B. (1998). Activin A stimulates meiotic maturation of human oocytes and modulates granulosa cell steroidogenesis *in vitro*. *Fertility and Sterility*, **70**, 1126–30.

Aleem, F.A., Omar, R.A. & El Tabbakh, G.H. (1986). Immunoreactive β-endorphin in human ovaries. *Fertility and Sterility*, **45**, 507–12.

Alpizar, E. & Spicer, L.J. (1994). Effects of interleukin-6 on proliferation and follicle-stimulating hormone-induced estradiol production by bovine granulosa cells *in vitro*: dependence on size of follicle. *Biology of Reproduction*, **50**, 38–43.

Andreani, C.L., Payne, D.W., Packman, J.N., Resnick, C.E., Hurwitz, A. & Adashi, E.Y. (1991). Cytokine-mediated regulation of ovarian function. Tumor necrosis factor alpha inhibits gonadotropin-supported ovarian androgen biosynthesis. *Journal of Biological Chemistry*, **266**, 6761–6.

Anteby, E.Y., Hurwitz, A., Korach, O., *et al.* (1996). Human follicular nitric oxide pathway: relationship to follicular size, oestradiol concentrations and ovarian blood flow. *Human Reproduction*, **11**, 1947–51.

Arakane, F., Sugawara, T., Kiriakidou, M., Kallen, C.B., *et al.* (1997). Molecular insights into the regulation of steroidogenesis: from laboratory to clinic and back. *Human Reproduction*, **12**, National Supplement, 46–50.

Araki, M., Fukumatsu, Y., Katabuchi, H., Schultz, L.D., Takahashi, K. & Okamura, H. (1996). Follicular development and ovulation in macrophage colony-stimulating factor-deficient mice homozygous for the osteopetrosis (*op*) mutation. *Biology of Reproduction*, **54**, 478–84.

Arici, A., Oral, E., Bahtiyar, O., Engin, O., Seli, E. & Jones, E.E. (1997). Leukaemia inhibitory factor expression in human follicular fluid and ovarian cells. *Human Reproduction*, **12**, 1233–9.

Armstrong, D.G. & Webb, R. (1997). Ovarian follicular dominance: the role of intra-ovarian growth factors and novel proteins. *Reviews of Reproduction*, **2**, 139–46.

Aschheim, S. & Zondek, B. (1927). Hypophysenvorderlappenhormon und Ovarialhormon im Harn von Schwangeren. *Klinische Wochenschrift*, **6**, 1322.

Baird, D.T. (1977). Evidence *in vivo* for the two-cell hypothesis of oestrogen synthesis by the sheep Graafian follicle. *Journal of Reproduction and Fertility*, **50**, 183–5.

Baird, D.T. (1984). The ovary. In *Reproduction in Mammals*, 2nd edn, ed. C.R. Austin & R.V. Short, pp. 91–114. Cambridge: Cambridge University Press.

Baird, D.T. & Smith, K. (1993). Inhibin and related peptides in the regulation of reproduction. *Oxford Reviews of Reproductive Biology*, **15**, 191–232.

Bao, B., Calder, M.D., Xie, S., *et al.* (1998). Expression of steroidogenic acute regulatory protein messenger ribonucleic acid is limited to theca of healthy bovine follicles collected during recruitment, selection and dominance of follicles of the first follicular wave. *Biology of Reproduction*, **59**, 953–9.

Ben-Shlomo, I., Kokia, E., Jackson, M., Adashi, E. & Payne, D.W. (1994). Interleukin-1β stimulates nitrite production in the rat ovary: evidence for heterologous cell–cell interaction and for insulin-mediated regulation of the inducible isoform of nitric oxide synthase. *Biology of Reproduction*, **51**, 310–18.

Benyo, D.F. & Pate, J.L. (1992). Tumor necrosis factor-α alters bovine luteal cell synthetic capacity and viability. *Endocrinology*, **130**, 854–60.

Bhatia, B. & Price, C.A. (2001). Insulin alters the effects of follicle stimulating hormone on aromatase in bovine granulosa cells *in vitro*. *Steroids*, **66**, 511–19.

Bodensteiner, K.J., Clay, C.M., Moeller, C.L. & Sawyer, H.R. (1999). Molecular cloning of the ovine growth/differentiation factor-9 gene and expression of growth/differentiation factor-9 in ovine and bovine ovaries. *Biology of Reproduction*, **60**, 381–6.

Bonello, N., McKie, K., Jasper, M., *et al.* (1996). Inhibition of nitric oxide: effects on interleukin-1β-enhanced ovulation rate, steroid hormones and ovarian leukocyte distribution at ovulation in the rat. *Biology of Reproduction*, **54**, 436–45.

Brännström, M. (1997). Intra-ovarian immune mechanisms in ovulation. In *Microscopy of Reproduction and Development: A Dynamic Approach*, ed. P.M. Motta, pp. 163–8. Rome: Antonio Delfino Editore.

Brännström, M., Bonello, N., Wang, L. & Norman, R.J. (1995). Effects of tumor necrosis factor α (TNFα) on ovulation in the rat ovary. *Reproduction, Fertility and Development*, **7**, 67–73.

Brännström, M., Böstrom, E.K., Encrantz, L. & Runesson, E. (1996). Chemotactic cytokines in the human follicle at ovulation. *Biology of Reproduction*, **54**, Supplement 1, 68 (Abstract).

Brännström, M., Friden, B.E., Jasper, M. & Norman, R.J. (1999). Variations in peripheral blood levels of immunoreactive tumor necrosis factor alpha (TNFα) throughout the menstrual cycle and secretion of TNFα from the human corpus luteum. *European Journal of Obstetrics, Gynaecology and Reproductive Biology*, **83**, 213–17.

Brännström, M., Mayerhofer, G. & Robertson, S.A. (1993a). Localisation of leukocyte subsets in the rat ovary during the periovulatory period. *Biology of Reproduction*, **48**, 277–86.

Brännström, M., Mikuni, M. & Hedin, L. (1997). Intra-ovarian events during follicular development and ovulation. *Human Reproduction*, **12**, National Supplement, 51–7.

Brännström, M., Norman, R.J., Seamark, R.F. & Robertson, S.A. (1994b). Rat ovary produces cytokines during ovulation. *Biology of Reproduction*, **50**, 88–94.

Brännström, M., Pascoe, V., Norman, R.J. & McClure, N. (1994a). Localisation of leukocyte subsets in the human follicle wall and in the corpus luteum throughout the menstrual cycle. *Fertility and Sterility*, **61**, 488–95.

Brännström, M., Wang, L. & Norman, R.J. (1993b). Ovulatory effect of interleukin-1 beta on the perfused rat ovary. *Endocrinology*, **132**, 399–404.

Brännström, M., Wang, L. & Norman, R.J. (1993c). Effects of cytokines on prostaglandin production and steroidogenesis of incubated preovulatory follicles of the rat. *Biology of Reproduction*, **48**, 165–71.

Bukovsky, A., Presl, J., Krabec, R. & Bednarik, T. (1977). Ovarian function in adult rats treated with antithymocyte serum. *Experientia*, **33**, 280–1.

Bukulmez, O. & Arici, A. (2000). Leukocytes in ovarian function. *Human Reproduction Update*, **6**, 1–15.

Campbell, B.K., Gordon, B.M. & Scaramuzzi, R.J. (1994). The effect of ovarian arterial infusion of transforming growth factor-α on ovarian follicle populations and ovarian hormone secretion in ewes with an autotransplant ovary. *Journal of Endocrinology*, **143**, 13–24.

Campbell, B.K. & Webb, R. (1995). Evidence that inhibin has autocrine and paracrine actions in controlling ovarian function in sheep. *Journal of Reproduction and Fertility, Abstract Series*, **15** Abstract 140.

Caron, K.M., Soo, S.C., Wetsel, W.C., Stocco, D.M., Clark, B.J. & Parker, K.L. (1997). Targeted disruption of the mouse gene encoding steroidogenic acute regulatory protein provides insights into congenital lipoid adrenal hyperplasia. *Proceedings of the National Academy of Sciences, USA*, **94**, 11540–5.

Carswell, E.A., Old, L.J., Kassel, R.L., Green, S., Fiore, N. & Williamson, B. (1975). An endotoxin-induced serum factor that causes necrosis of tumors. *Proceedings of the National Academy of Sciences, USA*, **72**, 3666–70.

Chang, R.J., Gougeon, A. & Erickson, G.F. (1998). Evidence for a neutrophil–interleukin-8 system in human folliculogenesis. *American Journal of Obstetrics and Gynecology*, **178**, 650–7.

Chen, Y.J., Feng, Q. & Liu, Y.X. (1999). Expression of the steroidogenic acute regulatory protein and luteinising hormone receptor and their regulation by tumor necrosis factor alpha in rat corpora lutea. *Biology of Reproduction*, **60**, 419–27.

Chun, S.Y., Eisenhauer, K.M., Kubo, M. & Hsueh, A.J.W. (1995). Interleukin beta suppresses apoptosis in rat ovarian follicles by nitric oxide production. *Endocrinology*, **136**, 3120–7.

Corner, G.W. (1919). On the origin of the corpus luteum of the sow from both granulosa and theca interna. *American Journal of Anatomy*, **26**, 117–83.

Davis, B.J., Lennard, D.E., Lee, C.A., *et al.* (1999). Anovulation in cyclooxygenase-2-deficient mice is restored by prostaglandin E_2 and interleukin-1β. *Endocrinology*, **140**, 2685–95.

Derynck, R. (1988). Transforming growth factor α. *Cell*, **54**, 593–5.

Dong, J., Albertini, D.F., Nishimori, K., Kumar, R.T., Lu, N. & Matzuk, M.M. (1996). Growth differentiation factor-9 is required during early ovarian folliculogenesis. *Nature (London)*, **383**, 531–5.

Dorrington, J., Chuma, A.V. & Bendell, J.J. (1988). Transforming growth factor β and follicle stimulating hormone promote rat granulosa cell proliferation. *Endocrinology*, **123**, 353–9.

Douglass, J., Cox, B., Quinn, B., Civelli, O. & Herbert, E. (1987). Expression of the prodynorphin gene in male and female reproductive tissues. *Endocrinology*, **120**, 707–13.

Driancourt, M.A., Gougeon, A., Royère, D. & Thibault, C. (1993). Ovarian function. In *Reproduction in Mammals and Man*, ed. C. Thibault, M.C. Levasseur & R.H.F. Hunter, pp. 281–305. Paris: Ellipses.

Elvin, J.A., Clark, A.T., Wang, P., Wolfman, N.M. & Matzuk, M. (1999). Paracrine actions of growth differentiation factor-9 in the mammalian ovary. *Molecular Endocrinology*, **13**, 1035–48.

Erickson, G.F., Chung, D.G., Sit, A., DePaolo, L.V., Shimasaki, S. & Ling, N. (1995). Follistatin concentrations in follicular fluid of normal and polycystic ovaries. *Human Reproduction*, **10**, 2120–4.

Erickson, G.F. & Hsueh, A.J.W. (1978). Secretion of inhibin by rat granulosa cells *in vitro*. *Endocrinology*, **103**, 1960–3.

Ethier, J.-F. & Findlay, J.K. (2001). Roles of activin and its signal transduction mechanisms in reproductive tissues. *Reproduction*, **121**, 667–75.

Falck, B. (1959). Site of production of oestrogen in rat ovary as studied in microtransplants. *Acta Physiologica Scandinavica*, **47**, Supplement 163, 1–101.

Feng, P., Catt, K.J. & Knecht, M. (1988). Transforming growth factor-β stimulates meiotic maturation of the rat oocyte. *Endocrinology*, **122**, 181–6.

Ferrara, N. & Davis-Smyth, T. (1997). The biology of vascular endothelial growth factor. *Endocrine Reviews*, **18**, 4–25.

Findlay, J.K. (1986). Angiogenesis in reproductive tissues. *Journal of Endocrinology*, **111**, 357–66.

Findlay, J.K. (1993). An update on the roles of inhibin, activin and follistatin as local regulators of folliculogenesis. *Biology of Reproduction*, **48**, 15–23.

Fortune, J.E. (1986). Bovine theca and granulosa cells interact to promote androgen production. *Biology of Reproduction*, **35**, 292–9.

Gadsby, J.E., Lovdal, J.A., Samaras, S., Barber, J.S. & Hammond, J.M. (1996). Expression of the messenger ribonucleic acids for insulin-like growth factor-1 and insulin-like growth factor binding proteins in porcine corpora lutea. *Biology of Reproduction*, **54**, 339–46.

Giudice, L.C. (1992). Insulin-like growth factors and ovarian follicular development. *Endocrine Reviews*, **13**, 641–69.

Glister, C., Tannetta, D.S., Groome, N.P. & Knight, P.G. (2001). Interactions between follicle-stimulating hormone and growth factors in modulating secretion of steroids and inhibin-related peptides by nonluteinised bovine granulosa cells. *Biology of Reproduction*, **65**, 1020–8.

Gong, J.G., Baxter, G., Bramley, T.A. & Webb, R. (1997). Enhancement of ovarian follicle development in heifers by treatment with recombinant bovine somatotrophin: a dose–response study. *Journal of Reproduction and Fertility*, **110**, 91–7.

Gong, J.G., McBride, D., Bramley, T.A. & Webb, R. (1994). Effects of recombinant bovine somatotrophin, insulin-like growth factor-1 and insulin on bovine granulosa cell steroidogenesis *in vitro*. *Journal of Endocrinology*, **143**, 157–64.

Gore-Langton, R.E. & Armstrong, D.T. (1988). Follicular steroidogenesis and its control. In *The Physiology of Reproduction*, ed. E. Knobil & J.D. Neill, pp. 331–85. New York: Raven Press.

Gougeon, A. (1997). Kinetics of human ovarian follicular development. In *Microscopy of Reproduction and Development: A Dynamic Approach*, ed. P.M. Motta, pp. 67–77. Rome: Antonio Delfino Editore.

Greenwald, G.S. & Roy, S.K. (1994). Follicular development and its control. In *The Physiology of Reproduction*, 2nd edn, ed. E. Knobil & J.D. Neill, pp. 629–724. New York: Raven Press.

Groome, N.P., Illingworth, P.J. & O'Brien, M. (1994). Detection of dimeric inhibin throughout the human menstrual cycle by two-site enzyme immunoassay. *Clinical Endocrinology*, **4**, 717–23.

Groome, N.P., Illingworth, P.J., O'Brien, M., *et al.* (1996). Measurement of dimeric inhibin-B throughout the human menstrual cycle. *Journal of Clinical Endocrinology and Metabolism*, **81**, 1401–5.

Guthrie, H.D., Rohan, R.M., Rexroad, C.E. Jr & Cooper, B.S. (1992). Changes in concentrations of follicular inhibin α and β_A subunit messenger ribonucleic acids and inhibin immunoactivity during preovulatory maturation in the pig. *Biology of Reproduction*, 47, 1018–25.

Hammond, J.M., Samaras, S.E., Grimes, R., *et al.* (1993). The role of insulin-like growth factors and epidermal growth factor-related peptides in intraovarian regulation in the pig ovary. *Journal of Reproduction and Fertility*, **48**, 117–25.

Harper, K.M. & Brackett, B.G. (1992). Enhanced bovine oocyte quality after *in vitro* maturation (IVM) with insulin-like growth factor-1 (IGF-1) and gonadotropins. *Biology of Reproduction*, **46** (Suppl. 1), 67.

Harper, M.J.K., Skarnes, R.C., Chang, M.C. & McGaughey, R.W. (1970). Effects of antilymphocytic serum (ALS) on pregnancy in rodents. *Biology of Reproduction*, **2**, 335–45.

Hayashi, M., McGee, E.A., Min, G., *et al.* (1999). Recombinant growth differentiation factor-9 (GDF-9) enhances growth and differentiation of cultured early follicles. *Endocrinology*, **140**, 1236–44.

Henderson, K.M. & Franchimont, P. (1983). Inhibin production by bovine ovarian tissues *in vitro*, and its regulation by androgens. *Journal of Reproduction and Fertility*, **67**, 291–8.

Hillier, S.G. & Miro, F. (1993). Inhibin, activin, and follistatin. Potential roles in ovarian physiology. *Annals of the New York Academy of Sciences*, **687**, 29–38.

Hillier, S.G., Whitelaw, P.F. & Smyth, C.D. (1994). Follicular oestrogen synthesis: the 'two cell, two gonadotrophin' model revisited. *Molecular and Cellular Endocrinology*, **100**, 51–4.

Hillier, S.G., Wickings, E.J., Illingworth, P.J., *et al.* (1991a). Control of immunoactive inhibin production by human granulosa cells. *Clinical Endocrinology*, **35**, 71–8.

Hillier, S.G., Wickings, E.J., Saunders, P.T.K., *et al.* (1989). Control of inhibin production by primate granulosa cells. *Journal of Endocrinology*, **123**, 65–73.

Hillier, S.G., Yong, E.L., Illingworth, P.I., Baird, D.T., Schwall, R.H. & Mason, A.J. (1991b). Effect of recombinant activin on androgen synthesis in cultured human thecal cells. *Journal of Clinical Endocrinology and Metabolism*, **72**, 1206–11.

Hsueh, A.J.W., Dahl, K.D., Vaughan, J., *et al.* (1987). Heterodimers and homodimers of inhibin subunits have different paracrine action in modulation of luteinising hormone-stimulated androgen synthesis. *Proceedings of the National Academy of Sciences, USA*, **84**, 5082–6.

Hu, P.C., Datto, M.B. & Wang, X.F. (1998). Molecular mechanisms of transforming growth factor-β signaling. *Endocrine Reviews*, 19, 349–63.

Hunter, M.G., Biggs, C., Pickard, A.R. & Faillace, L.S. (1994). Differences in follicular aromatase activity between Meishan and Large-White hybrid pigs. *Journal of Reproduction and Fertility*, **101**, 139–44.

Hunter, R.H.F. (1980). *Physiology and Technology of Reproduction in Female Domestic Animals*. London & New York: Academic Press.

Hunter, R.H.F., Cook, B. & Baker, T.G. (1976). Dissociation of response to injected gonadotrophin between the Graafian follicle and oocyte in pigs. *Nature (London)*, **260**, 156–8.

Hunter, R.H.F. & Poyser, N.L. (1985). Ovarian follicular fluid concentrations of prostaglandins E_2, $F_{2\alpha}$ and I_2 during the pre-ovulatory period in pigs. *Reproduction, Nutrition et Développement*, **25**, 909–17.

Hurwitz, A., Dushnik, M., Solomon, H., *et al.* (1993). Cytokine-mediated regulation of rat ovarian function: interleukin-1 stimulates the accumulation of a 92-kilodalton gelatinase. *Endocrinology*, **132**, 2709–14.

Hurwitz, A., Loukides, J., Ricciarelli, E., *et al.* (1992). The human intraovarian interleukin-1 (IL-1) system: highly-compartmentalised and hormonally dependent regulation of the genes encoding IL-1, its receptor, and its receptor antagonist. *Journal of Clinical Investigation*, **89**, 1746–54.

Ivell, R., Tillmann, G., Wang, H., *et al.* (2000). Acute regulation of the bovine gene for the steroidogenic acute regulatory protein in ovarian theca and adrenocortical cells. *Journal of Molecular Endocrinology*, **24**, 109–18.

Izadyar, F., Dijkstra, G., Van Tol, H.T.A., *et al.* (1998). Immunohistochemical localisation of mRNA expression of activin, inhibin, follistatin, and activin receptor in bovine cumulus–oocyte complexes during *in vitro* maturation. *Molecular Reproduction and Development*, **49**, 186–95.

Jablonka-Shariff, A. & Olson, L.M. (1998). The role of nitric oxide in oocyte meiotic maturation and ovulation: meiotic abnormalities of endothelial nitric oxide synthase knock-out mouse oocytes. *Endocrinology*, **139**, 2944–54.

Jablonka-Shariff, A. & Olson, L.M. (2000). Nitric oxide is essential for optimal meiotic maturation of murine cumulus–oocyte complexes *in vitro*. *Molecular Reproduction and Development*, **55**, 412–21.

Janson, P.O. (1975). Effects of the luteinising hormone on blood flow in the follicular rabbit ovary, as measured by radioactive microspheres. *Acta Endocrinologica (Copenhagen)*, **79**, 122–33.

Jasper, M.J., Brännström, M., Olofsson, J.I., *et al.* (1996). Granulocyte–macrophage colony-stimulating factor: presence in human follicular fluid, protein secretion and mRNA expression by ovarian cells. *Molecular Human Reproduction*, **2**, 555–62.

Jin, D.F., Muffly, K.E., Okulicz, W.C. & Kilpatrick, D.L. (1988). Estrous cycle- and pregnancy-related differences in expression of the proenkephalin and proopiomelanocortin genes in the ovary and uterus. *Endocrinology*, **122**, 1466–71.

Jones, J.I. & Clemmons, D.R. (1995). Insulin-like growth factors and their binding proteins: biological actions. *Endocrine Reviews*, **16**, 3–34.

Jooste, S.V. (1968). The effects of heterologous antilymphocytic antiserum in the late prenatal and early postnatal period in mice. *Transplantation*, **6**, 277–86.

Kaminski, T., Siawrys, G., Bogacka, I. & Przala, J. (2000). The physiological role of β-endorphin in porcine ovarian follicles. *Reproduction, Nutrition and Development*, **40**, 63–75.

Kato, T., Kumai, A. & Okamoto, R. (1993). Effect of β-endorphin on cAMP and progesterone accumulation in rat luteal cells. *Journal of Endocrinology*, **40**, 323–8.

Kerban, A., Boerboom, D. & Sirois, J. (1999). Human chorionic gonadotrophin induces an inverse regulation of steroidogenic acute regulatory protein messenger ribonucleic acid in theca interna and granulosa cells of equine preovulatory follicles. *Endocrinology*, **140**, 667–74.

Kim, J.J. & Fazleabas, A.T. (1999). Growth factors. In *Encyclopedia of Reproduction*, vol. 2, ed. E. Knobil & J.D. Neill, pp. 573–83. San Diego & London: Academic Press.

Kirby, D.R.S. (1967). Transplantation and pregnancy. In *Human Transplantation*, ed. F. Rapaport & J. Dausset, pp. 565–86. New York: Grune & Stratton.

Kiriakidou, M., McAllister, J.M., Sugawara, T., *et al.* (1996). Expression of steroidogenic acute regulatory protein (StAR) in the human ovary. *Journal of Clinical Endocrinology and Metabolism*, **81**, 4122–8.

Knight, P.G. (1996). Roles of inhibins, activins and follistatin in the female reproductive system. *Frontiers in Neuroendocrinology*, **17**, 476–509.

Knight, P.G. & Glister, C. (2001). Potential local regulatory functions of inhibins, activins and follistatin in the ovary. *Reproduction*, **121**, 503–12.

Lau, C.P., Ledger, W.L., Groome, N.P., Barlow, D.H. & Muttukrishna, S. (1999). Dimeric inhibins and activin A in human follicular fluid and oocyte–cumulus culture medium. *Human Reproduction*, **14**, 2525–30.

Lee, P.D., Conover, C.A. & Powell, D.R. (1993). Regulation and function of insulin-like growth factor-binding protein 1. *Proceedings of the Society for Experimental Biology and Medicine*, **204**, 4–29.

Levey, R.H. & Medawar, P.B. (1966). Nature and mode of action of antilymphocytic antiserum. *Proceedings of the National Academy of Sciences, USA*, **56**, 1130–7.

Li, M.D., DePaolo, L.V. & Ford, J.J. (1997). Expression of follistatin and inhibin/activin subunit genes in porcine follicles. *Biology of Reproduction*, **57**, 112–18.

Li, W.I., Wu, H. & Kumar, A.M. (1991). Synthesis and secretion of immunoreactive-methionine-enkephalin from rabbit reproductive tissues *in vivo* and *in vitro*. *Biology of Reproduction*, **45**, 691–7.

Lockwood, G.M., Muttukrishna, S. & Groome, N.P. (1996). Circulating inhibins and activin A during GnRH-analogue down-regulation and ovarian hyperstimulation with recombinant FSH for *in vitro* fertilisation and embryo transfer. *Clinical Endocrinology*, **45**, 741–8.

Lolait, S.J., Autelitano, D.J., Lim, A.T., Smith, A.I., Toh, B.H. & Funder, J.W. (1985). Ovarian immunoreactive β-endorphin and oestrous cycle in the rat. *Endocrinology*, **117**, 161–8.

Lolait, S.J., Autelitano, D.J., Markwick, A.J., Toh, B.H. & Funder, J.W. (1986). Co-expression of vasopressin with β-endorphin and dynorphin in individual cells from the ovaries of Brattleboro and Long–Evans rats: immunocytochemical studies. *Peptides*, **7**, 267–76.

Lovegren, E.S., Ziminski, S.J. & Puett, D. (1991). Ovarian contents of immunoreactive β-endorphin and α-N-acetylated opioid peptides in rats. *Journal of Reproduction and Fertility*, **91**, 91–100.

Maccio, A., Mantovani, G., Turnu, E., *et al.* (1993). Evidence that granulosa cells inhibit interleukin-1 alpha and interleukin-2 production from follicular lymphomonocytes. *Journal of Assisted Reproduction and Genetics*, **10**, 517–22.

Magoffin, D.A. & Jakimiuk, A.J. (1997). Inhibin A, inhibin B and activin A in the follicular fluid of regularly cycling women. *Human Reproduction*, **12**, 1714–19.

Marcinkiewicz, J.L., Krishna, A., Cheung, C.M. & Terranova, P.F. (1994). Oocytic tumor necrosis factor alpha: localisation in the neonatal ovary and throughout follicular development in the adult rat. *Biology of Reproduction*, **50**, 1251–60.

Mather, J.P., Moore, A. & Li, R.H. (1997). Activins, inhibins, and follistatins: further thoughts on a growing family of regulators. *Proceedings of the Society for Experimental Biology and Medicine*, **215**, 209–22.

Matzuk, M.M., Finegold, M.J., Su, J.G., Hsueh, A.J. & Bradley, A. (1992). Alpha-inhibin is a tumour-suppressor gene with gonadal specificity in mice. *Nature (London)*, **360**, 313–19.

Matzuk, M.M., Kumar, T.R., Shou, W., *et al.* (1996). Transgenic models to study the roles of inhibins and activins in reproduction, oncogenesis and development. *Recent Progress in Hormone Research*, **51**, 123–54.

Matzuk, M.M., Lu, N., Vogel, H., Sellheyer, K., Roop, D.R. & Bradley, A. (1995). Multiple defects and perinatal death in mice deficient in follistatin. *Nature (London)*, **374**, 360–3.

McGrath, S.A., Esquela, A.F. & Lee, S.J. (1995). Oocyte-specific expression of growth/differentiation factor-9. *Molecular Endocrinology*, **9**, 131–6.

Melner, M.H., Young, S.L., Czerwiec, F.S., *et al.* (1986). The regulation of granulosa cell proopiomelanocortin mRNA by androgens and gonadotropins. *Endocrinology*, **119**, 2082–8.

Miller, W.L. (1997). Congenital lipoid adrenal hyperplasia: the human gene knockout for the steroidogenic acute regulatory protein. *Journal of Molecular Endocrinology*, **19**, 227–40.

Minegishi, T., Hirakawa, T., Kishi, H., *et al.* (2000). A role of insulin-like growth factor 1 for follicle-stimulating hormone receptor expression in rat granulosa cells. *Biology of Reproduction*, **62**, 325–33.

Miro, F. & Hillier, S.G. (1992). Relative effects of activin and inhibin on steroid hormone synthesis in primate granulosa cells. *Journal of Clinical Endocrinology and Metabolism*, **75**, 1556–61.

Miro, F., Smyth, C.D. & Hillier, S.G. (1991). Development-related effects of recombinant activin on steroid synthesis in rat granulosa cells. *Endocrinology*, **129**, 3388–94.

Muttukrishna, S., Fowler, P.A., George, L., Groome, N.P. & Knight, P.G. (1996). Changes in peripheral serum concentrations of total activin A during the human menstrual cycle and pregnancy. *Journal of Clinical Endocrinology and Metabolism*, **81**, 3328–34.

Muttukrishna, S., Fowler, P.A., Groome, N.P., Mitchell, G.G., Robertson, W.R. & Knight, P.G. (1994). Serum concentrations of dimeric inhibin during the spontaneous human menstrual cycle and after treatment with exogenous gonadotrophin. *Human Reproduction*, **9**, 1634–42.

Muttukrishna, S., Groome, N. & Ledger, W. (1997). Gonadotropic control of secretion of dimeric inhibins and activin A by human granulosa-luteal cells *in vitro*. *Journal of Assisted Reproduction and Genetics*, **14**, 566–74.

Neufield, G., Ferrara, N., Schweigerer, L., Mitchell, R. & Gospodarowicz, D. (1987). Bovine granulosa cells produce basic fibroblast growth factor. *Endocrinology*, **121**, 597–603.

Nishimori, K. & Matzuk, M.M. (1996). Transgenic mice in the analysis of reproductive development and function. *Reviews of Reproduction*, **1**, 203–12.

Nishizuka, Y. & Sakakura, T. (1969). Thymus and reproduction: sex-linked dysgenesis of the gonad after neonatal thymectomy in mice. *Science*, **166**, 753–5.

Norman, R.J. & Brännström, M. (1996). Cytokines in the ovary: pathophysiology and potential for pharmacological intervention. *Pharmacology and Therapeutics*, **69**, 219–36.

Okuda, K., Sakumoto, R., Uenoyama, Y., Berisha, B., Miyamoto, A. & Schams, D. (1999). Tumor necrosis factor α receptors in microvascular endothelial cells from bovine corpus luteum. *Biology of Reproduction*, **61**, 1017–22.

Olson, L.M., Jones-Burton, C.M. & Joblonka-Shariff, A. (1996). Nitric oxide decreases estradiol synthesis in rat luteal cells *in vitro*: possible role for nitric oxide in functional luteal regression. *Endocrinology*, **137**, 3531–9.

Perks, C.M., Denning-Kendall, P.A., Gilmour, R.S. & Wathes, D.C. (1995). Localisation of messenger ribonucleic acids for insulin-like growth factor 1 (IGF-1), IGF-II and the type 1 IGF receptor in the ovine ovary throughout the estrous cycle. *Endocrinology*, **136**, 5266–73.

Pescador, N., Soumano, K., Stocco, D.M., Price, C.A. & Murphy, B.D. (1996). Steroidogenic acute regulatory protein in bovine corpora lutea. *Biology of Reproduction*, **55**, 485–91.

Peterson, C.M., Hales, H.A., Hatasaka, H.H., Mitchell, M.D., Rittenhouse, L. & Jones, K.P. (1993). Interleukin-1β (IL-1β) modulates prostaglandin production and the natural IL-1 receptor antagonist inhibits ovulation in the optimally stimulated rat ovarian perfusion model. *Endocrinology*, **133**, 2301–6.

Petraglia, F., Segre, A., Facchinetti, F., Campanini, D., Ruspa, M. & Genazzani, A.R. (1985). β-Endorphin and met-enkephalin in peritoneal and ovarian follicular fluids of fertile and post-menopausal women. *Fertility and Sterility*, **44**, 615–21.

Pitzel, L., Jarry, H. & Wuttke, W. (1993). Effects and interactions of prostaglandin $F_{2\alpha}$, oxytocin and cytokines on steroidogenesis of porcine luteal cells. *Endocrinology*, **132**, 751–6.

Qu, J., Godin, P.A., Nisolle, M. & Donnez, J. (2000). Expression of receptors for insulin-like growth factor-1 and transforming growth factor-β in human follicles. *Molecular Human Reproduction*, **6**, 137–45.

Rajaniemi, H. & Vanha-Perttula, T. (1972). Specific receptor for LH in the ovary: evidence by autoradiography and tissue fractionation. *Endocrinology*, **90**, 1–9.

Redmer, D.A. & Reynolds, L.P. (1996). Angiogenesis in the ovary. *Reviews of Reproduction*, **1**, 182–92.

Reynolds, L.P., Killilea, S.D. & Redmer, D.A. (1992). Angiogenesis in the female reproductive system. *FASEB Journal*, **6**, 886–92.

Roberts, A.J. & Skinner, M.K. (1991). Transforming growth factor-α and -β differentially regulate growth and steroidogenesis of bovine thecal cells during antral follicle development. *Endocrinology*, **129**, 2041–8.

Roby, K.F., Weed, J., Lyles, R. & Terranova, P.F. (1990). Immunological evidence for a human ovarian tumor necrosis factor-alpha. *Journal of Clinical Endocrinology and Metabolism*, **71**, 1096–102.

Rosselli, M., Imthurn, B., Magas, E., *et al.* (1994). Circulating nitrite/nitrate increase with follicular development: indirect evidence for estradiol mediated release. *Biochemical and Biophysical Research Communication*, **202**, 1543–52.

Rosselli, M., Keller, P.J. & Dubey, R.K. (1998). Role of nitric oxide in the biology, physiology and pathophysiology of reproduction. *Human Reproduction Update*, **4**, 3–24.

Runesson, E., Boström, E.K., Janson, P.O. & Brännström, M. (1996). The human pre-ovulatory follicle is a source of the chemotactic cytokine interleukin-8. *Molecular Human Reproduction*, **2**, 245–50.

Sakumoto, R., Berisha, B., Kawate, N., Schams, D. & Okuda, K. (2000). Tumor necrosis factor-α and its receptor in bovine corpus luteum throughout the estrous cycle. *Biology of Reproduction*, **62**, 192–9.

Sancho-Tello, M., Perez-Roger, I., Imakawa, K., Tilzer, L. & Terranova, P.F. (1992). Expression of tumor necrosis factor-alpha in the rat ovary. *Endocrinology*, **130**, 1359–64.

Sanders, S.L., Melner, M.H. & Curry, T.E. Jr (1990). Cellular localisation of ovarian proopiomelanocortin mRNA during follicular and luteal development in the rat. *Molecular Endocrinology*, **4**, 1311–19.

Savard, K., Marsh, J.M. & Rice, B.F. (1965). Gonadotrophins and ovarian steroidogenesis. *Recent Progress in Hormone Research*, **21**, 285–356.

Schwall, R.H. (1999a). Activin and activin receptors. In *Encyclopedia of Reproduction*, vol. 1, ed. E. Knobil & J.D. Neill, pp. 26–35. San Diego & London: Academic Press.

Schwall, R.H. (1999b). Inhibin. In *Encyclopedia of Reproduction*, vol. 2, ed. E. Knobil & J.D. Neill, pp. 832–9. San Diego & London: Academic Press.

Shimasaki, S., Koga, M., Esch, F., *et al.* (1988). Primary structure of the human follistatin precursor and its genomic organisation. *Proceedings of the National Academy of Sciences, USA*, **85**, 4218–22.

Shores, E.M. & Hunter, M.G. (1999). Immunohistochemical localisation of steroidogenic enzymes and comparison with hormone production during follicle development in the pig. *Reproduction, Fertility and Development*, **11**, 337–44.

Short, R.V. (1962). Steroids in the follicular fluid and the corpus luteum of the mare. A 'two-cell type' theory of ovarian steroid synthesis. *Journal of Endocrinology*, **24**, 59–63.

Shukovski, L. & Tsafriri, A. (1994). The involvement of nitric oxide in the ovulatory process in the rat. *Endocrinology*, **135**, 2287–90.

Shukovski, L., Zhang, Z.W., Michel, U. & Findlay, J.K. (1992). Expression of mRNA for follicle stimulating hormone suppressing protein in ovarian tissues of cows. *Journal of Reproduction and Fertility*, **95**, 861–7.

Silva, C.C. & Knight, P.G. (1998). Modulatory actions of activin-A and follistatin on the developmental competence of *in vitro*-matured bovine oocytes. *Biology of Reproduction*, **58**, 558–65.

Silva, C.C., Groome, N.P. & Knight, P.G. (1999). Demonstration of a suppressive effect of inhibin α-subunit on the developmental competence of *in vitro* matured bovine oocytes. *Journal of Reproduction and Fertility*, **115**, 381–8.

Simón, C., Frances, A., Piquette, G. & Polan, M.L. (1994a). Immunohistochemical localisation of the interleukin-1 system in the mouse ovary during follicular growth, ovulation and luteinisation. *Biology of Reproduction*, **50**, 449–57.

Simón, C., Tsafriri, A., Chun, S.Y., *et al.* (1994b). Interleukin-1 receptor antagonist suppresses human chorionic gonadotrophin-induced ovulation in the rat. *Biology of Reproduction*, **51**, 662–7.

Singh, J. & Adams, G.P. (1998). Immunohistochemical distribution of follistatin in dominant and subordinate follicles and the corpus luteum of cattle. *Biology of Reproduction*, **59**, 561–70.

Slomczynska, M., Pierzchala-Koziec, K., Gregoraszczuk, E., Maderspach, K. & Wierzchos, E. (1997). The kappa-opioid receptor is present in porcine ovaries: localization in granulosa cells. *Cytobios*, **92**, 195–202.

Smith, P.E. & Engle, E.T. (1927). Experimental evidence regarding the rôle of the anterior pituitary in the development and regulation of the genital system. *American Journal of Anatomy*, **40**, 159–217.

Soumano, K. & Price, C.A. (1997). Ovarian follicular steroidogenic acute regulatory protein, low-density lipoprotein receptor, and cytochrome P450 side-chain cleavage messenger ribonucleic acids in cattle undergoing superovulation. *Biology of Reproduction*, **56**, 516–22.

Spaczynski, R.Z., Arici, A. & Duleba, A.J. (1999). Tumor necrosis factor-α stimulates proliferation of rat ovarian theca-interstitial cells. *Biology of Reproduction*, **61**, 993–8.

Spicer, L.J. (1998). Tumor necrosis factor-α (TNF-α) inhibits steroidogenesis of bovine ovarian granulosa and theca cells *in vitro*. Involvement of TNF-α receptors. *Endocrine*, **8**, 109–15.

Stocco, D.M. & Clark, B.J. (1996). Regulation of the acute steroidogenic production of steroids in steroidogenic cells. *Endocrine Reviews*, **17**, 221–43.

Stock, A.E., Woodruff, T.K. & Smith, L.C. (1997). Effects of inhibin A and activin A during *in vitro* maturation of bovine oocytes in hormone- and serum-free medium. *Biology of Reproduction*, **56**, 1559–64.

Sturgis, S.H. (1961). Factors influencing ovulation and atresia of ovarian follicles. In *Control of Ovulation*, ed. C.A. Villee, pp. 213–18. Oxford & New York: Pergamon Press.

Sugawara, T., Holt, J.A., Driscoll, D. *et al.* (1995). Human steroidogenic acute regulatory protein: functional activity in COS-1 cells, tissue-specific expression, and mapping of the structural gene to 8p11.2 and a pseudogene to chromosome 13. *Proceedings of the National Academy of Sciences, USA*, **92**, 4778–82.

Suzuki, T., Miyamoto, K., Hasegawa, Y., *et al.* (1987). Regulation of inhibin production by rat granulosa cells. *Molecular and Cellular Endocrinology*, **54**, 185–95.

Takehara, Y., Dharmarajan, A., Kaufman, G. & Wallach, E. (1994). Effect of interleukin-1β on ovulation in the *in vitro* perfused rabbit ovary. *Endocrinology*, **134**, 1788–93.

Tarlatzis, B.C. & Bili, H. (1997). Steroid and cytokine production by human cumulus cells. In *Microscopy of Reproduction and Development: A Dynamic Approach*, ed. P.M. Motta, pp. 151–3. Rome: Antonio Delfino Editore.

Tekpetey, F., Engelhardt, H. & Armstrong, D. (1993). Differential modulation of porcine theca, granulosa, and luteal cell steroidogenesis *in vitro* by tumor necrosis factor. *Biology of Reproduction*, **48**, 936–43.

Terranova, P.F. & Rice, V.M. (1997). Review: cytokine involvement in ovarian processes. *American Journal of Reproductive Immunology*, **37**, 50–63.

Thissen, J.P., Ketelslegers, J.M. & Underwood, L.E. (1994). Nutritional regulation of the insulin-like growth factors. *Endocrine Reviews*, **15**, 80–101.

Tsafriri, A., Veljkovic, M.V., Pomerantz, S.H. & Ling, N. (1991). The action of transforming growth factors and inhibin-related proteins on oocyte maturation. *Bulletin of the Association of Anatomists*, **75**, 109–13.

Ueno, N., Ling, N., Ying, S.Y., Esch, F., Shimasaki, S. & Guillemin, R. (1987). Isolation and partial characterisation of follistatin: a single chain M_r 35,000 monomeric protein that inhibits the release of follicle-stimulating hormone. *Proceedings of the National Academy of Sciences, USA*, **84**, 8282–6.

Van Voorhis, B.J. (1999). Follicular development. In *Encyclopedia of Reproduction*, vol. 2, ed. E. Knobil & J.D. Neill, pp. 376–89. San Diego & London: Academic Press.

Van Voorhis, B.J., Dunn, M.S., Snyder, G.D. & Weiner, C.P. (1994). Nitric oxide: an autocrine regulator of human granulosa-luteal cell steroidogenesis. *Endocrinology*, **135**, 1799–806.

Van Voorhis, B.J., Moore, K., Strijbos, P.L.M., *et al.* (1995). Expression and localisation of inducible and endothelial nitric oxide synthase in the rat ovary. *Journal of Clinical Investigation*, **96**, 2719–26.

Vitt, U.A., Hayashi, M., Klein, C. & Hsueh, A.J.W. (2000). Growth differentiation factor-9 stimulates proliferation but suppresses the follicle stimulating hormone-induced differentiation of cultured granulosa cells from small antral and preovulatory rat follicles. *Biology of Reproduction*, **62**, 370–7.

Viveiros, M.M. & Liptrap, R.M. (1999). Glucocorticoid influence on porcine granulosa cell IGF-1 and steroid hormone production *in vitro*. *Theriogenology*, **51**, 1027–43.

Viveiros, M.M. & Liptrap, R.M. (2000). ACTH treatment disrupts ovarian IGF-1 and steroid hormone production. *Journal of Endocrinology*, **164**, 255–64.

Wang, L.J., Brännström, M., Cui, K.H., *et al.* (1995). Localisation of gene expression of interleukin-1 receptor and interleukin-1 receptor antagonist in the rat ovary. *Proceedings of the Australian Society for Reproductive Biology*, Abstract 133.

Wang, L.J., Brännström, M., Robertson, S.A. & Norman, R.J. (1992). Tumor necrosis factor α in the human ovary: presence in follicular fluid and effects on cell proliferation and prostaglandin production. *Fertility and Sterility*, **58**, 934–40.

Wang, L.J., Robertson, S., Seamark, R.F. & Norman, R.J. (1991). Lymphokines, including interleukin-2, alter gonadotropin-stimulated progesterone production and proliferation of human granulosa-luteal cells *in vitro*. *Journal of Clinical Endocrinology and Metabolism*, **72**, 824–31.

Watson, E.D., Thomson, S.R.M. & Howie, A.F. (2000). Detection of steroidogenic acute regulatory protein in equine ovaries. *Journal of Reproduction and Fertility*, **119**, 187–92.

Westman, A. (1934). Über die hormonale Funktion des unbefruchteten Eies. *Archiv für Gynaekologie*, **156**, 550–65.

Wilken, C., Van Kirk, E.A., Slaughter, R.G., Ji, T.H. & Murdoch, W.J. (1990). Increased production of ovarian thromboxane in gonadotropin-primed immature rats: relationship with the ovulatory process. *Prostaglandins*, **40**, 637–45.

Woodruff, T.K. & Mather, J.P. (1995). Inhibin, activin and the female reproductive axis. *Annual Review of Physiology*, **57**, 219–44.

Ying, S.-Y. & Zhang, Z. (1999). Ovarian hormones, overview. In *Encyclopedia of Reproduction*, vol. 3, ed. E. Knobil & J.D. Neill, pp. 578–82. San Diego & London: Academic Press.

Zachow, R.J., Tash, J.S. & Terranova, P.F. (1992). Tumor necrosis factor-alpha induces clustering in ovarian theca-interstitial cells *in vitro*. *Endocrinology*, **131**, 2503–13.

Zachow, R.J., Tash, J.S. & Terranova, P.F. (1993). Tumor necrosis factor-alpha attenuation of luteinising hormone-stimulated androstenedione production by ovarian theca-interstitial cells: inhibition at loci within the adenosine 3′,5′-monophosphate-dependent signaling pathway. *Endocrinology*, **133**, 2269–76.

Zachow, R.J. & Terranova, P.F. (1993). Involvement of protein kinase C and protein tyrosine kinase pathways in tumor necrosis factor-alpha-induced clustering of ovarian theca-interstitial cells. *Molecular and Cellular Endocrinology*, **97**, 37–49.

Zachow, R.J. & Terranova, P.F. (1994). The effects of tumor necrosis factor-α on luteinising hormone (LH)/-insulin-like growth factor-1 (IGF-1)-regulated androstenedione biosynthesis and IGF-1 directed LH receptor number in cultured ovarian theca-interstitial cells. *Endocrine*, **2**, 1145–50.

Zachow, R.J., Weitsman, S.R. & Magoffin, D.A. (1999). Leptin impairs the synergistic stimulation by transforming growth factor-β of follicle stimulating hormone-dependent aromatase activity and messenger ribonucleic acid expression in rat ovarian granulosa cells. *Biology of Reproduction*, **61**, 1104–9.

Zackrisson, U., Mikuni, M., Wallin, A., *et al.* (1996). Cell specific localisation of nitric oxide synthases (NOS) in the rat ovary during follicular development, ovulation and luteal formation. *Human Reproduction*, **11**, 2667–73.

Zeleznik, A.J. (1999). Luteinisation. In *Encyclopedia of Reproduction*, vol. 2, ed. E. Knobil & J.D. Neill, pp. 1076–83. San Diego & London: Academic Press.

Zhou, J., Kumar, T.R., Matzuk, M.M. & Bondy, C. (1997). Insulin-like growth factor 1 regulates gonadotropin responsiveness in the murine ovary. *Molecular Endocrinology*, **11**, 1924–33.

VI

Follicular recruitment, growth and development: selection – or atresia

Introduction

A topic of research that has remained conspicuous in the field of ovarian physiology for more than 25 years is that of follicular selection and dominance, and the underlying pattern of follicular growth and atresia. Bearing in mind the very extensive reserves of primordial follicles in the mammalian ovary (see Chapter II), one cannot escape the question of precisely what factors trigger (1) recruitment of follicles from the pool, (2) their further growth and development, and (3) the subsequent emergence of a follicle(s) that will dominate its contemporaries until the time of ovulation in that particular oestrous or menstrual cycle. In a sense, one is asking what cues identify those follicles that are selected to grow and emerge, and whether actual selection should be viewed primarily as a permissive or even chance event. Considered in an alternative light, even if phrased with less than elegance, are some primordial follicles better than others? The thought behind the question is whether interactions between oocytes and follicular cells produce a germinal unit that is more responsive to physiological cues in some follicles than in others. Or is there a chronological sequence – a predetermined programme – that prompts the first follicles to be formed to become the first to emerge from the pool (see Henderson & Edwards, 1968)? A substantial list of related questions could be offered.

Overall, there lingers the tantalising fact that although vast numbers of follicles are generated in mammalian ovaries, only a minute proportion of oocytes will ever leave an ovary by means of spontaneous ovulation. Atresia represents the destiny of an overwhelming majority of oocytes and follicles. Hence, there is the outstanding question of why so many primordial follicles are formed, and especially the question of whether there are subtle processes of genetic selection at play. Are follicles that emerge from the pool in some manner superior (e.g. possessing cells with fewer damaging mutations or a more supportive

186

vascular bed), or are they privileged by circumstances of timing, being at an appropriate stage to respond to particular patterns of gonadotrophin secretion? And is the vast number of oocytes and follicles generated necessary for some key physiological strategy, perhaps to facilitate waves of development and regression, or is it merely a vestige of external systems of fertilisation in aquatic ancestors?

As a final question by way of introduction, and one which arises repeatedly in this monograph, is there a specific rôle for the autonomic nervous system and its product of catecholamines in the processes of follicular selection and development? The question is posed despite the well-known fact that ovarian cyclicity and ovulation may continue after transplantation of the organ or portions thereof – a fact that has led all too many experienced researchers to disregard a potential contribution of the nervous system in the current context. To the present author, this is all the more surprising since many such physiologists readily concede that catecholamines could be involved in the growth and maturation of Graafian follicles.

Recruitment of follicles, selection, dominance

The concepts of recruitment, selection and dominance of ovarian follicles have been applied across species in mammals, and offer a useful framework within which attempts have been made to analyse physiological processes. Recruitment is a term used to indicate entrance into the gonadotrophin-dependent phase for a group or pool of follicles. It is a function of follicle size and clearly also of the availability of gonadotrophin receptors on somatic cells of the follicle. Selection indicates emergence of one or more ovulatory follicles for further growth from the cohort of recruited follicles, and involves a reduction in number of follicles potentially available for ovulation to that characteristic of the species. Dominance is a consequence of selection expressed as an identifiable follicle or number of follicles typical of the ovulation rate in that species and breed as, for example, in sheep. The phase of dominance is associated with rapid follicular development and, as a corollary, a wave of suppression and then atresia in other follicles of the cohort together with an apparent absence of any further immediate recruitment (for reviews, see Cahill, 1981; Hodgen *et al.*, 1983; Baird, 1990; Fortune, 1991, 1993; Hunter *et al.*, 1992; Driancourt *et al.*, 1993; Scaramuzzi *et al.*, 1993; Greenwald & Roy, 1994; Gougeon, 1996, 1997; Roche, 1996; Armstrong & Webb, 1997; Campbell *et al.*, 1999, 2000; Webb *et al.*, 1999a,b; Ginther, 2000; and McGee & Hsueh, 2000).

It follows from the above that a follicle destined to ovulate is derived from a cohort of growing follicles drawn, in turn, from a pool of non-proliferating primordial follicles formed during development of the fetus (Hodgen *et al.*, 1983).

Despite a wealth of detail, especially concerning endocrine profiles, the physiological regulation of recruitment, selection and dominance is not fully understood. This is in part because of the large number of follicular molecules, especially growth factors and other peptide secretions, that act in an autocrine or paracrine manner. Not only are various sub-populations of somatic cells involved in this regard, but the oocyte itself may have diverse influences according to the stage of follicular maturity, perhaps even orchestrating or fine-tuning events within the dominant follicle(s) as ovulation approaches. To counterbalance an emphasis on local influences in these remarks, it must be kept in mind that in our own species, as in cattle, horses and monovular breeds of sheep, the ovaries function asymmetrically in generating a dominant follicle. This could suggest a sensitive and precise means of inter-ovarian communication, which may involve hormones, peptide molecules, nerves, white blood cell traffic, peritoneal fluid components – or a contribution from each of these as potential vectors.

Waves of follicular development

Folliculogenesis is associated with the proliferation of a group of follicles at differing stages of development from which a species-specific number of follicles will emerge for continued growth. In small laboratory rodents, patterns of follicular development have been followed by means of classical histology whereas in large domestic ruminants, a traditional approach has been ovarian palpation *per rectum* to monitor follicular growth and surface morphology (Rajakoski, 1960). In the latter regard, a major advance was the introduction of real-time, trans-rectal ultrasonography – ultrasonic scanning – with a suitably positioned probe enabling accurate imaging and measurement of follicles in animals such as cow, sheep, horse, pig, goat and in our own species (e.g. Ginther, 1998). Indeed, trans-vaginal ultrasonic scanning of follicular development in human ovaries, usually as a sequel to treatment with gonadotrophic hormones, is an essential management tool in fertility (IVF) clinics. It enables oocyte retrieval by means of needle puncture and aspiration at an optimum stage of follicular development, judged primarily by size but frequently also by morphology of the cumulus oophorus and in conjunction with measurement of oestradiol concentrations in peripheral blood (see Bomsel-Helmreich, Bessis & Vu Gnoc Huyen, 1981; Bomsel-Helmreich, 1983).

Ultrasonic scanning combined with the measurement of circulating steroid hormone values provided sound evidence for the existence of waves of follicular

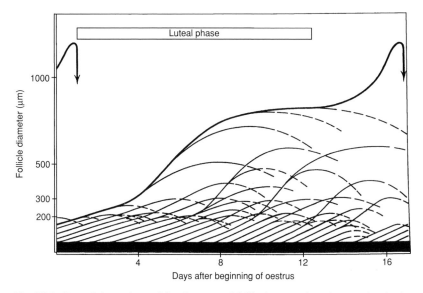

Fig. VI.1. One of the early models of curves of follicular growth and regression in the mammalian ovary that has subsequently been developed in detail – if not in concept – by various authors. (After Everett, 1961.)

growth in domestic animals (Figs. VI.1 and VI.2). These are seen as development of one or more follicles of conspicuously larger diameter during the early luteal phase of the oestrous cycle – typically several groups of follicles growing in parallel – followed by waning and regression. Then development of a second such wave of follicles during the mid-to-late luteal phase, again with regression before emergence of a fully dominant follicle(s) that grows to maturity and ovulation during the follicular phase of the oestrous cycle. A positive relationship exists between the number of follicular waves and duration of the oestrous cycle. All follicles detected initially in a wave are considered to be potentially capable of becoming dominant, and there is reasonably sound evidence for this view from experimental intervention (for reviews, see Hirshfield, 1991; Greenwald & Roy, 1994; Zeleznik & Benyo, 1994; Gougeon, 1996; McGee & Hsueh, 2000).

Human

Growth of follicles in the human ovary from the pre-antral stage to one at which they can be selected requires only basic (tonic) levels of gonadotrophins, and during this phase their granulosa cells exhibit low proliferative activity (Baird, 1983; Gougeon, 1997). Large numbers of follicles leave the pool of resting follicles to grow during each menstrual cycle, as many as fifty reaching a

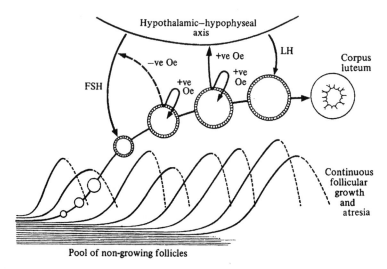

Fig. VI.2. A more detailed model of follicular growth up to the stage of ovulation indicating decline into atresia of all but the ovulatory follicle(s). The influence of pituitary gonadotrophic stimulation and of negative (−ve) and positive (+ve) feedback signals from the Graafian follicle – especially oestradiol (Oe) – is also depicted. LH, luteinising hormone; FSH, follicle stimulating hormone. (After Peters & McNatty, 1980.)

diameter of 1 mm or more in the follicular phase of the cycle. Selectable follicles of 2–5 mm diameter are available throughout the cycle, but only three or fewer of these attain a diameter of 8 mm by mid cycle, and the smaller follicles of 2–3 mm diameter succumb to atresia (Sturgis, 1961; Gougeon, 1997). By contrast, the follicle destined to ovulate grows to a diameter of 20 mm or more. As judged from ultrasonic scanning at the equator of a follicle rather than measurement of the protruding surface, fully grown follicles may reach a mean diameter of 22–23 mm or more shortly before ovulation (Table VI.1). Reference here is to a single ovulation, for twin or multiple ovulation is exceptional. Although the follicle destined to ovulate is well advanced in its growth by the mid-follicular phase, several waves of growing follicles progress at particular times during the menstrual cycle (Sturgis, 1961; Baird, 1983; Gougeon, 1997; McGee & Hsueh, 2000).

Dependence on FSH support and stimulation increases once a diameter of 2 mm has been achieved, and granulosa cell proliferation within selectable follicles increases in parallel with circulating concentrations of FSH (Baird, 1983; Hodgen *et al.*, 1983; Gougeon, 1997). In terms of steroid biosynthesis, the selected or dominant follicle is highly active due to the presence of appropriate enzymes in both the theca and granulosa cell layers; indeed the dominant follicle

Table VI.1. *Representative values for the pre-ovulatory diameter of Graafian follicles*

Species	Common mature diameter (mm)	Comment on diameter
Rat	0.6–0.9	0.7 mm is the usual minimum
Rabbit	1.8–2.0[a]	Coitally induced final growth
Sheep	6–8[a]	5 mm or less in prolific breeds
Pig	8–10[a]	9–12 mm sometimes recorded
Cow	18–20[a]	15–21 mm not infrequent
Human	20–23	18–25 mm not exceptional
Horse	40–50[a]	33–55 mm would include most breeds

[a] Depending especially on breed and maturity.
Adapted from various sources but principally Hunter (1980).

is the principal source of ovarian oestrogen – oestradiol-17β secretion. One of the most important events occurring during follicular selection is the activation of aromatase enzyme in granulosa cells. Levels of mRNA for P450 aromatase are low on days 8 and 9 but increase strongly on day 13 (Doody *et al.*, 1990). Such mRNA changes clearly underlie reports of enhanced aromatase activity during the pre-ovulatory phase, increases reaching 200 times the aromatase values in 'selectable' follicles (Hillier, Reichert & Van Hall, 1981). Degenerating follicles within a given cohort show a decline in steroidogenic enzyme activity.

Cow

Application of trans-rectal ultrasonography in cattle endorsed the long-held view (Rajakoski, 1960) that there were either two or three prominent waves of follicular growth during the 21 day oestrous cycle (see Ireland & Roche, 1987; Pierson & Ginther, 1988; Sirois & Fortune, 1988; Driancourt, 1991). Such waves of growth involved groups of three to six follicles of ≥4 mm diameter, emergence of successive waves reflecting increases in the circulating concentration of FSH that precede appearance of a wave by 1 day (Webb *et al.*, 1999a). Several days after the emergence of a wave, one of these follicles becomes dominant, enlarges to >10 mm diameter and continues growing while subordinate follicles in the wave or cohort slow down, then stop growing and become atretic (Sunderland *et al.*, 1994; Ginther *et al.*, 1996). The initial phase is sometimes referred to as deviation or divergence between growth of the dominant follicle and that of subordinate follicles. The diameter of a dominant bovine follicle at deviation is 8.5 mm (Ginther, 2000).

In a wave not culminating in ovulation, the dominant follicle is seen as the largest follicle for several days, even though the succeeding wave has already commenced. This led Fortune (1991) to conclude that a follicle may remain dominant in size for a period longer than it retains functional dominance. After regression of the corpus luteum, only the dominant follicle seen on entry into the follicular phase of the oestrous cycle continues development to the stage of ovulation. If deviation did not occur, the second largest follicle could become as large as the largest, a scenario that could culminate in twin ovulations.

Gong *et al.* (1996) have suggested that growth of bovine follicles to a diameter of 4.0 mm may not require stimulation by FSH or LH. Follicles of 5.0 mm diameter have the ability to suppress circulating concentrations of FSH, and after stimulation of a wave of follicular growth the peripheral FSH concentrations decline. At least in part, this is a function of a follicle's ability to produce significant levels of steroid and protein hormones. Experimental manipulation of the number of follicles of 5.0 mm diameter showed that the greater the number of such follicles, the more rapidly did mean FSH concentrations decline (Ginther *et al.*, 2000). One current proposal is that when the future dominant follicle reaches a threshold diameter, other follicles of the wave are suppressed by diminished concentrations of FSH before they can reach a similar diameter. The largest follicle has a greater ability to suppress FSH concentrations, driving these below the threshold requirement of all subordinate follicles. Appropriate treatment with FSH or anti-inhibin can prevent deviation of a dominant follicle (Ginther, 2000). Oestradiol is a specific deviation factor from a dominant follicle that supplements the suppressive influence of inhibin on FSH through its negative feedback ability. However, it is still not clear to what extent the dominant follicle brings about regression of subordinate follicles by means of local or systemic pathways, or both. If follicles regress more rapidly in the ovary containing the dominant follicle than follicles in the contralateral ovary, this would provide evidence for a local effect. Analysis of physiological changes in subordinate follicles immediately adjacent to the dominant follicle would then become important for distinguishing local influences (Sarty, Adams & Pierson, 2000). Despite the emphasis throughout this paragraph, the present author retains a strong view that involvement of the autonomic nervous system and its secretion of catecholamines cannot be discounted in these events.

Clarification is also required as to whether the dominant follicle in cattle requires continuing stimulation from the decreased circulating concentrations of FSH or whether LH can provide part of its gonadotrophic support (Webb *et al.*, 1999a). If there is not a progressive switch from FSH to LH dependence, then the perplexing question arises as to how the dominant follicle can thrive and develop on lower peripheral concentrations of FSH. Has it already achieved a sufficient growth impetus, or does it use large numbers of FSH receptors more

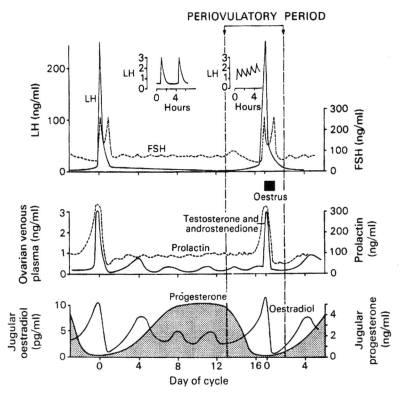

Fig. VI.3. A portrayal of hormonal changes throughout the oestrous cycle of the ewe, with some indication of curves of oestradiol secretion in step with waves of follicular development. (After Baird & McNeilly, 1981.)

effectively, or do reserves of FSH in the follicular fluid provide supplementary support? Further considerations might be that circulating FSH is sequestered by the dominant follicle more efficiently and/or that local support of follicular development comes from peptide growth factors, possibly in the form of synergistic influences with diminished gonadotrophins. Even so, the notion of a progressive switch from FSH to LH dependence is prominent in current interpretations (Adams, 1999).

Sheep

As is the case in cattle, there are waves of follicular development during the luteal phase of the oestrous cycle in sheep, preceded by increases in peripheral plasma concentrations of FSH and expressed as waves of oestradiol secretion (Fig. VI.3). Such peaks of FSH occur at intervals of approximately 6 days

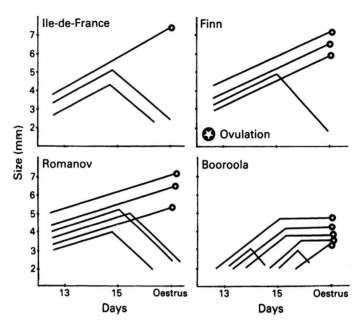

Fig. VI.4. Graphs of ovarian follicular development to indicate the diameter of mature follicles as influenced by the prolificacy of differing breeds of sheep. (After Driancourt *et al.*, 1986.)

(Campbell *et al.*, 1991a). Within a few hours of the onset of luteolysis, peripheral FSH concentrations commence to decrease, reaching a nadir within 24–36 hours (Scaramuzzi *et al.*, 1993), increasing once more only as part of the pre-ovulatory gonadotrophin surge. Oestradiol secretion commences a progressive increase with regression of the corpus luteum and in response to an increasing frequency of pulsatile LH secretion, achieving a threshold that triggers the pre-ovulatory gonadotrophin surge.

As to size, follicles commonly progress to a pre-ovulatory diameter of 7–8 mm in sheep, 9–11 mm in pigs, and 15–20 mm in cattle (Table VI.1). Among the domestic farm species, the horse is remarkable in terms of mature follicle size, the single Graafian follicle reaching a diameter of 35–55 mm. Measurements can be made with precision at the equator of a follicle by means of ultrasonic scanning. The breed of animal and indeed its maturity can have an important influence on follicular diameter in these farm species. For example, individual follicles in the more prolific breeds of sheep attain smaller diameters (Fig. VI.4).

Selection of the ovulatory follicle(s) in sheep occurs during the follicular phase of the oestrous cycle when the circulating concentrations of FSH are falling and the frequency of LH pulses increasing (Baird & McNeilly, 1981; Baird, 1983). Once again, a critical feature of follicle selection and dominance may be the ability of a maturing follicle to withstand the fall in circulating FSH concentrations that is seen during the follicular phase (Baird, 1983), although the antral fluid reserves of FSH may maintain a local threshold in this regard. The vital compensation, however, comes in the form of LH pulses expressed as a series of high frequency pulses of low amplitude that are seemingly essential for physiological levels of oestradiol secretion. It is clearly no coincidence that receptors for LH on granulosa cells become available during the final stages of follicular maturation. Even so, if FSH stimulates aromatase activity in granulosa cells, then an LH substitution for FSH in the dominant follicle might be expected to diminish oestradiol secretion and prompt that of progesterone.

Experimental evidence derived from an ovarian autotransplant model in sheep indicated that the ability of a follicle to transfer its gonadotrophic dependence from FSH to LH is fundamental to the mechanism of follicle selection (Campbell *et al.*, 1999). Although LH undoubtedly plays a central rôle in this selection in monovulatory species, such transfer from FSH to LH dependence does not clarify precisely how the dominant follicle emerges; rather, it describes what happens in the follicle that becomes dominant. An ability to develop LH receptors on granulosa cells could equally be invoked as a vital part of follicle selection, as could genes regulating the expression of aromatase. As to follicles not selected for dominance, the peak of FSH seen 24–36 hours after the pre-ovulatory gonadotrophin surge is associated with atresia of large non-ovulatory follicles (Scaramuzzi *et al.*, 1993).

Pig

A good indication of follicular growth in this species is given in Fig. VI.5, taken from McKenzie (1926). There is a marked contrast in the time course of follicular development and selection between the monovular cow and polyovular pig, even though their oestrous cycles are of comparable duration (see Hunter *et al.*, 1992). Despite this cyclic similarity, the follicular phase is somewhat shorter in cattle, for example of 3.5–4.0 days' duration in contrast to 5–6 days in pigs. In the latter species, only a first wave of follicle recruitment can be followed after ovulation, since follicle selection and dominance are suppressed or at least cannot be detected until the onset of the follicular phase of the cycle. In terms of the large number of 2–3 mm diameter follicles visible on the ovarian surface at days 3 and 5 of the cycle, fewer than 5% are compromised by atresia but by day 7 approximately 50% of the follicles have become atretic (Garrett & Guthrie,

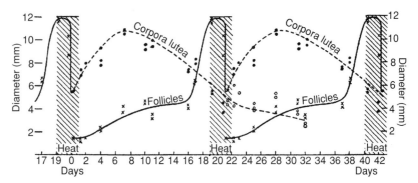

Fig. VI.5. A classical depiction of curves of follicular growth throughout the oestrous cycle of domestic pigs, suggesting pre-ovulatory follicular diameters of 10–12 mm in sows as distinct from gilts. (Adapted from McKenzie, 1926.)

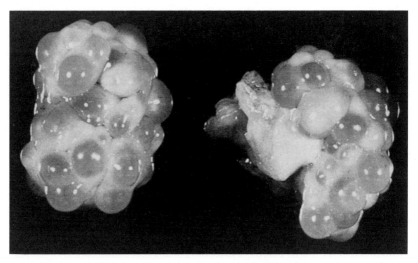

Fig. VI.6. The ovaries of a virgin pig (gilt) photographed on day 17 of the oestrous cycle to illustrate the large number of Graafian follicles of 6–7 mm in diameter. There would be a progressive and significant reduction in number during the remaining days of the follicular phase of the oestrous cycle. (After Hunter & Baker, 1975.)

1996). During the follicular phase of the oestrous cycle in pigs (Fig. VI.6), the group of follicles that will ovulate gradually emerges whereas growth of small- and medium-sized follicles and the secretion of FSH remain suppressed until the pre-ovulatory LH surge (Grant, Hunter & Foxcroft, 1989; Garrett *et al.*, 2000).

Endocrine activity associated with dominance

The follicular response to gonadotrophin stimulation is, in part, a function of the number of receptors available on somatic cells. Human follicles show a slowly increasing number of LH receptors on cells of the theca interna during basal growth whereas FSH receptors on the granulosa cells are little changed (Gougeon, 1997). However, FSH receptor activity or its mRNA level is not in itself a sufficient guide to trophic events, since downstream transducer coupling or modified adenylyl cyclase activity may change with developmental stage (Richards, 1994; Richards *et al.*, 1995). In addition, there is the complication that diverse factors present in serum or follicular fluid can modify the response to a given pattern of FSH stimulation, rendering interpretation of systemic gonadotrophin values a somewhat naïve approach to analysis of follicular development.

Conversion of androstenedione to oestradiol by granulosa cells from human follicles increases slowly as the diameter exceeds 3 mm, which suggests that suitable follicles are becoming more responsive to gonadotrophins, even though their FSH-induced aromatase remains poorly expressed. Local factors within a follicle may be acting to inhibit aromatisation of androgens. Nonetheless, between diameters of 5.5 and 8.2 mm, the newly selected human follicle expresses substantial aromatase activity in the granulosa layer, thought to be due to IGF and perhaps other factors synergising with FSH (Gougeon, 1997). There are marked changes in the steroidogenic activity of the maturing pre-ovulatory follicle. During the early to late follicular phase but before the gonadotrophin surge, oestradiol concentrations in human follicular fluid increase from approximately 650 to 2400 ng/ml; other values appear in Table V.1. Granulosa cells of the mature follicle, which are fully responsive to FSH, commence to express receptors for LH (Richards, 1994). A dramatic increase in LH binding occurs in human granulosa cells during pre-ovulatory maturation, such cells then being able to respond to the mid-cycle surge of gonadotrophins. Following the gonadotrophin surge, the concentrations of oestradiol and androstenedione in follicular fluid decrease whereas those of progesterone and 17α-hydroxyprogesterone increase markedly. In effect, the pre-ovulatory follicle has been reprogrammed primarily to produce progestins, in a steroidogenic sense thereby already functioning as a corpus luteum.

As noted in Chapter V, peptide growth factors are involved in local modulation, stimulating or attenuating cellular responses to gonadotrophins. Both EGF and IGF stimulate proliferation of FSH-responsive human granulosa cells. On the basis of detecting receptors for these growth factors in pre-antral and small antral follicles, Gougeon (1997) suggested an involvement in supporting development in human ovaries, as well as in regulating steroidogenesis during basal growth. Clearly, an understanding of the molecular control of growth factor availability is central to an understanding of folliculogenesis. Remaining with primates but turning to studies in marmoset monkeys, there is the further suggestion that activin may stimulate aromatase in both immature and pre-ovulatory tissues, acting in concert with IGFs at the time of follicle selection (Miro & Hillier, 1992). Indeed, aromatase activity in the granulosa cells of marmoset monkey pre-ovulatory follicles is at least ten times more sensitive to FSH than in the granulosa cells of small follicles (Harlow *et al.*, 1986).

Binding proteins for growth factors also have relevance to this discussion. A reduced intra-follicular concentration of IGFBPs should increase the availability of locally produced IGF to increase the response to gonadotrophins. As judged from *in vitro* studies with bovine follicles, a decrease in the concentrations of IGFBP-2, -4 and -5 in follicular fluid during the development of dominance supports this view. The decreased IGFBP-2 was associated with loss of expression of mRNA encoding the protein in granulosa cells of dominant follicles, an FSH-dependent inhibition (Armstrong & Webb, 1997). As a corollary, increased IGF bioactivity in potentially dominant follicles should increase the responsiveness of their granulosa cells to FSH.

A more detailed consideration of intra-follicular molecules associated with development has been presented in Chapter V. And, although focusing rather more closely on the oocyte, the review of Driancourt & Thuel (1998) gives a useful assessment of follicular dominance. As suggested in Fig. VI.7, taken from that review, LH is the primary gonadotrophic hormone during the dominance stage and IGF-1 together with inhibin constitute the principal locally acting compounds.

Models for follicular selection and dominance

Various attempts have been made in recent years to summarise the endocrine programme that facilitates selection of an ovulatory follicle, most notably in domestic animals such as cows and sheep (Hunter *et al.*, 1992; Scaramuzzi *et al.*, 1993) but also in humans (McGee & Hsueh, 2000). In the model described by Armstrong & Webb (1997), derived from extensive studies in cattle and set against a background of detailed information obtained in sheep, selection of a recruited follicle depends on increasing concentrations of FSH as measured in the systemic circulation. As already noted, however, the selected follicle continues development against a backcloth of declining FSH values, values

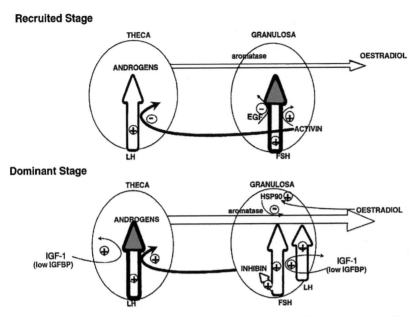

Fig. VI.7. Regulation of follicular maturation by both endocrine and local mechanisms. The top of the figure describes the follicle at the recruited stage while the bottom describes it during its dominant stage. During the recruited stage, FSH is the main gonadotrophin and activin the main locally acting compound. During the dominant stage, LH is the main gonadotrophin and IGF-1 together with inhibin are the main locally acting compounds. (After Driancourt & Thuel, 1998.)

considered to be below a threshold for other follicles in that particular wave or cohort (Fig. VI.8). In part at least, the depression in systemic FSH reflects a negative feedback influence of an increasing secretion of oestradiol and inhibin by granulosa cells of the selected follicle (Fig. VI.9). A shift in dependence of the selected follicle from FSH to LH support underlies the establishment of dominance.

The dominant follicle secretes increasing amounts of oestradiol-17β and the concentrations in systemic blood rise progressively. By means of a positive feedback loop influencing hypophyseal activity, the increasing concentration of oestradiol enhances the frequency of pulsatile LH secretion during the follicular phase of the cycle. It is worth underlining the point that growth of the pre-ovulatory follicle takes place even though the concentration of FSH monitored in the peripheral circulation is falling (Scaramuzzi *et al.*, 1993; Campbell *et al.*, 1999). Peak concentrations of oestradiol are usually recorded during the growth phase of the dominant follicle, although values may have commenced to fall

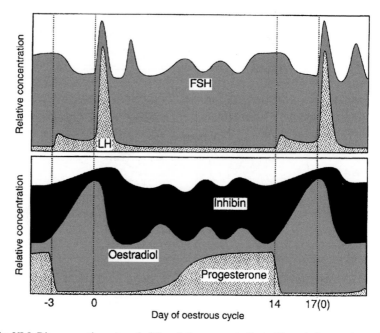

Fig. VI.8. Diagrammatic portrayal of the relative concentrations of key pituitary and gonadal hormones in the peripheral circulation of sheep during a 17-day oestrous cycle. Note the declining concentration of FSH during the follicular phase of the cycle. (After Scaramuzzi *et al.*, 1993.)

before maximum follicular diameter is achieved. Failure of a dominant follicle to ovulate, for example in a post-partum situation in cattle, may reflect an inadequate LH pulse frequency, which, in turn, leads to a low level of androgen substrate synthesis in the follicle and thereby to an inadequate positive feedback of oestradiol to trigger a pre-ovulatory gonadotrophin surge.

Gonadotrophic regulation of follicular oestradiol secretion depends on the location of the respective gonadotrophin binding sites. FSH receptors are exclusive to granulosa cells whereas LH receptors are expressed principally on cells of the theca interna, although also on granulosa cells in the final stages of follicular maturation and under the influence of FSH (see Chapter V). Follicular development is associated with changes in the expression of mRNA encoding gonadotrophin receptors (Armstrong & Webb, 1997). In their model, Armstrong & Webb (1997) emphasise the rôle of ovarian growth factors and the increasing vascularisation of the dominant follicle, offering greater access to systemic molecules.

In an earlier detailed analysis of follicle selection in the ewe that still retains much validity, Scaramuzzi *et al.* (1993) proposed five stages of follicular growth

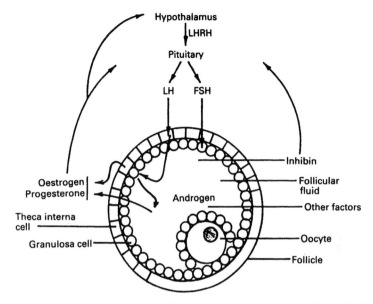

Fig. VI.9. Diagrammatic representation of a Graafian follicle to suggest diverse 'feedback influences' on the hypothalamus and pituitary gland of molecules from the theca and granulosa cells. Negative feedback influences of oestradiol and inhibin are prominent during much of an oestrous cycle. (Adapted from Christenson, Ford & Redmer, 1985.)

and development, culminating in ovulation of the dominant follicle (Fig. VI.10). Failure to respond to endocrine cues at any stage diverted such follicles into a pathway to atresia. The Scaramuzzi model sets out with a number of major assumptions. Primordial follicles were considered as largely quiescent until committed to growth. Once committed, they grow until the alternative destinations of atresia or, exceptionally, ovulation. The model proposes that only FSH and LH are essential for growth and development.

Primordial follicles progress to the committed stage, possibly in an ordered sequence reflecting a hierarchy, a suggestion stemming from Henderson & Edwards (1968). Hypophyseal gonadotrophic support appears not to be essential for this transition, but committed follicles nonetheless express gonadotrophin receptors on their theca and granulosa cells. Such follicles progress into gonadotrophin-responsive structures, showing an increasing ability for progesterone and androgen synthesis, but aromatase enzyme expression permitting oestradiol secretion requires further maturation. Aromatase activity per cell increases in parallel with the increasing sensitivity of granulosa cells to FSH (Henderson *et al.*, 1987). In terms of receptor activity, most gonadotrophin-responsive follicles – a pool of perhaps twenty-five or more follicles in the

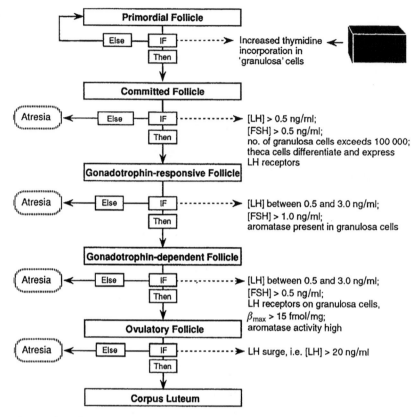

Fig. VI.10. A flow chart representing follicular growth and development in sheep, based on the dependence and sensitivity to gonadotrophic hormones. Approximately 180 days are required for progression from the primordial stage to that of pre-ovulatory follicle. The values given are approximations. (After Scaramuzzi *et al.*, 1993.)

ewe – are too immature to become gonadotrophin dependent. Dependency needs further growth and an absolute requirement for FSH support, as is also the case for LH to generate androgen substrate in the thecal cells; oestradiol secretion continues.

As already noted, a selected ovulatory (dominant) follicle prevents further development of gonadotrophin-dependent follicles by means of a negative feedback influence of oestradiol and inhibin to reduce secretion of FSH below a threshold value, thereby prompting a wave of atresia. The ovulatory follicle requires a low but critical concentration of FSH to support the final maturation of granulosa cells (Henderson *et al.*, 1988). In this regard, large numbers of LH receptors develop on the granulosa cells (Webb & England, 1982), making

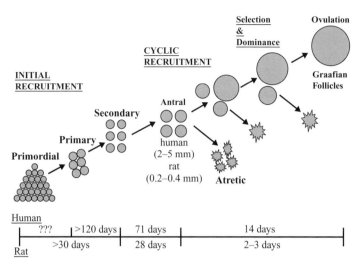

Fig. VI.11. A model for selection and dominance of a human Graafian follicle leading, in due course, to ovulation. A distinction has been drawn in this description between initial recruitment of follicles and the subsequent cyclic recruitment. (Adapted from McGee & Hsueh, 2000.)

them highly responsive to both gonadotrophic hormones. In terms of cAMP production in granulosa cells of the ovulatory follicle, LH progressively substitutes for FSH support. Local production of peptide factors may facilitate this change in sensitivity, acting by means of either autocrine or paracrine modulation.

A recent honing of a general model for selection of follicles comes from McGee & Hsueh (2000), primate ovaries being the special focus in this clinically orientated essay. They place a strong emphasis on the phase of recruitment that precedes selection, and argue for a distinction between initial recruitment and cyclic recruitment (Fig. VI.11). That being said, the factors underlying such ovarian events remain those already discussed by other authors, except in one regard. Based at least in part on the studies of Vanderhyden & Macdonald (1998), a molecular contribution of the oocyte to recruitment and selection of follicles is given more prominence than is customary by endocrinologists. Two molecules produced by the mammalian oocyte are highlighted, these being growth differentiation factor-9 (GDF-9) and GDF-9B or bone morphogenetic protein-15 (BMP-15), but their precise involvement in the processes of follicle recruitment and selection remains uncertain.

Follicle growth inhibitory factors

Growth factors and other bioactive protein factors produced by the dominant follicle can function to maintain dominance and suppress further growth or development of subordinate follicles. However, follicular dominance almost certainly depends on more than one means of physiological control and suppression, and elements of the autonomic nervous system may contribute at an ovarian level in this regard. Such involvement might stem from an influence of catecholamine molecules, but a rôle for nitric oxide also needs to be considered.

There is increasing evidence for follicle growth inhibitory factors (FGIFs) from dominant follicles being able to inhibit growth of subordinate follicles. For example, steroid- and inhibin-depleted follicular fluid can still act to inhibit follicular development in sheep and cattle (Campbell *et al.*, 1991a; Law *et al.*, 1992; Wood *et al.*, 1993). Partially purified follicular fluid fractions can inhibit cell proliferation and aromatase activity in cultured ovine, bovine and porcine granulosa cells. FGIFs are thought to function at least in part via systemic mechanisms, since a dominant follicle can influence the contralateral ovary in monovular species but, once again, a nervous contribution might be invoked in this regard.

The topic of factors that inhibit growth of follicles in human ovaries is touched on in the review by McGee & Hsueh (2000). Derived from ultrasound imaging of patients coupled with computer analysis and modelling, the dominant follicle has been suggested to exert a suppressive influence on neighbouring subordinate ones (Gore, Nayudu & Vlaisavljevic, 1997). Whilst this could very well be so for a variety of physiological reasons – and not just of an endocrine or paracrine nature – obtaining sound evidence remains a remarkably difficult practical problem. Experimental intervention will invariably perturb the physiological backcloth.

Atresia of follicles and germ cells

Embryonic ovaries undergo differentiation and development so that production of potential gametes – the primary oocytes – is essentially complete by the time of birth in most eutherian mammals so far studied (see Chapter II). This is due to waves of mitosis causing enormous proliferation in the stock of oogonia, followed by the first phase of meiosis and arrest of the nucleus at the dictyate (diplotene) stage until puberty. Oocytes at diplotene are then surrounded by follicle cells. Extensive atresia of oocytes and follicles is seen as a major feature of ovarian histology even before the time of birth (for reviews, see Baker, 1972, 1982; Peters & McNatty, 1980; Hirshfield, 1991; Greenwald & Roy,

1994; Hsueh, Billig & Tsafriri, 1994; Kapia & Hsueh, 1997), and an apparent resumption of meiosis may be a prominent feature of atresia. In the human fetus, there are six or seven million oocytes by the fifth month of pregnancy followed by a major wave of atresia (Baker, 1963; Fig. II.3). As suggested, the number of oocytes ovulated spontaneously is only a minute proportion of the hundreds of thousands of oocytes in the ovaries at birth.

Whether the potential for growth and development is equally distributed amongst the population of oocytes remains an unresolved question. In other words, must the great majority of oocytes within antral follicles be abandoned to the atretic process, as occurs in the biological situation, or could an increased proportion of oocytes be liberated by means of appropriate treatment? In fact, techniques such as multiple superovulation or repeated aspiration of oocytes from vesicular follicles make little overall inroad on the proportions, although they may increase the total number of available oocytes significantly. If most oocytes must indeed be abandoned to atresia, this may indicate qualitative deficiencies in germ cells or in their associated somatic cells. In this context, one suggestion invokes an accumulation of mutations in oocyte mitochondria (Krakauer & Mira, 1999), but such an interpretation is unsatisfactory on quantitative grounds: defective mitochondrial genomes could scarcely account for the massive scale of oocyte wastage (Perez *et al.*, 2000).

In women, the oft-repeated calculation of one egg ovulated per month for 40 years (assuming regular menstrual cycles and no pregnancies), would account for fewer than 500 oocytes, yet the ovaries are essentially depleted of germ cells by the time of menopause. Placing this 500 in comparison with an initial complement of 500 000 or more, then pre-, peri- and post-natal atresia must account for a devastating toll of oogonia, oocytes and developing follicles. Such wastage commonly exceeds 99% of the initial complement. Although atresia can occur at any stage of the lifespan, attention has focused on phases of degeneration seemingly linked to waves of follicular development. Moreover, statements from well-known authorities indicate that the highest rates of follicular atresia occur amongst pre-antral and antral follicles (Strömstedt & Byskov, 1999). Once follicles have become dependent on gonadotrophic support, then a dearth or inappropriate pattern of gonadotrophins – especially FSH – reaching individual follicles or a downgrading of receptors would lead to loss of integrity and wastage. Competition between growing follicles could result in a shortage of trophic support in the less advanced members of a wave. In addition, dominant follicles might be exerting paracrine influences to reduce or suppress competition from neighbouring follicles.

When considering the onset of atresia, it is important to view follicular vulnerability in terms of interactions between an oocyte and granulosa cells.

Germ and somatic cells need to have a functional metabolic relationship for the exchange of substrates and toxins if they are to avoid the destiny of atresia. It is essential to recall that there is no blood supply to the granulosa cells of a developing follicle, so gradients of diffusion and/or active transfer are of paramount importance. Inadequate or imbalanced exchange could be lethal.

To expand on this theme, although the reserves of primordial follicles in a mammalian ovary rest in a stable condition, they are certainly not dormant and will require an inflow of oxygen and nutrients from the somatic cells. In the other direction waste products will need to be eliminated from the oocytes and, once again, somatic cells will play a critical rôle. Therefore, not only is adequate functional contact with follicular cells required to sustain the viability of oocytes, but the somatic cells themselves will be vulnerable to inadequate support and/or to metabolic insult. Once the stage of gonadotrophin dependence is reached, specific patterns and absolute levels of endocrine support may become essential for follicle viability. Maintenance of such inputs and exchanges at a physiologically appropriate level in compartments not receiving vascular support will be open to error. This may be especially so in the highly competitive situation caused by the close proximity of oocytes and primordial follicles. Indeed, dying granulosa cells or oocytes may release breakdown products, in effect cytotoxic signals, whose influences damage neighbouring cells. One of the reasons why *in vitro* culture of oocytes or primordial follicles is relatively successful could be because intra-ovarian competition is reduced and availability of oxygen and substrate enhanced.

Turning to characteristics of follicular atresia, detachment of granulosa cells from the basement membrane, hypertrophy of the theca layer, modifications to its vascular bed and passive infiltration of inflammatory cells are all well-known anatomical pointers (see Greenwald & Roy, 1994). An absence of mitotic activity is a classical description, but some granulosa cells may still be proliferating in the early stages of atresia. According to a recent study, the presence of dying cells at all stages of follicle maturation indicates that the population of granulosa cells during follicular growth and maturation is in continuous exchange, the old cells being discarded by apoptosis (Rosales-Torrès *et al.*, 2000). Perturbations or physiological modification of this delicate equilibrium between proliferation and death of granulosa cells could promote the onset of atresia. Progression of atresia will involve an increase in the proportion of both necrotic and apoptotic granulosa cells. As atresia continues, fragmentation of the basal lamina and the appearance of pycnotic nuclei within degenerating granulosa cells are typical. Advanced stages of atresia demonstrate a markedly reduced vascular supply or specific follicular ischaemia followed by extensive disorganisation, death and dissolution of the granulosa cell layers. These steps lead eventually to the formation of scar tissue.

Even in its early stages, modifications to the theca vascular supply and a consequential diminished oxygen diffusion to the innermost granulosa cells may accelerate the process of atresia in these cell layers (see Hirshfield, 1989, 1991, 1997). Larger follicles with greater numbers of granulosa cells would seem more susceptible to this gaseous constraint, but there can be little doubt that more than one mode of cell death contributes to the process of atresia. As already noted, pseudomaturation of the oocyte may occur during the atretic process, as seen in breakdown of the germinal vesicle and an apparently organised resumption of meiosis. This may be due primarily to loss of intimate contact (gap junctions) between granulosa cells and the oocyte plasma membrane (oolemma) upon withdrawal of corona cell foot processes (Szöllösi, 1978, 1993; Szöllösi *et al.*, 1978). Such functional contacts would normally be acting to impose meiotic arrest.

In primate ovaries, atresia continues throughout the menstrual cycle, affecting follicles at different stages of development, even though waves of atresia may be related to certain times of the cycle. This observation led Sturgis (1961) to believe that there must be a functional element underlying the dozens of follicles undergoing atresia in the adult ovary at any one time. On the basis of observations in rhesus monkeys, he noted that atresia 'wipes out all follicles of second rank in each cycle, the ones that ordinarily would be most likely next to mature and ovulate'; in other words, induction of atresia in all but one of the ripening follicles each month. Sturgis (1961) further added that such atresia could not be simply a local phenomenon 'because second rank follicles are wiped out equally just prior to ovulation in the contralateral ovary as well'.

Over and above referring to dissolution of the granulosa layer as the first sign of impending atresia, emphasis was given to a short-lived period of hypertrophy of the theca interna in primates (Sturgis, 1961). Such cells were regarded as secretory before the follicles of second rank collapsed rapidly within the next 2 or 3 days. Moreover, Sturgis offered the speculation – some 40 years ago – that factors causing hypertrophy of the theca interna in primates might in some way be responsible for the limitation of ovulation to a single follicle each month. The fate of theca cells is of interest, since species differences have been highlighted. Whereas rat theca cells also undergo conspicuous hypertrophy during follicular atresia (Braw *et al.*, 1976), sheep theca cells from atretic follicles show degeneration and nuclear condensation comparable to that found in granulosa cells (Hay, Cran & Moor, 1976; O'Shea, Hay & Cran, 1978). Hypertrophy of theca cells may be associated with their progressive incorporation into interstitial tissue and the potential to exert both local and systemic endocrine influences; secretion of steroid hormones is a major consideration in this regard.

The process of atresia is reflected in an increased production of progesterone and a concomitant decline in the synthesis of oestrogens in all species examined. Two explanations appear relevant – a loss of aromatase activity (Maxson, Haney & Schomberg, 1985), with decreased aromatase mRNA, and a decreased C17,20-lyase activity causing a loss of substrate for aromatisation by granulosa cells. Decreased synthesis of oestrogens may lead to an accumulation of androgens in the antral fluid of atretic follicles, for McNatty *et al.* (1979) noted high levels of androgens in human follicular fluid correlated with morphological features of atresia. Nonetheless, there remains the question of cause and effect. Can androgens act to promote follicular atresia, and if so how, or do they reflect the onset and progress of such degeneration? Overall, the steroid hormone profile of atretic follicles, especially the high androstenedione to oestradiol ratio, stands in contrast to the relatively high concentrations of oestradiol and low concentrations of progesterone in healthy Graafian follicles. Interruption of oestradiol synthesis at any stage in a given Graafian follicle will result in atresia (Baird, 1990).

In the early stages of atresia, administration of appropriate doses of FSH is sufficient to rescue potentially atretic follicles, at least in sheep (Hay *et al.*, 1979). However, as degeneration proceeds, the follicle loses its ability to be rescued even though the oocyte may be retrieved and subjected successfully to *in vitro* maturation and fertilisation. Ultimately, the somatic cells lose their ability to bind FSH and LH, and an oocyte remaining *in situ* shows progressive shrinkage and loss of viability. The presence, extent and function of gonadotrophin receptors on theca and granulosa cells are clearly critical determining factors underlying the destiny of a given follicle.

Among the domestic species, a high proportion of follicles visible on the surface of cow ovaries have, to some degree, embarked on the process of atresia (Kruip & Dieleman, 1982). The concentration of steroid hormones in follicular fluid is one of the criteria used in this species for determining atresia. The pattern of IGFBP expression can be another (Monget *et al.*, 1993). In pig ovaries, changes in the pattern of atresia, follicle replenishment and steroidogenesis in non-ovulatory follicles during the pre- and post-ovulatory period are associated with changes in secretion of FSH that may be regulated by inhibin secretion (see Chapter V). Inhibin is present in both atretic and non-atretic follicles, but the predominant forms are less abundant in atretic follicles (Garrett *et al.*, 2000).

During the past 15 years or so, there has been an ever-increasing consideration of the relative contributions of the processes of necrosis and apoptosis to the phenomenon of atresia. However, instead of the debate resting polarised and

Table VI.2. *A summary of important cellular and molecular features that enable a distinction to be drawn between apoptosis and necrosis*

Apoptosis	Necrosis
Affects scattered individual cells	Affects tracts of contiguous cells
Chromatin and cytoplasmic condensation, cell shrinkage	Cell swelling and rupture of plasma membrane
May require mRNA and protein synthesis	Not dependent upon new mRNA or protein synthesis
Normal ATP level	Decreased ATP level
Endonuclease activation and internucleosomal DNA cleavage (ladder pattern)	Activation of non-specific DNases with generalized DNA breakdown (smearing)

After Tilly & Hsueh, 1992.

somewhat sterile, there has been a gradual appreciation that both processes may be contributing simultaneously to follicular atresia (Rosales-Torrès *et al.*, 2000), even though they have characteristic distinguishing features. Some of these are summarised in Table VI.2. As will be described below, apoptosis is an active and orderly physiological process in single cells whereas necrosis is characterised by a haphazard loss of structure in groups of cells, with swelling and rupture leading to specific responses from the immune system.

Apoptosis within ovarian follicles

As already emphasised, the fate of the great majority of ovarian oocytes and follicles is the degenerative process of atresia. Ovulation is the destiny of a privileged few, in fact only a tiny percentage of the original stock. Whereas the precise nature of such privilege awaits full definition, useful steps have been made in understanding atresia itself. The process could be viewed as passive in the sense of straightforward cellular lysis and degeneration, but evidence is accumulating in favour of intrinsic gene activity leading to the phenomenon of apoptosis, usually taken to mean programmed cell death. The term apoptosis describes a mode of cell death that is responsible for the selective deletion of cells in normal tissue (Wyllie, Kerr & Currie, 1980); it generally affects individual cells, causing them to detach from their neighbours without damaging the latter. Apoptosis is believed to constitute the major pathway to follicular degeneration, and such a form of cell death is certainly implicated in atresia of subordinate follicles (see Palumbo & Yeh, 1994; Hsueh *et al.*, 1996). Nonetheless, the extent to which apoptosis is representative of a selection process during differentiation of ovarian tissues remains to be resolved. It seemingly plays a major rôle in the depletion of oocytes during ageing of ovarian tissues (Tilly, 1996; Morita & Tilly, 1999).

As to the relative contributions of necrosis and apoptosis to follicular atresia, a useful standpoint derived from studies in domestic farm animals may be that necrosis constitutes the dominating mechanism of cell death in granulosa cells taken from atretic large follicles. By contrast, its participation in the death of small follicles tends to be negligible (Rosales-Torrès *et al.*, 2000). In other words, there is a prevalence of necrosis over apoptosis in late atretic follicles, which could explain instances of inflammatory reaction observed during atresia. In terms of distinguishing the two modes of cell death, a quantifiable feature of early necrosis might be the release of lysosomal enzymes into the cytoplasmic compartment of the cell or, in specific cases of atresia, into the cell-free fraction of follicular fluid.

The initial focus of apoptosis leading to atresia is on the oocyte in primordial, primary and pre-antral stages of follicular development. This stands in marked contrast to the initiation of apoptosis in the granulosa cells of follicles destined to undergo atresia at later stages of development (Murdoch, 1995; Perez *et al.*, 2000). Concerning somatic cells, apoptosis is predominant in granulosa cells and seen only occasionally in theca cells, but it has been observed in avian and porcine theca cells of atretic follicles (Tilly *et al.*, 1991). One review suggests that apoptosis of theca interna cells may be species-specific (Terranova & Taylor, 1999). In the mouse, degradation of DNA characteristic of apoptosis in the granulosa cells indicates that follicles at an early antral stage are the most vulnerable to degeneration whereas pre-antral follicles show minimal degeneration (Danilovich, Bartke & Winters, 2000). In our own species, the incidence of apoptosis was significantly higher in granulosa cells from an ovary without a dominant follicle compared with granulosa cells from an ovary with a dominant follicle (Mikkelsen, Høst & Lindenberg, 2001). Although diverse details of the apoptotic cascade are now available for ovarian cells, such detail should not divert attention from the fact that specific cues first rendering an oocyte or follicle susceptible to apoptosis remain unknown. It would seem highly probable that they are multi-factorial. One of the signals that may trigger apoptosis is detachment of cells from the extracellular matrix, and this phenomenon has been termed anoikis (Frisch & Francis, 1994).

A family of proteins termed cadherins may have relevance in the above regard. These are cell adhesion molecules localised to the plasma membrane. Epithelial cadherin (E-cadherin) is expressed in rat granulosa cells *in vitro* (Farookhi & Blaschuk, 1989) and its mRNA noted in mouse ovaries *in vivo* (MacCalman

Farookhi & Blaschuk, 1994). E-cadherin may be involved as an anti-apoptotic molecule, since it was reported in the granulosa cells of healthy ovine follicles whereas expression was not detected in atretic follicles (Ryan, Valentine & Bagnell, 1996), although the time of monitoring would doubtless be critical. Further support for an involvement of cadherins in the regulation of apoptosis comes from studies involving neuronal cadherin (N-cadherin). Apoptosis of rat granulosa cells *in vitro* is inhibited by N-cadherin (Peluso, Pappalardo & Trolice, 1996).

The notion that cell death may be associated with characteristic morphological changes in individual organelles has a substantial pedigree. Long before publication of the now classical paper of Kerr, Wyllie & Currie (1972), features of impending cell death were identifiable in a reasonably systematic way, not least by means of electron microscopy. Cells undergoing apoptosis show decreased cell size with shrinkage of the cell membrane, condensation of nuclear chromatin and breakdown of the nuclear envelope, and compacted cytoplasmic organelles. Apoptotic bodies are then formed – vesicular membranous structures containing nuclear fragments. Apoptotic cells and the apoptotic bodies are rapidly engulfed by neighbouring cells, thereby avoiding an inflammatory reaction leading to tissue damage. This is why the phenomenon is sometimes referred to as silent death (see Nagata, 1997). Apoptosis is thus distinct from necrotic cell death, as it is also on morphological grounds. Recent evidence suggests that Mer receptor tyrosine kinase is essential for the engulfment and efficient clearance of apoptotic cells (Scott *et al.*, 2001).

Ingestion of apoptotic cells by macrophages provokes the release of anti-inflammatory cytokines and suppression of pro-inflammatory cytokines (Green & Beere, 2000). Recognition of cells in the late stages of apoptosis involves exposure of the lipid phosphatidylserine on the outer membrane leaflet of dying cells. This feature of apoptotic cells during their removal by phagocytes has been highly conserved throughout phylogeny (Fadok *et al.*, 2000). Even though the products of apoptotic cell death are engulfed by phagocytic cells, effectively eliminating the physical evidence of death, there are post mortem consequences expressed as an altered behaviour of the phagocytes (Fadok *et al.*, 2000; Green & Beere, 2000). One such response is the induced release of anti-inflammatory cytokines such as TGF-β.

In conjunction with mitosis, it is this active or physiological and orderly form of cell death that regulates the number of cells in a given tissue, and is usually considered as an important aspect of tissue homeostasis (Wyllie, 1992). In the words of a concise review entitled 'When cells die', physiologically occurring cell death is presented as a very important aspect of healthy life (Bergmann, 2000). In an ovarian context, there is regulation by the two opposing processes of follicular development and atresia. That being said, it is scarcely surprising that apoptosis in the gonad appears to be under endocrine, and possibly paracrine, regulation (Yuan & Giudice, 1997; Chun & Hsueh, 1998). Indeed, it would seem that expression of genes that encode for apoptosis-related proteins is regulated by sex steroid hormones (Critchley *et al.*, 1999). Indirectly, the incidence and extent of apoptosis can be influenced by season of the year, nutritional backcloth, and genetic status of the animal in question. *In vitro* studies suggest an involvement of diverse growth factors, not least since absence of particular growth factors in culture systems can prompt apoptosis.

Expression of apoptosis in various cell systems is associated with activation of calcium–magnesium-dependent endonuclease, an enzyme that cleaves genomic DNA into approximately 185–200 base-pair fragments at internucleosomal sites (Jewgenow, Wood & Wildt, 1997; Ejima *et al.*, 2000). The base-pair fragments can be visualised as a distinct ladder of DNA bands upon agarose gel electrophoresis and staining with ethidium bromide, and are a hallmark of the early stages of apoptotic cell death. A dependence on the expression of nuclear and cytosolic proteins is seemingly required in the cleavage process. Endonuclease activity has been detected in granulosa cells, but there are very few reports of endonuclease activity in germ cells (Arends, Morris & Wyllie, 1990; Terranova & Taylor, 1999).

Apoptosis is brought about due to the activity of a death-promoting family of proteases called caspases (cysteine-aspartate proteases) (Fig. VI.12). Perhaps the best characterised apoptosis-inducing signals are those represented by the cell surface receptor for Fas. Fas antigen is a transmembrane receptor that triggers apoptosis in appropriate cells when bound to Fas ligand (FasL). Upon activation, receptors aggregate as trimers and the receptor death domains interact with assembling protein–protein molecules. Hence, Fas activation indirectly triggers a proteolytic cascade leading to apoptosis (Terranova & Taylor, 1999). Fas antigen-mediated cell death may be implicated in follicular atresia, for Fas antigen and FasL are both expressed in the ovary (Porter *et al.*, 2000; Vickers *et al.*, 2000).

Human, mouse and cow ovarian cells are sensitive to Fas antigen-mediated death in the presence of cytokines such as interferon (IFN) γ. IFN-γ will induce Fas antigen expression and enhances Fas antigen-mediated death in many cell types. In a study on bovine ovarian tissue, Fas antigen mRNA levels were highest in granulosa cells from subordinate as compared to other follicles (Porter *et al.*, 2000). FasL alone had no effect on the viability of granulosa or theca cells but became cytotoxic in the presence of IFN-γ. However, granulosa cells from pre-ovulatory follicles exposed to the LH surge *in vivo* were completely resistant to FasL-induced killing.

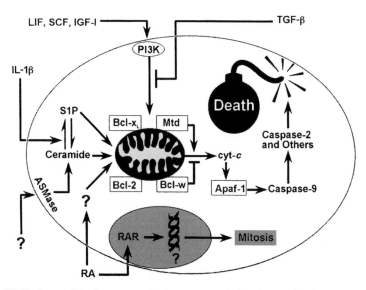

Fig. VI.12. Some of the factors contributing to apoptosis in mice. During fetal ovarian development, the survival or death of oogonia and oocytes is modulated by many extracellular stimuli, including survival factors (e.g. SCF, LIF, IGF-1, IL-1β), retinoic acid (RA), and death factors (e.g. TGF-β). Survival factors utilise, among other messengers, phosphatidylinositol 3'-kinase (PI3K) to prevent apoptosis, possibly involving c-Akt phosphorylation and activation. Whatever the initiating stimulus, signalling into mitochondria via Bcl family members (e.g. Bcl-2, Bcl-w, Bcl-x_L, Mtd/Bok, Bax) is probably central to the final decision made by the female germ cell to commit to or repress apoptosis execution. If apoptosis is the outcome, mitochondrial cytochrome c (cyt-c) release probably triggers the caspase cascade through Apaf-1 oligomerisation. ASMase, acid sphingomyelinase; SAP, sphingosine 1-phosphate; RAR, retinoic acid receptor; LIF, leukaemia inhibitory factor; SCF, stem cell factor; IGF, insulin-like growth factor; IL, interleukin. (After Morita & Tilly, 1999.)

Stem cell or Steel factor (SCF) is a major regulator of growth in the embryonic and fetal ovary (see Chapter II), and could be involved in the events of cell death. Mice lacking either the SCF receptor (Kit) or SCF (Kit ligand) had defective gametogenesis and impaired fertility (Terranova & Taylor, 1999). Both SCF and LIF suppressed apoptosis among mouse primordial germ cells in culture (Pesce *et al.*, 1993), thereby promoting germ cell survival.

As of writing, there is more direct evidence for the Bcl-2 family of proteins having an important influence as apoptotic and anti-apoptotic agents. Bcl-2 protein is the prototype of the gene family that is associated with the inhibition of apoptosis (Wyllie, 1994). Bcl-2 family members are membrane pore-forming proteins, Bcl-2 itself regulating release of cytochrome c from mitochondria and this may then promote activation of caspases. Mitochondria are thought to play a key rôle in integrating the function of Bcl-2 proteins (Green & Reed, 1998).

Some members of the family prompt apoptosis whilst others act as negative regulators. Regulation is complex, and the ratio of promoters to suppressors will determine whether a cell succumbs to apoptosis. Bax protein, also a Bcl-2 family member promotes cell death, possibly antagonising the action of Bcl-2 through heterodimer interaction (Oltavi, Milliman & Korsmeyer, 1993). Hence, the Bcl-2 (anti-apoptotic) to Bax (apoptotic) ratio may be critical in regulating apoptosis. Bcl-2-deficient mice have fewer primordial follicles as compared with controls whereas Bax-deficient animals have a greater number of primordial follicles (Terranova & Taylor, 1999). Such findings suggest that Bcl-2 and Bax interact to influence the final number of oogonia available for recruitment into the pool of primordial follicles.

The factors regulating expression of Bcl-2 and Bax are uncertain, but the protein p53 may enhance apoptosis by increasing expression of Bax. Gonadotrophic hormones reduce apoptosis by reducing p53 expression and Bax expression. This suggests that treatment with gonadotrophic hormones would favour Bcl-2 acting as an anti-apoptotic factor (Terranova & Taylor, 1999). Nonetheless, the extent to which the population of ovarian oocytes and follicles can be protected from degenerative events *in vivo* by manipulation of apoptosis-inducing mechanisms remains uncharted territory, as therefore do its long-term influences on fertility (see Morita & Tilly, 1999).

As pointed out in a recent editorial comment (Anonymous, 2001), dysregulation of apoptosis could result in cancer. Hence, in order to avoid inappropriate cell death, death-receptor signals must be tightly regulated. The FLICE (FADD-like IL-1β-converting enzyme)/caspase-8 inhibitory proteins, or FLIPs, are a recently identified group of inhibitors that control the Fas death receptor signalling pathway. Viral and cellular FLIPs are recruited into death receptor pathways where they inhibit the activation of caspase-8 at the receptor level. FLIP expression is tightly regulated and might be involved in the control of cell activation and cell death.

Although considered classically in a context of inducing regression of the Müllerian ducts in embryonic males (Roberts *et al.*, 1999), a rôle for Müllerian-inhibiting substance (MIS) in apoptotic events in the female gonad should not be overlooked, at least in the mature ovary (see Hunter, 1995). Nor should a contribution from gradients in temperature (see Chapter III) which might be acting to influence the ratio of inducers or inhibitors within the Bcl-2 family.

One other aspect of apoptosis deserves mention, preliminary though the observations are. Nitric oxide may play a rôle in inhibiting apoptosis in pre-ovulatory rat follicles and in stimulating vasodilation and increased blood flow to developing follicles. In one sense, therefore, nitric oxide could have a regulatory influence on programmed cell death (Chun *et al.*, 1995; Rosselli *et al.*,

1998). More specifically, nitric oxide has been shown to mediate IL-1 induced anti-apoptotic effects in culture (Jablonka-Shariff & Olson, 2000).

Concluding remarks

This chapter has ranged widely over a series of major topics, and each has had a bearing on the development of an ovulatory follicle(s) and the processes that underlie this cyclic event – cyclic, that is, in the absence of seasonal breeding, pregnancy or pathology. No specifically novel features have been adduced to enable a satisfactory answer as to why so many oocytes and follicles are generated in the mammalian ovary, with the inevitable destiny of atresia for the great majority. However, the possibility was raised that this ancestral trait, stemming from species with external (aquatic) fertilisation, may have been retained and exploited because of the dual rôle of the mammalian ovary. Not only is there a vital germinal function but there is also a powerful endocrine capability associated ultimately with the establishment and maintenance of pregnancy. Developmental events underlying recruitment of follicles followed by waves of growth, a phase of selection and then dominance would reflect this endocrine aspect of gonadal function, culminating in the release of a viable gamete. Indeed, treatment with appropriate gonadotrophins can demonstrate unequivocally that endocrine regulation is a key to follicular selection and dominance, although, of course, such a statement does not preclude a sensitive contribution of both the autonomic nervous system and the vascular bed.

A vital influence also of the oocyte within a developing follicle needs to be appreciated, and similarly the range of molecules that can be secreted by this impressively large cell – a cell moreover traditionally regarded as essentially quiescent until shortly before ovulation. The latter is certainly not the case. In fact, as suggested in Chapter V, the oocyte could usefully be viewed as a centrally placed computer, coordinating and regulating diverse developmental events culminating in selection of an ovulatory follicle. Nonetheless, the extent to which an oocyte can influence episodes of apoptosis and dictate the onset of follicular atresia remains to be resolved. The author's own inclination is that a germ cell is more likely to be exerting the primary controlling influence on the destiny of associated somatic cells than vice versa, despite the intimate and reciprocal relationship between the two. Viewed quantitatively, as a single cell, the oocyte would appear more vulnerable than follicular cells and, for various reasons, is more likely to have a stronger voice – not least in coordinating granulosa cell renewal and proliferation versus granulosa cell apoptosis.

References

Adams, G.P. (1999). Comparative patterns of follicle development and selection in ruminants. *Journal of Reproduction and Fertility, Supplement*, **54**, 17–32.

Anonymous (2001). Controlling apoptosis. *Nature (London)*, **413**, xi.

Arends, M.J., Morris, R.G. & Wyllie, A.H. (1990). Apoptosis: the role of the endonuclease. *American Journal of Pathology*, **136**, 593–608.

Armstrong, D.G. & Webb, R. (1997). Ovarian follicular dominance: the role of intraovarian growth factors and novel proteins. *Reviews of Reproduction*, **2**, 139–46.

Baird, D.T. (1983). Factors regulating the growth of the preovulatory follicle in the sheep and human. *Journal of Reproduction and Fertility*, **69**, 343–52.

Baird, D.T. (1990). The selection of the follicle of the month. In *From Ovulation to Implantation*, Proceedings of the VIIth Reinier de Graaf Symposium, Maastricht, ed. J.L.H. Evers & M.J. Heineman, Excerpta Medica International Congress Series 917, pp. 3–19. Amsterdam: Elsevier Science Publishers.

Baird, D.T. & McNeilly, A.S. (1981). Gonadotrophic control of follicular development and function during the oestrous cycle of the ewe. *Journal of Reproduction and Fertility, Supplement*, **30**, 119–33.

Baker, T.G. (1963). A quantitative and cytological study of germ cells in human ovaries. *Proceedings of the Royal Society of London, series B*, **158**, 417–33.

Baker, T.G. (1972). Oogenesis and ovulation. In *Reproduction in Mammals*, ed. C.R. Austin & R.V. Short, pp. 14–45. Cambridge: Cambridge University Press.

Baker, T.G. (1982). Oogenesis and ovulation. In *Reproduction in Mammals*, 2nd edn, ed. C.R. Austin & R.V. Short, pp. 17–45. Cambridge: Cambridge University Press.

Bergmann, A. (2000). When cells die. *Trends in Cell Biology*, **10**, 82–3.

Bomsel-Helmreich, O. (1983). The preovulatory human oocyte and its microenvironment. In *Fertilisation of the Human Egg In Vitro*, ed. H.M. Beier & H.R. Lindner, pp. 19–34. Berlin & Heidelberg: Springer-Verlag.

Bomsel-Helmreich, O., Bessis, R. & Vu Gnoc Huyen, L. (1981). Cumulus oophorus of the pre-ovulatory follicle assessed by ultrasound and histology. In *Ultrasound and Infertility*, ed. A.D. Christie, pp. 105–20. Secaucus, NJ: Chartwell and Bratt.

Braw, R.H., Byskov, A.G., Peters, H. & Faber, M. (1976). Follicular atresia in the infant human ovary. *Journal of Reproduction and Fertility*, **46**, 55–9.

Cahill, L.P. (1981). Folliculogenesis in the sheep as influenced by breed, season and oestrous cycle. *Journal of Reproduction and Fertility, Supplement*, **30**, 135–42.

Campbell, B.K., Dobson, H., Baird, D.T. & Scaramuzzi, R.J. (1999). Examination of the relative role of FSH and LH in the mechanism of ovulatory follicle selection in sheep. *Journal of Reproduction and Fertility*, **117**, 355–67.

Campbell, B.K., Picton, H.M., Mann, G.E., McNeilly, A.S. & Baird, D.T. (1991a). The effect of steroid- and inhibin-free ovine, follicular fluid on ovarian follicles and ovarian hormone secretion. *Journal of Reproduction and Fertility*, **93**, 81–96.

Campbell, B.K., Scaramuzzi, R.J., Evans, G. & Downing, J.A. (1991b). Increased ovulation rate in androstenedione-immune ewes is not due to elevated plasma concentrations of FSH. *Journal of Reproduction and Fertility*, **91**, 655–66.

Campbell, B.K., Telfer, E.E., Webb, R. & Baird, D.T. (2000). Ovarian autografts in sheep as a model for studying folliculogenesis. *Molecular and Cellular Endocrinology*, **163**, 131–9.

Christenson, R.K., Ford, J.J. & Redmer, D.A. (1985). Maturation of ovarian follicles in the prepubertal gilt. *Journal of Reproduction and Fertility, Supplement,* **33,** 21–36.

Chun, S.Y., Eisenhauer, K.M., Kubo, M. & Hsueh, A.J.W. (1995). Interleukin-1β suppresses apoptosis in rat ovarian follicles by increasing nitric oxide production. *Endocrinology,* **136,** 3120–7.

Chun, S.Y. & Hsueh, A.J.W. (1998). Paracrine mechanisms of ovarian follicle apoptosis. *Journal of Reproductive Immunology,* **39,** 63–75.

Critchley, H.O.D., Tong, S., Cameron, S.T., Drudy, T.A., Kelly, R.W. & Baird, D.T. (1999). Regulation of *bcl-2* gene family members in human endometrium by antiprogestin administration *in vivo. Journal of Reproduction and Fertility,* **115,** 389–95.

Danilovich, N.A., Bartke, A. & Winters, T.A. (2000). Ovarian follicle apoptosis in bovine growth hormone transgenic mice. *Biology of Reproduction,* **62,** 103–7.

Doody, K.J., Lorence, M.C., Mason, J.I. & Simpson, E.R. (1990). Expression of messenger ribonucleic acid species encoding steroidogenic enzymes in human follicles and corpora lutea throughout the menstrual cycle. *Journal of Clinical Endocrinology and Metabolism,* **70,** 1041–5.

Driancourt, M.A. (1991). Follicular dynamics in sheep and cattle. *Theriogenology,* **35,** 55–79.

Driancourt, M.A., Gauld, I.K., Terqui, M. & Webb, R. (1986). Variations in patterns of follicle development in prolific breeds of sheep. *Journal of Reproduction and Fertility,* **78,** 565–75.

Driancourt, M.A., Gougeon, A., Royère, D. & Thibault, C. (1993). Ovarian function. In *Reproduction in Mammals and Man,* ed. C. Thibault, M.C. Levasseur & R.H.F. Hunter, pp. 281–305. Paris: Ellipses.

Driancourt, M.A. & Thuel, B. (1998). Control of oocyte growth and maturation by follicular cells and molecules present in follicular fluid. *Reproduction, Nutrition, Development,* **38,** 345–62.

Ejima, K., Koji, T., Tsuruta, D., Nanri, H., Kashimura, M. & Ikeda, M. (2000). Induction of apoptosis in placentas of pregnant mice exposed to lipopolysaccharides: possible involvement of Fas/Fas ligand system. *Biology of Reproduction,* **62,** 178–85.

Everett, J.W. (1961). The mammalian female reproductive cycle and its controlling mechanisms. In *Sex and Internal Secretions,* 3rd edn, ed. W.C. Young, pp. 497–555. Baltimore, MD: Williams & Wilkins.

Fadok, V.A., Bratton, D.L., Rose, D.M., Pearson, A., Ezekewitz, R.A.B. & Henson, P.M. (2000). A receptor for phosphatidylserine-specific clearance of apoptotic cells. *Nature (London),* **405,** 85–90.

Farookhi, R. & Blaschuk, O.W. (1989). E-cadherin may be involved in FSH-stimulated responses in rat granulosa cells. In *Growth Factors and the Ovary,* ed. A.N. Hirshfield, pp. 257–65. New York: Plenum Press.

Fortune, J.E. (1991). Ovarian follicular growth and development in mammals. *Biology of Reproduction,* **50,** 225–32.

Fortune, J.E. (1993). Follicular dynamics during the bovine oestrous cycle: a limiting factor in improvement of fertility? *Animal Reproduction Science,* **33,** 111–25.

Frisch, S.M. & Francis, H. (1994). Disruption of epithelial cell–matrix interactions induces apoptosis. *Journal of Cell Biology,* **124,** 619–26.

Garrett, W.M. & Guthrie, H.D. (1996). Expression of androgen receptors and steroido-genic enzymes in relation to follicular growth and atresia following ovulation in pigs. *Biology of Reproduction*, **55**, 949–55.

Garrett, W.M., Mack, S.O., Rohan, R.M. & Guthrie, H.D. (2000). *In situ* analysis of the changes in expression of ovarian inhibin subunit mRNAs during follicle recruitment after ovulation in pigs. *Journal of Reproduction and Fertility*, **118**, 235–42.

Ginther, O.J. (1998). *Ultrasonic Imaging and Animal Reproduction*, Book 3, *Cattle*. Cross Plains, WI: Equiservices Publishing.

Ginther, O.J. (2000). Selection of the dominant follicle in cattle and horses. *Animal Reproduction Science*, **60/61**, 61–79.

Ginther, O.J., Bergfelt, D.R., Kulick, L.J. & Kot, K. (2000). Selection of the dominant follicle in cattle: role of two-way functional coupling between follicle-stimulating hormone and the follicles. *Biology of Reproduction*, **62**, 920–7.

Ginther, O.J., Wiltbank, M.C., Fricke, P.M., Gibbons, J.R. & Kot, K. (1996). Selection of the dominant follicle in cattle. *Biology of Reproduction*, **55**, 1187–94.

Gong, J.G., Campbell, B.K., Bramley, T.A., Gutierrez, C.G., Peters, A.R. & Webb, R. (1996). Suppression in the secretion of follicle-stimulating hormone and lutein-ising hormone and ovarian follicle development in heifers continuously infused with a gonadotrophin-releasing hormone agonist. *Biology of Reproduction*, **55**, 68–74.

Gore, M.A., Nayudu, P.L. & Vlaisavljevic, V. (1997). Attaining dominance *in vivo*: distinguishing dominant from challenger follicles in humans. *Human Reproduction*, **12**, 2741–7.

Gougeon, A. (1996). Regulation of ovarian follicular development in primates: facts and hypotheses. *Endocrine Reviews*, **17**, 121–55.

Gougeon, A. (1997). Kinetics of human ovarian follicular development. In *Microscopy of Reproduction and Development: A Dynamic Approach*, ed. P.M. Motta, pp. 67–77. Rome: Antonio Delfino Editore.

Grant, S.A., Hunter, M.G. & Foxcroft, G.R. (1989). Morphological and biochemical characteristics during ovarian follicular development in the pig. *Journal of Reproduction and Fertility*, **86**, 171–83.

Green, D.R. & Beere, H.M. (2000). Gone but not forgotten. *Nature (London)*, **405**, 28–9.

Green, D.R. & Reed, J.C. (1998). Mitochondria and apoptosis. *Science*, **281**, 1309–12.

Greenwald, G.S. & Roy, S.K. (1994). Follicular development and its control. In *The Physiology of Reproduction*, ed. E. Knobil & J. Neill, pp. 629–724. New York: Raven Press.

Harlow, C.R., Shaw, H.J., Hillier, S.G. & Hodges, J.K. (1986). Factors influencing follicle-stimulating hormone-responsive steroidogenesis in marmoset granulosa cells: effects of androgens and the stage of follicular maturity. *Endocrinology*, **122**, 2780–7.

Hay, M.F., Cran, D.G. & Moor, R.M. (1976). Structural changes occurring during atresia in sheep ovarian follicles. *Cell and Tissue Research*, **169**, 515–29.

Hay, M.F., Moor, R.M., Cran, D.G. & Dott, H.M. (1979). Regeneration of atretic sheep ovarian follicles *in vitro*. *Journal of Reproduction and Fertility*, **55**, 195–207.

Henderson, K.M., McNatty, K.P., O'Keefe, L.E., Lun, S., Heath, D.A. & Prisk, M.D. (1987). Differences in gonadotropin-stimulated cyclic AMP production by granulosa cells from Booroola × Merino ewes which were homozygous, heterozygous or non-carriers of a fecundity gene influencing their ovulation rate. *Journal of Reproduction and Fertility*, **81**, 395–402.

Henderson, K.M., Savage, L.C., Ellen, R.L., Ball, K. & McNatty, K.P. (1988). Consequences of increasing or decreasing plasma FSH concentrations during the preovulatory period in Romney ewes. *Journal of Reproduction and Fertility*, **84**, 187–96.

Henderson, S.A. & Edwards, R.G. (1968). Chiasma frequency and maternal age in mammals. *Nature (London)*, **218**, 22–8.

Hillier, S.G., Reichert, L.E. & Van Hall, E.V. (1981). Control of preovulatory estrogen biosynthesis in the human ovary. *Journal of Clinical Endocrinology and Metabolism*, **52**, 847–56.

Hirshfield, A.N. (1988). Size-frequency analysis of atresia in cycling rats. *Biology of Reproduction*, **38**, 1181–8.

Hirshfield, A.N. (1989). Rescue of atretic follicles *in vitro* and *in vivo*. *Biology of Reproduction*, **40**, 181–90.

Hirshfield, A.N. (1991). Development of follicles in the mammalian ovary. *International Review of Cytology*, **124**, 43–101.

Hodgen, G.D., Goodman, A.L., Stouffer, R.L., *et al.* (1983). Selection and maturation of the dominant follicle and its ovum in the menstrual cycle. In *Fertilisation of the Human Egg In Vitro*, ed. H.M. Beier & H.R. Lindner, pp. 57–69. Berlin, Heidelberg & New York: Springer-Verlag.

Hsueh, A.J.W., Billig, H. & Tsafriri, A. (1994). Ovarian follicle atresia: a hormonally controlled apoptotic process. *Endocrine Reviews*, **15**, 707–24.

Hsueh, A.J.W., Eisenhauer, K., Chun, S.Y., Hsu, S.Y. & Billig, H. (1996). Gonadal cell apoptosis. *Recent Progress in Hormone Research*, **51**, 433–56.

Hunter, M.G., Biggs, C., Faillace, L.S. & Picton, H.M. (1992). Current concepts of folliculogenesis in monovular and polyovular farm species. *Journal of Reproduction and Fertility, Supplement*, **45**, 21–38.

Hunter, R.H.F. (1980). *Physiology and Technology of Reproduction in Female Domestic Animals*. London & New York: Academic Press.

Hunter, R.H.F. (1995). *Sex Determination, Differentiation and Intersexuality in Placental Mammals*. Cambridge: Cambridge University Press.

Hunter, R.H.F. & Baker, T.G. (1975). Development and fate of porcine Graafian follicles identified at different stages of the oestrous cycle. *Journal of Reproduction and Fertility*, **43**, 193–6.

Ireland, J.J. & Roche, J.F. (1987). Hypotheses regarding development of dominant follicles during a bovine oestrous cycle. In *Follicular Growth and Ovulation Rate in Farm Animals*, ed. J.F. Roche & D. O'Callaghan, pp. 1–18. Dordrecht & Boston: Martinus Nijhoff Publishers.

Jablonka-Shariff, A. & Olson, L.M. (2000). Nitric oxide is essential for optimal meiotic maturation of murine cumulus-oocyte complexes *in vitro*. *Molecular Reproduction and Development*, **55**, 412–21.

Jewgenow, K., Wood, T.C. & Wildt, D.E. (1997). DNA degradation in mural granulosa cells of non- and slightly atretic follicles of fresh and cold-stored domestic cat ovaries. *Molecular Reproduction and Development*, **48**, 350–5.

Kapia, A. & Hsueh, A.J.W. (1997). Regulation of ovarian follicle atresia. *Annual Review of Physiology*, **59**, 349–63.

Kerr, J.F.R., Wyllie, A.H. & Currie, A.R. (1972). Apoptosis: a basic biological phenomenon with wide-ranging implications in tissue kinetics. *British Journal of Cancer*, **26**, 239–57.

Krakauer, D.C. & Mira, A. (1999). Mitochondria and germ-cell death. *Nature (London)*, **400**, 125–6.

Kruip, T.A.M. & Dieleman, S.J. (1982). Macroscopic classification of bovine follicles and its validation by micromorphological and steroid biochemical procedures. *Reproduction, Nutrition and Development*, **22**, 465–73.

Law, A.S., Baxter, G., Logue, D.N., O'Shea, T. & Webb, R. (1992). Evidence for the action of a bovine follicular fluid factor(s) other than inhibin in suppressing follicular development and delaying oestrus in heifers. *Journal of Reproduction and Fertility*, **96**, 603–16.

MacCalman, C.D., Farookhi, R. & Blaschuk, O.W. (1994). Oestradiol regulates E-cadherin mRNA levels in the surface epithelium of the mouse ovary. *Clinical and Experimental Metastasis*, **12**, 276–82.

McGee, E.A. & Hsueh, A.J.W. (2000). Initial and cyclic recruitment of ovarian follicles. *Endocrine Reviews*, **21**, 200–14.

McKenzie, F.F. (1926). *The Normal Oestrous Cycle in the Sow*. University of Missouri, College of Agriculture, Agricultural Experiment Station, Research Bulletin no. 86.

McNatty, K.P., Smith, D.M., Makris, A., Osathanondh, R. & Ryan, K.J. (1979). The microenvironment of the human antral follicle: interrelationships among the steroid levels in antral fluid, the population of granulosa cells, and the status of the oocyte *in vivo* and *in vitro*. *Journal of Clinical Endocrinology and Metabolism*, **49**, 851–60.

Maxson, W.S., Haney, A.F. & Schomberg, D.W. (1985). Steroidogenesis in porcine atretic follicles: loss of aromatase activity in isolated granulosa and theca. *Biology of Reproduction*, **33**, 495–501.

Mikkelsen, A.L., Høst, E. & Lindenberg, S. (2001). Incidence of apoptosis in granulosa cells from immature human follicles. *Reproduction*, **122**, 481–6.

Miro, F. & Hillier, S.G. (1992). Relative effects of activin and inhibin on steroid hormone synthesis in primate granulosa cells. *Journal of Clinical Endocrinology and Metabolism*, **75**, 1556–61.

Monget, P., Monniaux, D., Pisselet, C. & Durand, P. (1993). Changes in insulin-like growth factor-1 (IGF1), IGF-II, and their binding proteins during growth and atresia of ovine ovarian follicles. *Endocrinology*, **132**, 1438–46.

Morita, Y. & Tilly, J.L. (1999). Oocyte apoptosis: like sand through an hourglass. *Developmental Biology*, **213**, 1–17.

Murdoch, W.J. (1995). Programmed cell death in pre-ovulatory ovine follicles. *Biology of Reproduction*, **53**, 8–12.

Nagata, S. (1997). Apoptosis by death factor. *Cell*, **88**, 355–65.

Oltavi, Z.N., Milliman, C.L. & Korsmeyer, S.J. (1993). Bcl-2 heterodimerises *in vivo* with a conserved homologue Bax that accelerates programmed cell death. *Cell*, **74**, 609–19.

O'Shea, J.D., Hay, M.F. & Cran, D.G. (1978). Ultrastructural changes in the theca interna during follicular atresia in sheep. *Journal of Reproduction and Fertility*, **54**, 183–7.

Palumbo, A. & Yeh, J. (1994). In situ localisation of apoptosis in the rat ovary during follicular atresia. *Biology of Reproduction*, **51**, 888–95.

Peluso, J.J., Pappalardo, A. & Trolice, M.P. (1996). N-cadherin-mediated cell contact inhibits granulosa cell apoptosis in a progesterone-independent manner. *Endocrinology*, **137**, 1196–203.

Perez, G.I., Trbovich, A.M., Gosden, R.G. & Tilly, J.L. (2000). Mitochondria and the death of oocytes. *Nature (London)*, **403**, 500–1.

Pesce, M., Farrace, M.G., Piacentini, M., Dolci, S. & De Felici, M. (1993). Stem cell factor and leukemia inhibitory factor promote primordial germ cell survival by suppressing programmed cell death (apoptosis). *Development*, **118**, 1089–94.

Peters, H. & McNatty, K.P. (1980). *The Ovary*. London & Toronto: Paul Elek, Granada Publishing.

Pierson, R.A. & Ginther, O.J. (1988). Ultrasonic imaging of the ovaries and uterus in cattle. *Theriogenology*, **29**, 3–20.

Porter, D.A., Vickers, S.L., Cowan, R.G., Huber, S.C. & Quirk, S.M. (2000). Expression and function of FAS antigen vary in bovine granulosa and theca cells during ovarian follicular development and atresia. *Biology of Reproduction*, **62**, 62–6.

Rajakoski, E. (1960). The ovarian follicular system in sexually mature heifers with special reference to seasonal, cyclical and left-right variations. *Acta Endocrinologica, Supplement*, **52**, 1–68.

Richards, J.S. (1994). Hormonal control of gene expression in the ovary. *Endocrine Reviews*, **15**, 725–51.

Richards, J.S., Fitzpatrick, S.L., Clemens, J.W., Morris, J.K., Alliston, T. & Sirois, J. (1995). Ovarian cell differentiation: a cascade of multiple hormones, cellular signals and regulated genes. *Recent Progress in Hormone Research*, **50**, 223–54.

Roberts, L.M., Hirokawa, Y., Nachtigal, M.W. & Ingraham, H.A. (1999). Paracrine-mediated apoptosis in reproductive tract development. *Developmental Biology*, **208**, 110–22.

Roche, J.F. (1996). Control and regulation of folliculogenesis – a symposium in perspective. *Reviews of Reproduction*, **1**, 19–27.

Rosales-Torrès, A.M., Avalos-Rodríguez, A., Vergara-Onofre, M., *et al.* (2000). Multiparametric study of atresia in ewe antral follicles: histology, flow cytometry, internucleosomal DNA fragmentation and lysosomal enzyme activities in granulosa cells and follicular fluid. *Molecular Reproduction and Development*, **55**, 270–81.

Rosselli, M., Keller, P.J. & Dubey, R.K. (1998). Role of nitric oxide in the biology, physiology and pathophysiology of reproduction. *Human Reproduction Update*, **4**, 3–24.

Ryan, P.L., Valentine, A.F. & Bagnell, C.A. (1996). Expression of epithelial cadherin in the developing and adult pig ovary. *Biology of Reproduction*, **55**, 1091–7.

Sarty, G.E., Adams, G.P. & Pierson, R.A. (2000). Three-dimensional magnetic resonance imaging for the study of ovarian function in a bovine *in vitro* model. *Journal of Reproduction and Fertility*, **119**, 69–75.

Scaramuzzi, R.J., Adams, N.R., Baird, D.T., *et al.* (1993). A model for follicle selection and the determination of ovulation rate in the ewe. *Reproduction, Fertility and Development*, **5**, 459–78.

Scott, R.S., McMahon, E.J., Pop, S.M., *et al.* (2001). Phagocytosis and clearance of apoptotic cells is mediated by MER. *Nature (London)*, **411**, 207–11.

Sirois, J. & Fortune, J.E. (1988). Ovarian follicular dynamics during the estrous cycle in heifers monitored by real-time ultrasonography. *Biology of Reproduction*, **39**, 308–17.

Strömstedt, M. & Byskov, A.G. (1999). Oocyte, Mammalian. In *Encyclopedia of Reproduction*, vol. 3, ed. E. Knobil & J.D. Neill, pp. 468–80. San Diego & London: Academic Press.

Sturgis, S.H. (1961). Factors influencing ovulation and atresia of ovarian follicles. In *Control of Ovulation*, ed. C.A. Villee, pp. 213–18. Oxford, London & New York: Pergamon Press.

Sunderland, S.J., Crowe, M.A., Boland, M.P., Roche, J.F. & Ireland, J.J. (1994). Selection, dominance and atresia of follicles during the oestrous cycle of heifers. *Journal of Reproduction and Fertility*, **101**, 547–55.

Szöllösi, D. (1978). On the role of gap junctions between follicle cells and the oocyte in the mammalian ovary. *Research in Reproduction*, **10**, 3–4.

Szöllösi, D. (1993). Oocyte maturation. In *Reproduction in Mammals and Man*, ed. C. Thibault, M.C. Levasseur & R.H.F. Hunter, pp. 307–25. Paris: Ellipses.

Szöllösi, D., Gérard, M., Ménézo, Y. & Thibault, C. (1978). Permeability of ovarian follicle: corona cell–oocyte relationship in mammals. *Annales de Biologie Animale, Biochimie et Biophysique*, **18**, 511–21.

Terranova, P.F. & Taylor, C.C. (1999). Apoptosis (cell death). In *Encyclopedia of Reproduction*, vol. 1, ed. E. Knobil & J.D. Neill, pp. 261–73. San Diego & London: Academic Press.

Tilly, J.L. (1996). Apoptosis and ovarian function. *Reviews of Reproduction*, **1**, 162–72.

Tilly, J.L. & Hsueh, A.J.W. (1992). Apoptosis as the basis of ovarian follicular atresia. In *Gonadal Development and Function*, ed. S.G. Hillier, Serono Symposium Publications, no. 94, pp. 157–65. New York: Raven Press.

Tilly, J.L., Kowalski, K.I., Johnson, A.L. & Hsueh, A.J.W. (1991). Involvement of apoptosis in ovarian follicular atresia and postovulatory regression. *Endocrinology*, **129**, 2799–801.

Vanderhyden, B.C. & Macdonald, E.A. (1998). Mouse oocytes regulate granulosa cell steroidogenesis throughout follicular development. *Biology of Reproduction*, **59**, 1296–301.

Vickers, S.L., Cowan, R.G., Harman, R.M., Porter, D.A. & Quirk, S.M. (2000). Expression and activity of the Fas antigen in bovine ovarian follicle cells. *Biology of Reproduction*, **62**, 54–61.

Webb, R., Campbell, B.K., Garverick, N.A., Gong, J.G., Gutierrez, C.G. & Armstrong, D.G. (1999a). Molecular mechanisms regulating follicular recruitment and selection. *Journal of Reproduction and Fertility, Supplement*, **54**, 33–48.

Webb, R., Gosden, R.G., Telfer, E.E. & Moor, R.M. (1999b). Factors affecting folliculogenesis in ruminants. *Animal Science*, **68**, 257–84.

Webb, R. & England, B.G. (1982). Identification of the ovulatory follicle in the ewe: associated changes in follicular size, thecal and granulosa cell luteinising hormone receptors, antral fluid steroids, and circulating hormones during the preovulatory period. *Endocrinology*, **110**, 873–81.

Wood, S.C., Glencross, R.G., Bleach, E.C., Lovell, R., Beard, A.J. & Knight, P.G. (1993). The ability of steroid-free bovine follicular fluid to suppress FSH secretion and delay ovulation persists in heifers actively immunised against inhibin. *Journal of Endocrinology*, **136**, 137–48.

Wyllie, A.H. (1992). Apoptosis and the regulation of cell numbers in normal and neoplastic tissues: an overview. *Cancer Metastasis Review*, **11**, 95–103.

Wyllie, A.H. (1994). Death gets a break. *Nature (London)*, **369**, 272–3.

Wyllie, A.H., Kerr, J.F.R. & Currie, A.R. (1980). Cell death: the significance of apoptosis. *International Review of Cytology*, **68**, 251–306.

Yuan, W. & Giudice, L.C. (1997). Programmed cell death in human ovary is a function of follicle and corpus luteum status. *Journal of Clinical Endocrinology and Metabolism*, **82**, 3148–55.

Zeleznik, A.J. & Benyo, D.F. (1994). Control of follicular development, corpus luteum function, and the recognition of pregnancy in higher primates. In *The Physiology of Reproduction*, ed. E. Knobil & J.D. Neill, pp. 751–82. New York: Raven Press.

VII

Follicular responses to the pre-ovulatory surge of gonadotrophic hormones

Introduction

The ovaries respond in a variety of ways and in different cellular compartments to what has become known as the pre-ovulatory gonadotrophin surge. As a consequence of the positive feedback influence of oestradiol secretion from a mature follicle(s) on the hypothalamus and on the anterior lobe of the pituitary gland, these central structures are coordinated to release a massive surge of gonadotrophic hormones that is clearly detectable in the peripheral circulation. In reality, this consists of peaks of both LH and FSH, the latter being the smaller of the two (Fig. VII.1). The fact that these peaks of gonadotrophin discharge are closely coincident or superimposed was taken as one of the first lines of evidence for a GnRH common to both the LH and FSH activities of the anterior pituitary gland.

The adjective gonadotrophic of course indicates that LH and FSH have a stimulatory influence on the ovaries, and this is because circulating gonadotrophins are taken up preferentially by ovarian tissues. If we use a radio-labelled preparation of LH, for example, to trace its accumulation in target tissues, we find that the hormone is rapidly sequestered in the antral fluid of pre-ovulatory follicles in minutes where its concentration then increases over a more extended period. As a sequel to such accumulation in follicular fluid, labelled hormone can be demonstrated to bind to mural granulosa cells that have been suitably primed by FSH. Prior to the pre-ovulatory gonadotrophin surge, the predominant binding site for LH is on cells of the theca interna, with little or no binding detectable in the granulosa cell layers (see Chapters V and VI).

When viewed as an endocrine structure, the state of maturity of a Graafian follicle is largely expressed in terms of its synthesis and secretion of oestradiol-17β. Oestradiol not only acts locally to influence intra-follicular events such as the availability of hormone binding sites, the rate and distribution of blood

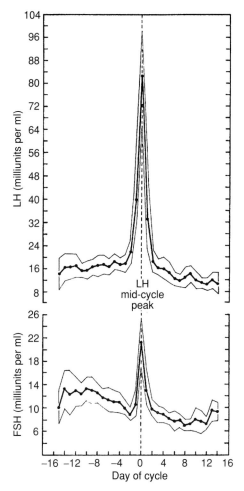

Fig. VII.1. Classical graphs to depict the coincident or closely coincident peaks of the two gonadotrophic hormones at the mid-cycle stage of a normal menstrual cycle. LH, luteinising hormone; FSH, follicle stimulating hormone. (After Ross *et al.*, 1970.)

flow and the nature and extent of autonomic nervous activity, but also acts centrally in terms of programming sexual behaviour and the positive feedback responses described. For these last two influences to be expressed, an oestradiol threshold must be reached in the concentration of steroid hormone released into the systemic circulation. In one sense, therefore, it is the maturity of a Graafian follicle that regulates and precipitates its own time of ovulation.

Table VII.1. *Examples of the approximate*
interval between the pre-ovulatory surge of
gonadotrophic hormones and ovulation of
mature Graafian follicles

Species	Interval until ovulation (hours)
Rabbit[a]	9–11
Rat	12–15
Mouse	12–15
Sheep	25–26
Cow	28–30
Pig	39–40
Human	38–40

[a]Ovulation in rabbits is normally induced by coitus.

Events underlying the pre-ovulatory surge

If a focus is placed upon the conspicuous or massive surge of pituitary hormones detected in the peripheral circulation a characteristic number of hours
before ovulation (Table VII.1), intensive blood sampling reveals this to be composed of stepwise increases in the frequency and amplitude of gonadotrophin
secretion (Fig. VII.2). Although drawn classically as a smooth curve, this con-
ceals the intermittent nature of the discharge from hypophyseal tissue, which,
in turn, reflects the pulsatile pattern of hypothalamic secretion of GnRH into
the capillary portal system. This hypothalamic secretion stimulates a group of
cells in the anterior lobe, the so-called gonadotrophs, to secrete LH and FSH
episodically. The responsiveness of gonadotrophs to GnRH has been increased
by the positive feedback influence of oestradiol.

As to whether the consequences of positive feedback by oestradiol involve increased synthesis of LH and FSH, observations have been usefully summarised
by Clarke (1996). Increased LHβ subunit mRNA levels are seen in rats before
the LH surge, in contrast to increased FSHβ mRNA levels after the surge. Turning to sheep, the level of α subunit mRNA doubles during the follicular phase
of the oestrous cycle, but with a non-significant increase in LHβ subunit mRNA
and a fall in FSHβ mRNA. An overall point of interest is the divergent regulation
of gonadotrophin subunit mRNAs, with that for LHβ tending to be upregulated.

Development of remarkably skilled surgical techniques in sheep enabled
simultaneous sampling of hypophyseal portal blood and of pituitary secretion;
the relationships between GnRH and gonadotrophin secretion could thereby
be examined. In summary, short-lived pulses of GnRH were noted to generate

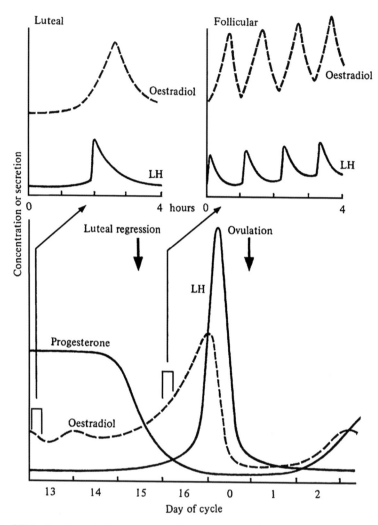

Fig. VII.2. Graphs to indicate the changing frequency and amplitude of gonadotrophin secretion leading to the pre-ovulatory peak during the 16- or 17-day oestrous cycle of sheep. (Adapted from Baird & McNeilly, 1981.)

correspondingly large pulses of LH secretion (Clarke, 1996). The sampling technique also revealed smaller pulses of GnRH not reflected in increased peripheral concentrations of LH. One rôle suggested for these minor pulses of GnRH was to sustain synthesis of gonadotrophins. In fact, small responding pulses of LH can be detected in the pituitary secretion collected directly in cavernous sinus blood of sheep (Clarke, 1996).

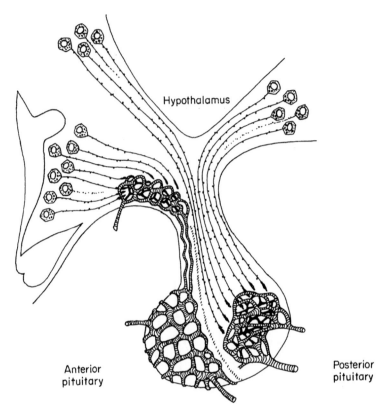

Fig. VII.3. To illustrate the functional anatomy underlying physiological relationships between the hypophysiotrophic area of the hypothalamus and the pituitary gland. Note particularly the capillary portal system to the anterior pituitary and direct innervation of the posterior pituitary. (After Hunter, 1980.)

The pre-ovulatory gonadotrophin surge is widely accepted to be due to a rise in GnRH secretion, although an increased responsiveness of the pituitary gonadotrophs would also contribute. Nonetheless, considerable variation between animals in the magnitude of the GnRH rise has been recorded. Frequent sampling of hypophyseal portal blood (Fig. VII.3) points to augmented GnRH secretion being a consequence of an increased frequency of GnRH pulses. However, focusing on cellular activity at the level of the gonadotrophs, secretory granules of hormone are mobilised towards the plasma membrane. One of the actions of positive oestradiol feedback can be seen as packaging of LH in the form of such secretory granules moving towards the cell membrane. A concomitant increased activation of protein kinase C as second messenger can be demonstrated in these cells. There is also a marked self-priming of GnRH during the pre-ovulatory interval, an action involving the upregulation of G-protein-coupled second messenger systems within the pituitary gonadotrophs.

A finding of great importance in the present context is that of multiple GnRH mRNA transcripts that are apparently differentially regulated (Clarke, 1996), clearly a route to sophisticated levels of control, flexibility and interaction. Readers seeking further detail on these endocrine, neuroendocrine and molecular events preceding the pre-ovulatory gonadotrophin surge are referred to Crowley *et al.* (1985), Moenter *et al.* (1991), Clarke (1993) and Evans *et al.* (1994).

Gonadotrophin surge attenuating factor

There is accumulating evidence that the ovary produces a substance that acts by negative feedback to attenuate the pre-ovulatory surge of gonadotrophic hormones. Although not yet fully characterised, and therefore not identified with certainty, it is referred to as a gonadotrophin surge inhibiting or attenuating factor (GnSAF). In women undergoing a superovulation protocol as part of fertility treatment (see Chapter XI), a supposed GnSAF has been shown to attenuate the anticipated endogenous surge of LH that would normally precede ovulation, thereby leading to failure of this vital event (Messinis & Templeton, 1989, 1990a,b, 1991). In experimental animals such as monkeys and rats, the putative factor will block the pre-ovulatory LH surge and was thus more appropriately referred to as an inhibiting factor – gonadotrophin surge inhibiting factor (GnSIF: Littman & Hodgen, 1984; Koppenaal, Tijssen & de Koning, 1991). As to the site of the negative feedback influence of this ovarian factor, a reduced pituitary sensitivity to GnRH indicates that it is primarily hypophyseal. Indeed, the reduced pituitary responsiveness could explain an attenuated pre-ovulatory LH surge (Messinis & Templeton, 1990a). Other evidence suggests that ovarian secretion of GnSAF can be stimulated by FSH (Koppenaal *et al.*, 1991; Messinis *et al.*, 1993; Fowler & Price, 1997) and that GnSAF antagonises the sensitising influence of oestradiol on the anterior pituitary, thereby regulating the amplitude of the mid-cycle LH surge (Messinis *et al.*, 1998).

The modulation of pituitary LH secretion achieved by GnSAF is seen in circumstances of GnRH-induced secretion of LH, not in basal levels of gonadotrophin secretion. Various studies have emphasised that it is a non-steroidal factor distinct from inhibin (e.g. Fowler, Messinis & Templeton, 1990), the latter being a gonadal peptide concerned with feedback regulation of FSH rather than LH (see Chapter V). One suggestion is that GnSAF represents an ovarian hormone especially involved with feedback control of LH secretion at the time of the pre-ovulatory LH surge (Fowler & Templeton, 1996).

In vitro bioassay models were very much to the fore in studies of GnSAF, notably a rat pituitary bioassay. Monitoring the change in response of pituitary LH to GnRH enabled values for ovarian follicular fluid GnSAF bioactivity to be estimated in monkeys (Schenken & Hodgen, 1986), pigs (Danforth *et al.*, 1987) and rats (Busbridge *et al.*, 1988), and more recently in women (Busbridge *et al.*, 1990; Fowler *et al.*, 1990; Knight *et al.*, 1990; Mroueh *et al.*, 1996). When GnSAF bioactivity was purified to homogeneity, a 69 kDa candidate protein could be isolated from bovine follicular fluid (Danforth & Cheng, 1995). Comparable studies with human follicular fluid purifying GnSAF to homogeneity isolated a candidate protein of 12.5 kDa, with sequence homology to the 12.5 kDa C-terminal fragment of human serum albumin (Pappa *et al.*, 1999); evidence suggested that GnSAF was a glycoprotein. As of writing, full sequencing of the GnSAF molecule is awaited, not least since this would remove uncertainty as to whether GnSAF really is a novel ovarian protein.

Resumption of meiosis

In a majority of mammalian species so far examined, oocytes enter the diplotene stage of the first meiotic prophase at approximately the time of birth (Brambell, 1956; Zuckerman, 1960; and see Chapter II). Thereafter, until beyond puberty and shortly before ovulation or the onset of atresia, the germinal material is arranged in a large vesicular nucleus and this prolonged state of arrest is referred to as the dictyate stage (Fig. VII.4). The stages of meiosis between diakinesis (i.e. late prophase with resumption of meiotic activity) and metaphase of the second meiotic division occur, in most species, during the relatively brief period between stimulation of the Graafian follicle by the surge of LH and ovulation some hours later. Under appropriate endocrinological conditions, the dictyate stage in the oocytes of adult mammals can be terminated and the resumption of meiosis induced (Pincus & Enzmann, 1935; Edwards, 1962, 1965), as, for example, following systemic injection of hCG or LH (Tables VII.2 and VII.3). Moreover, the stage of meiosis at a particular time after gonadotrophin injection can usually be predicted with considerable accuracy (Edwards & Gates, 1959; Dziuk, 1965; Hunter & Polge, 1966).

Specific responses to the pre-ovulatory surge of gonadotrophic hormones are noted in both the follicular oocyte and its surrounding somatic cells. After an interval during which morphological changes are not readily detectable at the

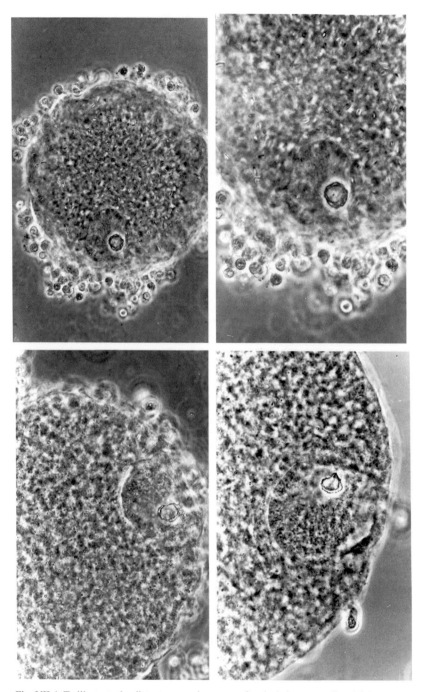

Fig. VII.4. To illustrate the dictyate or resting stage of meiosis in mammalian (pig) oocytes. There is a large peripherally placed germinal vesicle (nucleus) with the condensed chromatin arranged in the form of a ring or horseshoe around a single eccentrically located nucleolus – features characteristic of a primary oocyte.

Table VII.2. *The approximate duration of the maturation stages in follicular oocytes after injecting animals in pro-oestrus with 500 i.u. hCG*

Stage of maturation	Interval from hCG injection (h)	Approximate duration of stage (h)	% of oocytes conforming to timetable
Germinal vesicle	0–18	18	100
Germinal vesicle to prometaphase	18–22	4	100
Prometaphase	22–26	4	100
Metaphase 1	26–35	9	78.6
Anaphase 1, telophase 1, and abstriction of first polar body	35–37	2	71.3

i.u., international units; hCG, human chorionic gonadotrophin.
After Hunter & Polge, 1966.

Table VII.3. *Timing of the response of the oocyte nucleus after the pre-ovulatory gonadotrophin surge or following a systemic injection of a luteinising hormone preparation. All times are in hours*

Species	Latent period[a] after stimulus (h)	Stage of meiotic maturation			
		First metaphase	First anaphase	First telophase	Second metaphase
Cow	10–12	14–21	22	23	24
Sheep	10–11	12–20	21	22	24
Pig	17–18	26–34	35	36	37

[a]The oocyte nucleus remains as a germinal vesicle during this latent period.
After Hunter, 1980.

level of the light microscope, resumption of meiosis is seen as a progressive condensation of chromatin, undulation then breakdown and disappearance of the nuclear envelope that surrounds the so-called germinal vesicle, and gradual alignment of chromosomes as bivalents on a metaphase spindle. This is followed by anaphase, telophase and extrusion of the first polar body (Fig. II.4). A detailed ultrastructural description of these events has been given by Szöllösi (1993). Eggs are usually ovulated at this stage of secondary oocyte with the chromosomes once more assembled on a metaphase spindle. In this condition of second meiotic metaphase, an egg awaits a fertilising spermatozoon and is generally displaced rapidly to the ampullary–isthmic junction of the Fallopian tube, the site of fertilisation (see Chapter IX).

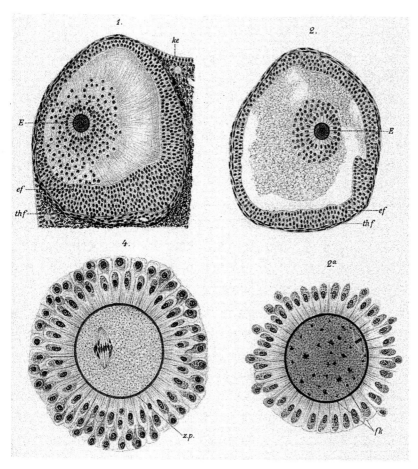

Fig. VII.5. Beautiful drawings of the late nineteenth century that illustrate the rearrangement of the cellular investment around an oocyte during the pre-ovulatory interval, i.e. following the surge of gonadotrophic hormones. The corona radiata is especially highlighted in the lower part of the plate. (Adapted from Sobotta, 1895.)

Turning to the somatic cells that encompass the oocyte, those of the cumulus oophorus and its innermost corona radiata, their most conspicuous responses to the pre-ovulatory gonadotrophin surge are (1) a specific expansion and loosening of the cell mass in association with a characteristic mucification of the intercellular spaces (see below and Bomsel-Helmreich, 1985), and (2) a rearrangement of the corona radiata in contact with both the zona pellucida and the oocyte plasma membrane (Fig. VII.5). The latter contact consists of long cytoplasmic processes with bulbous foot swellings that indent the vitelline membrane

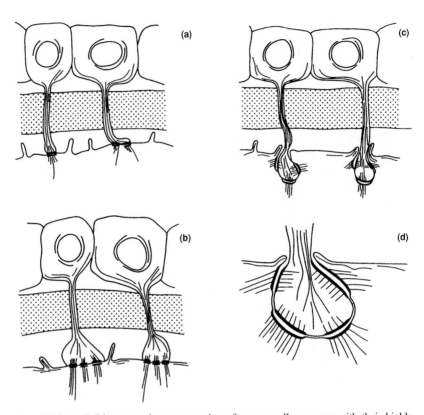

Fig. VII.6. (a–d) Diagrammatic representation of corona cell processes with their highly specialised bulbous feet that may conspicuously indent the egg plasma membrane (oolemma). According to species and the stage of the pre-ovulatory interval, the junctional complexes between somatic cell processes and oocyte show characteristic features. (Courtesy of the late D.G. Szöllösi, and after Szöllösi, 1993.)

(Fig. VII.6) and exhibit intercellular gap junctions. Specialised contact with the zona pellucida and oocyte membrane gradually diminishes during the post-gonadotrophin surge (Anderson & Albertini, 1976; Thibault, Szöllösi & Gérard, 1987), pre-ovulatory interval with undulation and then retraction of the cytoplasmic processes. These processes contain f-actin filaments and microtubules. Progressive loss of somatic cells continues after ovulation. Initial modifications to the cumulus cell mass increase its dimensions, improve the effectiveness of egg displacement from the gonad into the duct system, and enhance the chances of sperm–egg contact within the Fallopian tube (see Chapter IX). Loss of the cumulus investment is generally accelerated in mated animals, since sperm-associated or sperm-derived hyaluronidase acts to depolymerise the hyaluronic acid cement substance. However, in a major review of pre-fertilisation events, the kinetics of cumulus investment 'break-up' were surprisingly stated to be similar between mated and unmated animals (Harper, 1994).

If mating or coitus has preceded ovulation by several hours or more in normal healthy animals or patients with functional Fallopian tubes, then a secondary oocyte should commence to be fertilised soon after ovulation. Penetration and activation of an oocyte by the fertilising spermatozoon lead to anaphase and telophase of the second meiotic division, with extrusion of the second polar body (Fig. VII.7). A perspective worth emphasising here in the context of fertilisation is the protracted arrest of meiosis in ovarian oocytes from the time of birth until final maturation of a Graafian follicle, followed by completion of the first meiotic division shortly before ovulation and completion of the second meiotic division shortly after ovulation. This represents a chronologically compact resumption and completion of meiosis in oocytes undergoing normal fertilisation. The same overall perspective would apply to oocytes of the dog and fox, although the timescale of resumption to completion of meiosis is even more condensed. In these species, oocytes show no conspicuous morphological response of the nucleus to the pre-ovulatory increase of gonadotrophins in the interval leading up to ovulation. Rather, they are shed from the ovary with the nucleus still in first meiotic prophase and resume nuclear maturation in the lumen of the Fallopian tube either 'spontaneously' or after sperm penetration (Pearson & Enders, 1943; Farstad *et al.*, 1989).

Numerous reviews exist on the topic of pre-ovulatory maturation of oocytes (e.g. Thibault, 1977; Tsafriri, 1985, 1989; Dekel, 1999; Picton & Gosden, 1999; Strömstedt & Byskov, 1999). One of the introductory points invariably made in such reviews is that meiotic maturation is only possible after a long series of preparatory processes reflected in growth of the oocyte and increase in the volume of follicular antral fluid. As a generalisation, oocytes must achieve approximately 80% of their final mature diameter before becoming competent to resume meiosis: even then, many such oocytes will only develop until first metaphase (Szöllösi, 1993). Critical changes in nucleolar morphology and nucleolar RNA synthetic activity have taken place shortly before the stage of competence is reached (Szöllösi, 1993; Hyttel *et al.*, 1996, 1997). Such competence reflects not only transcription and the correct ordering of maternal transcripts expressed during first meiotic prophase but an increase in cytoplasmic volume with the deposition of 'yolk'.

The involvement of centrioles has also been discussed (Szöllösi, 1993; Schatten, 1994). They are replaced by a filamentous mass (pericentriolar material) from which microtubules originate for formation of the meiotic spindle(s). Because 'disassembly' of centrioles has been noted in the oocytes of diverse

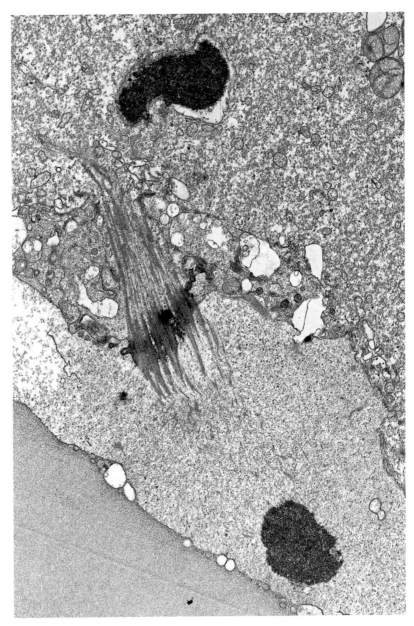

Fig. VII.7. An electron micrograph of a recently activated egg to show the microtubules of the spindle apparatus at completion of the second meiotic division. Note that a nuclear membrane fails to form around the chromatin of the second polar body. (After Szöllösi & Hunter, 1973.)

animal species, the suggestion arose that if the centrioles remained intact at the spindle poles and fully functional, then perhaps oocyte maturation would not be arrested at the second meiotic metaphase (Szöllösi, 1993).

A major question still awaiting a full response is precisely how the pre-ovulatory gonadotrophin surge acts to trigger resumption of meiosis in oocytes enclosed within follicles destined for ovulation. A general theme in most explanations is that the gonadotrophin surge either overcomes a molecular inhibitor that has maintained the meiotic arrest or, alternatively, generates local trophic signals within the follicle – or a combination of both such actions.

Long-standing evidence for an inhibitor comes from the now classical observation that liberating oocytes from pre-ovulatory follicles into a suitable culture medium prompts the spontaneous resumption of meiosis in a high proportion of instances across a wide range of species (Pincus & Enzmann, 1935; Chang, 1955; Edwards, 1962, 1965; Hunter, Lawson & Rowson, 1972). Anaphase, telophase and the formation of a first polar body follow timescales corresponding closely to the *in vivo* counterparts for most, if not all, of the oocytes examined. A potential shortcoming of the inhibitor hypothesis could be that the meiotic arrest is imposed on oocytes in the earliest (primordial) stages of follicle formation long before accumulation of follicular fluid and formation of an antrum (see Chapter II). This is not to argue that the inhibitory molecules would not be already expressed at such early stages, but the manner of their presentation to the cell surface and nucleus might differ from that in follicular fluid.

Cumulus cells or a sub-population thereof would seem to play a rôle in the arrest of meiosis. Early evidence for granulosa cell production of a meiotic inhibitor came from the studies of Tsafriri & Channing (1975) in the mouse. A putative inhibitor was said to be produced when a follicle is immature while another protein, an antagonist that nullifies the influence of the inhibitor, is seemingly produced in response to the gonadotrophin peak. These preliminary studies led in due course to isolation of a potential inhibitor from follicular fluid, a substance thought not to be species-specific (Tsafriri, 1978). Purification of an oocyte maturation inhibitor from pig follicular fluid and partial characterisation suggested a peptide of relative molecular mass less than 1000. It is trypsin sensitive but heat stable. Specific details of the characterisation procedures are given by Tsafriri (1985) and Tsafriri & Pomerantz (1986).

More recent work dealing with inhibitors of meiotic resumption in the oocyte concerns adenosine, a purine base, and hypoxanthine, a purine nucleotide. They were identified as low molecular weight substances in follicular fluid that inhibited resumption of meiosis upon isolation and culture of oocyte–cumulus cell complexes (Eppig & Downs, 1988), but they were not the only components of an inhibitory system (Tsafriri, 1988). A factor of overwhelming importance as an inhibitor appears to be the relative level of cAMP in the oocyte. cAMP or indeed other inhibitory molecules may enter the oocytes through gap junctions at the tip of the corona cell processes and function to maintain the oocyte in meiotic arrest. Reduction in the level of cAMP and that of the associated protein kinase provides a milieu for resumption of meiosis (Schultz, 1987; Dekel, 1999), whereas increase of cAMP in an oocyte by experimental means (e.g. dibutyryl cAMP, forskolin or injection of adenylyl cyclase) compromises or prevents resumption of meiosis *in vitro*. PKA mediates the action of cAMP due to its phosphorylation of target cells (Dekel, 1996). Only type I PKA has been detected in the oocyte whereas both types I and II are present in cumulus cells, a fact underlying the concentration sensitivity to cAMP and response to an LH peak (Strömstedt & Byskov, 1999). Protein kinase C may also be contributing to meiotic regulation. Within isolated mouse oocytes, direct activation of protein kinase C suppresses maturation, whereas stimulation within cumulus cells generates a positive milieu leading to resumption of meiosis (Downs, Cottom & Hunzicker-Dunn, 2001).

If the corona cells play a critical rôle in meiotic inhibition, in part by means of their bulbous foot processes in direct contact with the oolemma, then one function of the gonadotrophin surge could be to cause interruption of the gap junctional contacts between the corona cell processes and oocyte – a breakdown of the physiological syncytium (Szöllösi, 1978; Thibault *et al.*, 1987). A stimulated follicle generates increased amounts of cAMP which, in turn, influences the gap junction protein, connexin 43 (Cx43). Cell-to-cell communication is reduced owing to conformational changes in this protein, with the consequence that flow of cAMP into the oocyte is insufficient to maintain meiotic arrest (Dekel, 1999). However, passage of small molecules (< 1 kDa) from follicle cells into the oocyte continues well after resumption of meiosis, suggesting a persistence of at least some of the gap junctions between the corona radiata and oocyte (see Szöllösi, 1993). And there is evidence to suggest that withdrawal of the corona cell processes does not occur in primates and cattle following the gonadotrophin surge. Rather, this key endocrine event apparently serves either to maintain or enhance the stability and cytoskeletal composition of trans-zonal processes in such species (Can, Holmes & Albertini, 1997).

An alternative mode of response to the gonadotrophin surge might involve positive signals from the granulosa cells that have acquired LH receptors and to which LH or hCG will bind. Because the corona cell foot processes make intimate junctional contact with the oocyte membrane, passage of trophic molecules into the oocyte could be envisaged, leading to signal transduction, breakdown of the germinal vesicle and resumption of meiosis. Of course, the two proposed lines of explanation, removal or dissipation of an inhibitor or prompting by a positive signal, are not mutually exclusive: they could coexist or be separated chronologically in a pre-ovulatory follicle.

In fact, there is abundant evidence for a maturation promoting factor or metaphase promoting factor (MPF). Classical work in amphibian oocytes was summarised by Masui & Markert (1971) and Masui (1991), and some of the details concerning MPF in mammals were reviewed by Szöllösi (1993), Whitaker (1996) and Strömstedt & Byskov (1999). The activity first described in amphibian oocytes was found in the cytoplasm before breakdown of the germinal vesicle and prompted resumption of meiosis. Further studies indicated that MPF was a regulator of the cell cycle G_2 to M transition in organisms ranging from yeast to humans. In mammals, MPF is thought to have two protein components, a protein kinase and a cyclin, i.e.

1. a 34 kDa protein catalytic subunit, a homologue of the yeast cell division control gene, cdc_2/cdc_{28};
2. a 45 kDa protein regulatory subunit, cyclin B, a homologue of yeast cdc_{13} product.

Cyclins are synthesised during interphase and are phosphorylated, probably for stabilisation. In mouse oocytes, MPF is thought to be present in excess in the form of a precursor, a pro-MPF, probably activated at the post-translational level by the c-*mos* proto-oncogene protein, a component of cytostatic factor (CSF). Still in mice, c-*mos* mRNA is expressed in growing and fully grown oocytes, but the Mos protein is found only in fully grown oocytes and unfertilised eggs. Female c-*mos* deficient mice have substantially reduced fertility and their oocytes fail to arrest at second meiotic metaphase. The ovaries frequently reveal cysts and teratomas. One conclusion would be that c-*mos* is crucial for the checkpoint at metaphase II arrest (Strömstedt & Byskov, 1999). In other mammals, protein synthesis in the oocyte appears necessary before meiosis can be resumed; RNA synthesis in the cumulus cells seems essential (Galli & Moor, 1991).

The specific activity of MPF depends on the state of phosphorylation of the protein kinase (Fig. VII.8). MPF activation is thought to be related to dephosphorylation of the $p34^{cdc2}$ component on its tyrosine and threonine residues by a phosphatase that, in turn, is produced by an MPF kinase (Cdc25). And, as a kinase (Whitaker, 1996), active MPF itself initiates a cascade of phosphorylations important for (1) breakdown of the germinal vesicle, (2) condensation of chromosomes, (3) polymerisation of spindle microtubules and (4) reorganisation

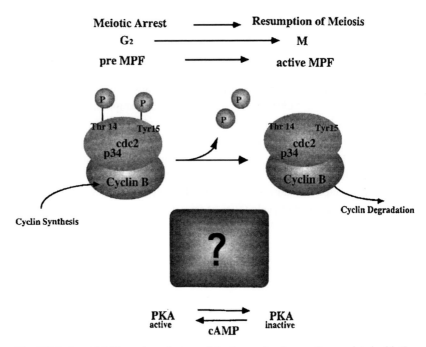

Fig. VII.8. A useful illustration of some of the key molecular events associated with the meiotic cell cycle in mammalian oocytes, a critical point being the state of phosphorylation of the protein kinase. See text for further description. P, phosphate; PKA, protein kinase A; MPF, metaphase promoting factor; cAMP, cyclic AMP. (After Dekel, 1999.)

of the vitellus. MPF activity achieves peak levels at the first and second meiotic metaphases, decreasing at anaphase I and II (Strömstedt & Byskov, 1999). On the basis of recent studies in *Xenopus* oocytes, the interplay between Cdc2 kinase and the c-Mos/mitogen-activated protein (MAP) kinase pathway is complex (Frank-Vaillant *et al.*, 2001).

One line of recent research has focused on a meiosis-activating substance (MAS) or, more specifically, meiosis-inducing sterols (Byskov *et al.*, 1999; Leonardsen *et al.*, 2000). Treatment with FSH, forskolin or dibutyryl cAMP induces mouse cumulus cells to secrete a MAS. This substance is diffusible and, although cumulus–oocyte connections are required for initiating the response, actual transfer from cumulus cells to the oocyte does not require gap junctions. MASs have been purified from human follicular fluid (FF-MAS) and from bull testes; they are different but closely related sterols, and both are intermediaries in the cholesterol biosynthetic pathway. Treatment with gonadotrophic hormones enhances the enzymatic activity leading to FF-MAS. Both sterols will induce resumption of meiosis in hypoxanthine-arrested naked oocytes as well as in

Table VII.4. *Results to illustrate the influence of precocious (gonadotrophin-induced) ovulation upon the maturity of pig oocytes and the incidence of polyspermic penetration in a total of 88 animals*

Day of oestrous cycle when injected with hCG	Proportion of primary oocytes shed (%)	Proportion of eggs polyspermic (%)	Evolution of sperm head[a] (%)
17	81.2	91.7	<10
18	17.4	38.0	–
19	2.1	2.4	–
20	1.4	2.8	>90

[a]Transformation into a male pronucleus.
Adapted from Hunter *et al.*, 1976.

cumulus-intact oocytes (Strömstedt & Byskov, 1999). As of writing, the precise mechanism of action of the sterols is not known, but preliminary results from human IVF clinics suggest a beneficial influence of MAS added to the culture system (Cavilla *et al.*, 2001).

In a context of LH or hCG-induced ovulation during a spontaneous oestrous cycle, there is one particular observation worth highlighting at this stage – that it is possible to dissociate the process of ovulation from resumption of meiosis in the oocyte being shed (Hunter, Cook & Baker, 1976; Baker & Hunter, 1978; Hunter, 1979). This has been emphasised as a dichotomy between follicle and oocyte not only of nuclear maturation but also of cytoplasmic and membranous components of oocyte maturation at a time when oestradiol concentrations in the antral fluid were low. The evidence for failure of nuclear maturation was morphological whereas failure of membranous and cytoplasmic components was indicated, respectively, by polyspermic penetration of the vitellus in surgically inseminated animals and failure of sperm heads to decondense and evolve into pronuclear structures (Table VII.4). The extent of the polyspermic condition in individual primary oocytes was frequently massive (Fig. VII.9), with as many as 60–80 unswollen sperm heads in the egg cytoplasm and still more spermatozoa in the perivitelline space (Hunter, 1967, 1976; Hunter *et al.*, 1976).

Extensive polyspermy in primary oocytes released after reprogramming of oestrous cycles with oral progestagens had previously been noted by Polge & Dziuk (1965), with the suggestion that a potential block to polyspermy develops in step with the acquisition of meiotic competence (Dziuk & Dickmann, 1965). Of course, it is now appreciated that maturation and migration of the cortical granules, organelles underlying a block to polyspermy, are occurring during resumption of meiosis in the pre-ovulatory interval. In order to become

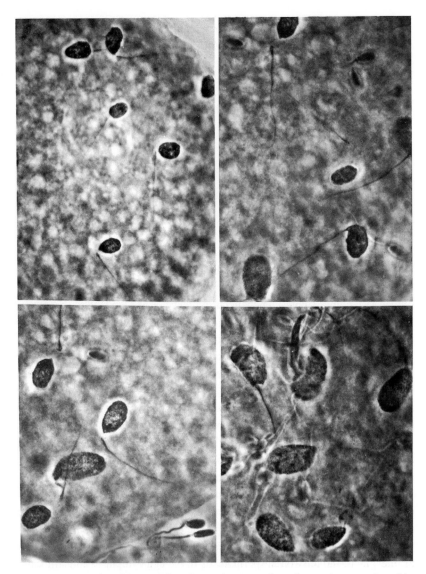

Fig. VII.9. Illustration of the condition of extensive or massive polyspermic penetration of the vitellus. Apart from a modest degree of enlargement, most of the sperm heads have undergone little transition towards pronuclear structures, even though these are not primary oocytes. (After Hunter, 1967.)

functional, cortical granules must have achieved a location just beneath the vitelline membrane, enabling fusion with the plasmalemma and exocytosis of their contents upon activation of the egg by a fertilising spermatozoon (Szöllösi, 1967; Fléchon, 1970; Gulyas, 1980). Similarly, vitelline factors such as MPGF

RESPONSE TO A SINGLE INJECTION OF GONADOTROPHIN
IN EARLY OR LATE FOLLICULAR PHASE

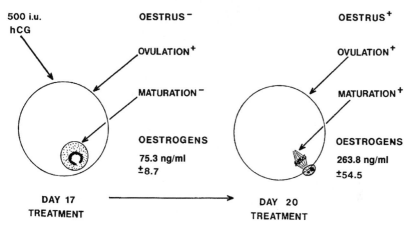

Fig. VII.10. Diagrammatic representation of the condition of oocytes shed into the Fallopian tubes of sexually mature pigs when ovulation is induced in the early follicular phase or at the onset of oestrus with a single injection of 500 i.u. human chorionic gonadotrophin (hCG). Minus signs indicate absence of an event; plus signs indicate its occurrence. (Adapted from Hunter *et al.*, 1976.)

that promote decondensation of the sperm head chromatin in spontaneously ovulated oocytes are maturing or being made available during resumption of meiosis (Thibault & Gérard, 1970, 1973).

Classical studies demonstrated receptors for LH on porcine granulosa cells, and also that hCG can bind to such granulosa cells (Schomberg & Tyrey, 1972; Channing & Kammerman, 1973). Accordingly, a proposition concerning the dichotomy discussed above between follicle and oocyte was that the increasing titres of intra-follicular oestradiol recorded during the latter part of the oestrous cycle mediate in the response of the oocyte to the gonadotrophin stimulus. This would be in part, at least, by increasing the binding sites for LH or hCG on the granulosa cells (Hunter *et al.*, 1976); these binding sites have been reported to be similar (Kammerman *et al.*, 1972; Lee, 1976). In both laboratory and farm animals, the proposed involvement of oestradiol within the ripening follicle was also seen in a broader biological context, since oocyte maturation would normally be occurring only in circumstances in which the animal was exhibiting behavioural oestrus and in which the reproductive tract was tonic and oedematous (Fig. VII.10). In other words, a gonadal steroid hormone would coordinate changes during the latter part of the follicular phase within the follicle, the reproductive tract and the hypothalamo-pituitary axis. After the LH

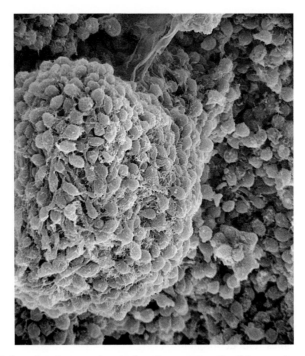

Fig. VII.11. Scanning electron micrograph to demonstrate cells of the cumulus oocyte complex. Expansion and mucification of the cellular mass become prominent during the pre-ovulatory interval as a response to the gonadotrophin surge in follicles selected for ovulation. (Adapted from Fléchon *et al.*, 1986b.)

surge, the availability of granulosa cell aromatase decreases rapidly, resulting in a shift from oestradiol to progesterone dominance during the phase when resumption of meiosis is becoming apparent. Indeed, progesterone titres within the follicular fluid may achieve values of 6400 ng/ml in domestic farm animals (Hunter *et al.*, 1976). The rôle of steroid signals in the maturation of oocytes was given some prominence by Osborn & Moor (1983).

Expansion and mucification of cumulus oophorus

The mass of cells arranged tightly around the follicular oocyte constitutes, with its enclosure, the cumulus–oocyte complex (Fig. VII.11). Those cells situated closest to the oocyte must be capable of responding to locally secreted molecules, even though a primary response of the cumulus oophorus is to systemic programming by a surge of pituitary hormones, in this instance especially FSH. The evidence here is that FSH will stimulate expansion of the cumulus cell mass *in vitro* when isolated from mice, rats and pigs (Dekel & Phillips, 1979; Dekel, Hillensjo & Kraicer, 1979; Eppig, 1980a). Specific expansion and loosening of the cumulus cell investment following the pre-ovulatory gonadotrophin surge have been referred to above (and see Bomsel-Helmreich, 1985), but the characteristic mucification of the intercellular spaces requires further comment.

Mucification of the cumulus mass within pre-ovulatory follicles is represented by deposition of an extracellular matrix that is sensitive to the action of hyaluronidase (Schuetz & Schwartz, 1979; Eppig, 1980b), being rich in hyaluronic acid. Hyaluronic acid is the major structural macromolecule of the matrix (Eppig, 1979). This large polyanionic polymer is part of the GAG family of molecules, is hydrophilic and thus able to attract water and expand as a coil – hence pre-ovulatory enlargement of the cumulus mass seen at mucification and an increase in volume of the follicular antrum. In fact, cumulus expansion is a good example of cellular movement and rearrangement on the extracellular matrix, the latter acting as a substrate (Sutovsky *et al.*, 1994; Sutovsky & Fléchon, 1997). A thin network of fibrils can be detected in the expanded cumulus cell mass, and this network extends into the follicular fluid (Salustri *et al.*, 1992). The cytoskeleton, especially filamentous actin and actin-associated proteins, enables remodelling, and is considered the principal force underlying gonadotrophin-induced expansion of the cumulus oophorus (Sutovsky & Fléchon, 1997). Molecular interactions within the matrix facilitate deformation but not disaggregation by shearing forces, a flexibility of significance during displacement of the cumulus cell mass through an aperture in the follicle wall at ovulation (Chen, Russell & Larsen, 1993).

Deposition of mucus between cells of the cumulus oophorus has been examined by scanning electron microscopy, with timescales available for morphological changes relative to ovulation in several species (Dekel *et al.*, 1978; Dekel & Phillips, 1979; Fléchon *et al.*, 1986a). Although an extracellular matrix appears to accumulate gradually during the pre-ovulatory interval, complete coverage of cumulus cells with a hyaluronidase-sensitive material is achieved only several hours before ovulation. Cumulus expansion in the rat progresses characteristically from the accumulation of secreted glycosaminoglycans in the peripheral portion of the cumulus mass and gradually involves the more central cells (Dekel *et al.*, 1981). Eventually, disruption of the cumulus attachment to the oocyte occurs, with cell–cell communication between cumulus-granulosa cells slowly distintegrating. In a study of mucification in pig granulosa cells (Fig. VII.11), the intercellular matrix was first observed largely on the stalk between the parietal granulosa and the cumulus oophorus, but it soon extended to the deeper layers of cumulus cells (Fléchon *et al.*, 1986b), as also noted in mice (Sato, Inoue & Toyoda, 1993). Mucification of the corona radiata occurred shortly before ovulation in pig oocytes when coupling between the oocyte and surrounding cell layer was minimal (Fléchon *et al.*, 1986b; Motlik, Fulka & Fléchon, 1986).

Expansion of the cumulus oophorus is associated with the interruption of gap junctions and a reduction in the physical integrity of the cumulus–oocyte complex (Anderson & Albertini, 1976), expressed as a decreased ionic coupling and metabolic cooperation between cumulus cells and the oocyte (Moor *et al.*, 1981; Sato, 1997). The sharpest decline in intercellular coupling between the cumulus oophorus and oocyte was correlated with modification of the corona radiata (Moor *et al.*, 1981; Eppig, 1982; Motlik *et al.*, 1986). The gonadotrophin-induced reorganisation of the cumulus cell cytoskeleton almost certainly results in altered metabolic and signalling pathways (Sutovsky & Fléchon, 1997).

Expansion of mouse cumulus complexes *in vitro* may be promoted by a serum factor that stabilises hyaluronic acid in the cumulus matrix (Eppig, 1980b). However, accumulation and organisation of hyaluronan molecules depends largely on interaction between inter-α-trypsin inhibitor, a factor present in serum, and macromolecules synthesised by the cumulus cells that bind tightly to hyaluronan (for a review, see Camaioni *et al.*, 1997). Proteoglycans and proteins that will bind specifically to hyaluronan are found in the extracellular matrix of various tissues (Knudson & Knudson, 1993). *In vitro* studies have suggested that three different types of proteoglycan can be synthesised by mouse cumulus cells, but only a dermatan sulphate proteoglycan had the same matrix distribution as hyaluronan (Camaioni *et al.*, 1997); it is primarily accumulated in the matrix in the presence of serum. Retention of hyaluronan in the intercellular spaces appears essential for maintenance of structural integrity in the expanding cumulus. The serum factor required for successful matrix formation may cooperate with structural proteins to establish cross-links between the hyaluronan strands. Cells present in inflammatory sites such as the pre-ovulatory follicle (see Chapter VIII) synthesise a glycoprotein, TNF-stimulated gene *6*, which binds specifically to hyaluronan (Lee *et al.*, 1993) and forms a stable complex with inter-α-trypsin inhibitor, enhancing its protease inhibitory action (Wisniewski *et al.*, 1996).

Addition of EGF or cAMP to the culture medium may also facilitate expansion of mouse cumulus complexes during *in vitro* maturation of oocytes (Downs, 1989; Downs & Hunzicker-Dunn, 1995). Mouse oocytes are known to secrete a cumulus expansion-enabling factor when maintained under suitable *in vitro* conditions, a factor with a molecular mass of between 30–100 kDa (Eppig *et al.*, 1993a). In a separate report, the mouse factor was found to be: (1) retained by 100 kDa but not by 300 kDa membranes, (2) heat labile (65 °C for 15 s), and

(3) lost by proteinase K digestion. The suggestion was thus that the factor is a protein (Eppig, Wigglesworth & Chesnel, 1993b). In experiments involving physical separation of the oocyte from its somatic cells, it was revealed that the oocyte secretes a soluble factor necessary for synthesis of hyaluronic acid by the cumulus cells (Buccione *et al.*, 1990; Salustri, Yanagishita & Hascall, 1990). Such physical removal of the oocyte also inhibits FSH-induced expansion of the cumulus *in vitro*, but conditioning of the culture medium with denuded oocytes will overcome the inhibition. This influence of the oocyte is developmentally regulated during meiotic maturation: in the case of mouse oocytes, the ability to secrete a cumulus expansion-enabling factor is acquired by the time the cell becomes competent to undergo germinal vesicle breakdown, but such ability is lost after fertilisation (Vanderhyden *et al.*, 1990). As to the nature of cumulus expansion-enabling factor in the mouse, one proposed candidate is GDF-9, a paracrine factor that seemingly can regulate both synthesis and retention of hyaluronic acid in the cell matrix (Elvin *et al.*, 1999).

As an overall perspective, it seems reasonable to conclude that an oocyte embarking on pre-ovulatory resumption of meiosis would itself be a significant source of molecules acting to influence and disperse its surrounding somatic cell investment – a theme that emerges from the reports of Buccione *et al.* (1990), Salustri *et al.* (1990), Vanderhyden, Telfer & Eppig (1992) and Vanderhyden (1993). However, the impressive results with mouse oocytes were not endorsed in comparable *in vitro* studies with pig and cow oocytes, in which expansion of the cumulus cell mass appeared not to depend on factors synthesised in the oocyte (Procházka *et al.*, 1991, 1998; Ralph, Telfer & Wilmut, 1995). A similar qualification applies to synthesis of hyaluronic acid by pig cumulus cells *in vitro* in response to FSH (Nagyová, Procházka & Vanderhyden, 1999), even though pig and cow oocytes produce a cumulus expansion-enabling factor that will promote FSH-stimulated expansion of oocyte-free mouse cumulus complexes (Singh, Zhang & Armstrong, 1993; Vanderhyden, 1993; Ralph *et al.*, 1995). Once again, the factor(s) is developmentally regulated and the stage of its production in pig oocytes differs from that in mice (Nagyová, Vanderhyden & Procházka, 2000). A reservation here would be one of interpretation: what can be demonstrated *in vitro* does not necessarily correspond precisely with the physiological situation, and because a component or molecule appears not to be essential *in vitro* does not mean that it is not contributing to events *in vivo*. Oocyte-secreted factors are known to influence functional differences between bovine mural granulosa cells and cumulus cells (Li *et al.*, 2000).

Although typically round in appearance, cumulus cells become polarised during the phase of expansion. They then form an anastomosing network interconnected by large cytoplasmic projections – projections filled with bundles of microfilaments, microtubules and intermediate filaments (Sutovsky & Fléchon, 1997). Proteins, glycosaminoglycans and proteoglycans are all implicated in such cell spreading. Whereas the basal lamina serves initially to support and orientate granulosa cells, significant changes in this membrane would be anticipated during mucification of the cumulus oophorus and final pre-ovulatory remodelling of a Graafian follicle, not least its lateral expansion. Nonetheless, one key function of the membrane at this mature stage may be to help to maintain large molecules such as proteoglycans in the follicular fluid (Rodgers *et al.*, 1998).

Extracellular matrix of follicle

Due to the constantly changing nature of its follicle population, the ovary represents a vigorously dynamic structure and the necessary remodelling is facilitated by the extracellular matrix (for a review, see McIntush & Smith, 1998). In particular, remodelling of this matrix underlies growth of a Graafian follicle and the major changes at ovulation or those involved in the process of atresia. Growth of an antral follicle is represented on the one hand by breakdown of the extracellular matrix and, on the other, by its reconstruction. Such breakdown and reconstruction lead to a changing composition of the extracellular matrix. Major components of the matrix include proteins and proteoglycans, and have been reviewed in depth by Luck (1994). In its diverse forms, collagen seemingly constitutes the most abundant protein of the extracellular matrix in the ovary. Other prominent proteins include fibronectin, laminin, keratin and connexin. Two members of the connexin family, connexin 37 (Cx37) and (Cx43), are critical for normal folliculogenesis (Nuttinck *et al.*, 2000). This can be highlighted in Cx37-deficient mice, which show profound defects in follicular growth to the extent that the Graafian stage is never achieved (Simon *et al.*, 1997).

In terms of follicular components, it should be recalled from Chapter II that the basement membrane is a specialised layer or sheet of the extracellular matrix, hence the name basal lamina. Major changes would be anticipated in the basal lamina during the remodelling of a Graafian follicle shortly before ovulation, as just mentioned in the previous section. Because the basal lamina supports and orientates cells of the granulosa, substantial changes would also be expected in the extracellular matrix of this epithelial layer, even though the matrix is relatively scant. The prominent extracellular matrix of the theca interna is likewise involved in a major phase of pre-ovulatory remodelling.

The extent and architecture of the extracellular matrix are influenced by members of the TGF-β family. These molecules can promote the deposition of proteins involved in cell adhesion, seen notably in the laying down of fibronectin, and they may also stimulate expression of integrin receptors in diverse types of ovarian cell. In one specific sense, TGF-βs can function to inhibit degradation of matrix proteins. By contrast, enzymatic degradation of the proteinaceous components of the extracellular matrix is undertaken by matrix metalloproteases (MMPs). Regulation of the activity of these proteases is achieved in several ways, not least by molecules termed tissue inhibitors of metalloproteases (TIMPs). There are four confirmed members of this inhibitor family. TIMP-1 and -2 are soluble in the extracellular matrix, TIMP-3 is bound to the matrix, whereas the state of TIMP-4 requires clarification. TIMP-1 is a 28 kDa polypeptide widely synthesised and secreted in the female gonad and duct system (Buhi *et al.*, 1997). Suggested involvements for TIMP-1 have included those of acting as a growth factor or regulation of growth factor bioavailability or stimulation of steroidogenesis (for a review, see Shores & Hunter, 2000). As of writing, physiological functions for TIMP-2 (21 kDa) and TIMP-3 (24 kDa) are uncertain.

Growth of the follicle and remodelling of the basement membrane depend on coordination in the activity of MMPs and TIMPs. It follows that the balance between extracellular matrix breakdown and construction would reflect the relative activities of MMPs and TIMPs. The earlier literature bearing on such a balance revealed uncertainty as to the follicular cells producing TIMPs, although granulosa cells were favoured in species such as rats, sheep, pigs and human (for a review, see Shores & Hunter, 2000). Shores & Hunter (2000) emphasised, however, that the potential of theca cells to generate TIMPs had not been specifically examined until their own study – which also examined the response of theca cells to treatment with TIMP *in vitro*. Working with pig ovaries, Shores & Hunter demonstrated the presence of TIMPs in granulosa cells, theca cells and in the follicular fluid of follicles ≥5 mm in diameter both before and after the pre-ovulatory gonadotrophin surge. In fact, TIMP concentrations were higher in the theca layer than in granulosa cells or follicular fluid, which would be in line with an anticipated greater MMP activity to degrade the substantial extracellular matrix of the theca (Shores & Hunter, 2000). As to given members of the TIMP family, TIMP-1, -2 and -3 were produced by both large and small pig follicles.

Proteolytic degradation of the follicle cell layers by MMPs is required for shedding of the oocyte. TIMPs are involved in the control of this process and prevent potentially wayward MMPs from degrading tissues elsewhere in the ovary. As a separate mode of involvement in the events leading up to ovulation, the extracellular matrix may regulate the activity of components of the IGF system and likewise the activity of other growth factors (for a review, see Armstrong & Webb, 1997). In other words, a specific interplay would seem to be occurring between diverse growth factors and a dynamic extracellular matrix. Such an interplay can be deduced from the report of Ingman, Owens &

Armstrong (2000), in which growth factor binding proteins are influenced by an oocyte in studies of the bovine extracellular matrix.

Consideration must also be given to fluid components of the extracellular matrix. The majority of granulosa cells in a Graafian follicle make contact neither with the basement membrane nor, as corona radiata cells, with the zona pellucida. Rather, they are exposed to, or suspended in, a matrix of follicular fluid. Characteristically, this forms a reservoir of hyaluronidase-digestible mucopolysaccharide material whose quantity and composition vary with development of a follicle. Nonetheless, there is heavy sulphation. Incubation evidence derived from ^{35}S tracer studies to examine GAGs of the granulosa and theca of antral follicles revealed a 60% association with heparan sulphate, 25% with dermatan sulphate and 15% in chondroitin sulphates. Further details were reviewed by Luck (1994) and Camaioni *et al.* (1997). Changes in proteoglycan composition can be detected during folliculogenesis, with an overall reduction in content as the follicle grows, although there is an increase in the ratio of chondroitin sulphate to hyaluronic acid. The data suggest that modified proteoglycan production during follicular maturation is under close gonadotrophic regulation, and that proteoglycan molecules contribute strongly to the relatively high viscosity of follicular fluid (Luck, 1994).

Remodelling of basement membrane

Subtle differences between the terms basement membrane and basal lamina have not been insisted upon in this monograph. They have been discussed elsewhere (Luck, 1994), and in essence concern the presence of collagen type IV, proteoglycans and laminin in the extracellular membrane correctly termed a basal lamina. In some contrast, a basement membrane is seen as a less-defined or more complex structure, frequently incorporating an additional collagenous matrix and associated cell layer. Rodgers *et al.* (1998) offered a stringent interpretation: a basal lamina is a specialised sheet of extracellular matrix that separates epithelial cell layers (e.g. stratum granulosum) from underlying mesenchyme in organs throughout the body. The basement membrane and basal lamina influence epithelial cell migration, proliferation and differentiation, and can selectively retard the passage of molecules from one side of a basal lamina to the other. There is thus a regulatory influence on the formation of antral fluid.

Although the basal lamina changes in composition during follicular development and is being continually remodelled, it nonetheless represents a lattice-type development of collagen IV closely intertwined with a network of laminin. Binding of entactin to the collagen and laminin confers stability. Fibronectin, heparan sulphate, proteoglycans and other molecules are associated with the collagen IV–laminin backbone (Rodgers *et al.*, 1998). During development of a follicle, synthesis and organisation of membrane must be proceeding at a high level. As frequently noted, the surface area of the membrane will increase as the square of the follicular radius. Moreover, in some species, the latter dimension may double on almost a daily basis as ovulation approaches (Gosden *et al.*, 1988). van Wezel & Rodgers (1996) have estimated that the surface area of a bovine follicle doubles nineteen times during follicular development, emphasising the requisite remodelling of the basal lamina.

At a molecular level, such remodelling may be facilitated by the existence of multiple forms of collagen, laminin and fibronectin. Each collagen IV molecule is composed of three α chains, although six different types of α chain have so far been discovered (Rodgers *et al.*, 1998). Similarly, each laminin molecule is composed of one α, one β and one γ chain, yet five different α chains, three β chains and two γ chains have been discovered. And, as to fibronectin, at least twenty different isoforms have been revealed. There is thus enormous scope for molecular modifications to the composition of the basal lamina. The nature of the control or programming mechanism for such modifications is far from being understood, but it would be prudent to anticipate important inputs from the oocyte itself. Diverse studies have indicated that granulosa cells are able to secrete many of the components of a basal lamina, but theca cells may also make a contribution in developing follicles (Rodgers *et al.*, 1998).

As a consequence of the lateral expansion and then degradation of the basal lamina preceding ovulation, there is invasion of the thecal capillary bed followed by massive angiogenesis amongst the granulosa cells in concert with clear evidence of luteinisation or incipient luteinisation (for a review, see Redmer & Reynolds, 1996). Pre-ovulatory follicular fluid is especially rich in angiogenic factors (e.g. VEGF). The oocyte and its somatic cell investment are no longer isolated from direct contact with the vascular system. In addition to the colonisation by capillaries and nerve cell processes, there is a gradual ingress of theca cells themselves whereas granulosa cells project in the opposing direction as dissolution of the basal lamina proceeds.

Concluding remarks

The contents of this chapter have centred on ovarian events set in train by the pre-ovulatory surge of gonadotrophic hormones that paves the way to ovulation. Events underlying the gonadotrophin outpouring could be traced back to episodes of neurosecretion in the basal region of the hypothalamus, with the pulsatile form of secretion receiving appropriate emphasis. However, there remains an enigma in terms of the so-called pulse generator, with the consideration that such a putative regulator may not be a specifically identifiable or discrete group of cells. Rather, it may represent the integrating functions of various groups of cells expressed first as neuronal (electrical) activity and then transduced as neuroendocrine secretion (GnRH). Whether the form of the surge is modified by a unique GnSAF has not been finally resolved, and awaits molecular analysis at critically timed intervals.

There has been significant progress in describing responses of the oocyte investment within pre-ovulatory follicles to the surge of gonadotrophic hormones, techniques involving scanning electron microscopy and fluorescent imaging of specific proteins. Although modifications in the cumulus oophorus are clearly an important preliminary to actual ovulation, they also contribute to the post-ovulatory destiny of an oocyte within the lumen of the Fallopian tube (see Chapters IX and X). There has been conspicuous progress in clarifying factors underlying resumption of meiosis in the oocyte itself, focusing especially on biochemical events. Modulation in the level of cAMP inputs from the surrounding corona cells is one of the key steps in generating an intra-oocyte cAMP level compatible with resumption of meiosis, but an involvement of MASs was also examined. The interplay of a spectrum of factors, both positive and negative, will doubtless be found to regulate the resumption of meiosis in pre-ovulatory follicles.

Finally, various structural changes in the extracellular matrix and particularly in the basal lamina were considered, with the outstanding phase of re-modelling required in these elements and in specific regions of the follicle wall (Chapter VIII) as ovulation approaches, involving not only shedding of an oocyte but also of a substantial population of follicular cells. A contribution of these somatic cells to physiological events within the Fallopian tube(s) is examined in Chapter IX.

References

Anderson, E. & Albertini, D.F. (1976). Gap junctions between the oocyte and companion follicle cells in the mammalian ovary. *Journal of Cell Biology*, **71**, 680–6.

Armstrong, D.G. & Webb, R. (1997). Ovarian follicular dominance: the role of intra-ovarian growth factors and novel proteins. *Reviews of Reproduction*, **2**, 139–46.

Baird, D.T. & McNeilly, A.S. (1981). Gonadotrophic control of follicular development and function during the oestrous cycle of the ewe. *Journal of Reproduction and Fertility, Supplement* **30**, 119–33.

Baker, T.G. & Hunter, R.H.F. (1978). Interrelationships between the oocyte and somatic cells within the Graafian follicle of mammals. *Annales de Biologie Animale, Biochimie, Biophysique*, **18**, 419–26.

Bomsel-Helmreich, O. (1985). Ultrasound and the preovulatory human follicle. *Oxford Reviews of Reproductive Biology*, **7**, 1–72.

Brambell, F.W.R. (1956). Development of the ovary and oogenesis. In *Marshall's Physiology of Reproduction*, 3rd edn, vol. 1, part 1, ed. A.S. Parkes, pp. 397–467. London: Longmans, Green & Co.

Buccione, R., Vanderhyden, B.C., Caron, P.J. & Eppig, J.J. (1990). FSH-induced expansion of the mouse cumulus oophorus *in vitro* is dependent upon a specific factor(s) secreted by the oocyte. *Developmental Biology*, **138**, 16–25.

Buhi, W.C., Alvarez, I.M., Pickard, A.R., *et al.* (1997). Expression of tissue inhibitor of metalloproteinase-1 protein and messenger ribonucleic acid by the oviduct of cyclic, early pregnant and ovariectomised steroid-treated gilts. *Biology of Reproduction*, **57**, 7–15.

Busbridge, N.J., Buckley, D.M., Cornish, M. & Whitehead, S.A. (1988). Effects of ovarian hyperstimulation and isolated pre-ovulatory follicles on LH responses to GnRH in rats. *Journal of Reproduction and Fertility*, **82**, 329–36.

Busbridge, N.J., Chamberlain, G.V.P., Griffiths, A. & Whitehead, S.A. (1990). Non-steroidal follicular factors attenuate the self-priming action of gonadotrophin-releasing hormone on the pituitary gonadotroph. *Neuroendocrinology*, **51**, 493–9.

Byskov, A.G., Andersen, C.Y., Leonardsen, L. & Baltsen, L. (1999). Meiosis activating sterols (MAS) and fertility in mammals and man. *Journal of Experimental Zoology*, **285**, 237–42.

Camaioni, A., Hascall, V.C., Yanagishita, M., D'Alessandris, C. & Salustri, A. (1997). Dynamics of the extracellular matrix of the cumulus oophorus. In *Microscopy of Reproduction and Development: A Dynamic Approach*, ed. P.M. Motta, pp. 137–41. Rome: Antonio Delfino Editore.

Can, A., Holmes, R.M. & Albertini, D.F. (1997). Analysis of the mammalian ovary by confocal microscopy. In *Microscopy of Reproduction and Development: A Dynamic Approach*, ed. P.M. Motta, pp. 101–8. Rome: Antonio Delfino Editore.

Cavilla, J.L., Kennedy, C.R., Baltsen, M., Klentzeris, L.D., Byskov, A.G. & Hartshorne, G.M. (2001). The effects of meiosis activating sterol on in-vitro maturation and fertilisation of human oocytes from stimulated and unstimulated ovaries. *Human Reproduction*, **16**, 547–55.

Chang, M.C. (1955). The maturation of rabbit oocytes in culture and their maturation, activation, fertilisation and subsequent development in the Fallopian tubes. *Journal of Experimental Zoology*, **128**, 379–405.

Channing, C.P. & Kammerman, S. (1973). Characteristics of gonadotropin receptors of porcine granulosa cells during follicle maturation. *Endocrinology*, **92**, 531–40.

Chen, L., Russell, P.T. & Larsen, W.J. (1993). Functional significance of cumulus expansion in the mouse: roles for the preovulatory synthesis of hyaluronic acid within the cumulus mass. *Molecular Reproduction and Development*, **34**, 87–93.

Clarke, I.J. (1993). Variable patterns of gonadotropin-releasing hormone secretion during the estrogen-induced luteinising hormone surge in ovariectomised ewes. *Endocrinology*, **133**, 1624–32.

Clarke, I.J. (1996). The hypothalamo-pituitary axis. In *Scientific Essentials of Reproductive Medicine*, ed. S.G. Hillier, H.C. Kitchener & J.P. Neilson, pp. 120–32. London & Philadelphia: W.B. Saunders Company Ltd.

Crowley, W.F., Filicori, M., Spratt, D.I. & Santoro, N.F. (1985). The physiology of gonadotropin-releasing hormone (GnRH) secretion in men and women. *Recent Progress in Hormone Research*, **41**, 473–531.

Daen, F.P., Sato, E., Naito, K. & Toyoda, Y. (1994). The effect of pig follicular fluid fractions on cumulus expansion and male pronucleus formation in porcine oocytes matured and fertilised *in vitro*. *Journal of Reproduction and Fertility*, **101**, 667–73.

Danforth, D.R. & Cheng, C.Y. (1995). Purification of a candidate gonadotropin surge inhibiting factor from porcine follicular fluid. *Endocrinology*, **136**, 1658–65.

Danforth, D.R., Sinosich, M.J., Anderson, T.L., *et al.* (1987). Identification of gonadotropin surge-inhibiting factor (GnSIF) in follicular fluid and its differentiation from inhibin. *Biology of Reproduction*, **37**, 1075–82.

Dekel, N. (1996). Protein phosphorylation/dephosphorylation in the meiotic cell cycle of mammalian oocytes. *Reviews of Reproduction*, **1**, 82–8.

Dekel, N. (1999). Meiotic cell cycle, oocytes. In *Encyclopedia of Reproduction*, vol. 3, ed. E. Knobil & J.D. Neill, pp. 168–76. San Diego & London: Academic Press.

Dekel, N., Hillensjo, Y. & Kraicer, P.F. (1979). Maturation effects of gonadotropins on the cumulus–oocyte complex of the rat. *Biology of Reproduction*, **20**, 191–7.

Dekel, N., Kraicer, P.F., Phillips, D.M., Sanchez, R.S. & Segal, S.J. (1978). Cellular association in the rat oocyte–cumulus cell complex: morphology and ovulatory changes. *Gamete Research*, **1**, 47–57.

Dekel, N., Lawrence, T.S., Gilula, N.B. & Beers, W.H. (1981). Modulation of cell-to-cell communication in the cumulus–oocyte complex and the regulation of oocyte maturation by LH. *Developmental Biology*, **80**, 356–62.

Dekel, N. & Phillips, D.M. (1979). Maturation of the rat cumulus oophorus: a scanning electron microscopic study. *Biology of Reproduction*, **21**, 9–18.

Downs, S.M. (1989). Specificity of epidermal growth factor action on maturation of the murine oocyte and cumulus oophorus *in vitro*. *Biology of Reproduction*, **41**, 371–9.

Downs, S.M., Cottom, J. & Hunzicker-Dunn, M. (2001). Protein kinase C and meiotic regulation in isolated mouse oocytes. *Molecular Reproduction and Development*, **58**, 101–15.

Downs, S.M. & Hunzicker-Dunn, M. (1995). Differential regulation of oocyte maturation and cumulus expansion in the mouse oocyte–cumulus cell complex by site-selective analogues of cyclic adenosine monophosphate. *Developmental Biology*, **172**, 72–85.

Dziuk, P.J. (1965). Timing of maturation and fertilisation of the sheep egg. *Anatomical Record*, **153**, 211–24.

Dziuk, P. & Dickmann, Z. (1965). Failure of the zona reaction in five pig eggs. *Nature (London)*, **208**, 502–3.

Edwards, R.G. (1962). Meiosis in ovarian oocytes of adult mammals. *Nature (London)*, **196**, 446–50.

Edwards, R.G. (1965). Maturation *in vitro* of mouse, sheep, cow, pig, rhesus monkey and human ovarian oocytes. *Nature (London)*, **208**, 349–51.

Edwards, R.G. & Gates, A.H. (1959). Timing of the stages of the maturation divisions, ovulation, fertilisation and the first cleavage of eggs of adult mice treated with gonadotrophins. *Journal of Endocrinology*, **18**, 292–304.

Elvin, J.A., Clark, A.T., Wang, P., Wolfman, N.M. & Matzuk, M.M. (1999). Paracrine actions of growth differentiation factor-9 in the mammalian ovary. *Molecular Endocrinology*, **13**, 1035–48.

Eppig, J.J. (1979). FSH stimulates hyaluronic acid synthesis by oocyte cumulus cell complex from mouse preovulatory follicles. *Nature (London)*, **281**, 483–4.

Eppig, J.J. (1980a). Role of serum in FSH stimulated expansion by mouse oocyte–cumulus cell complexes *in vitro*. *Biology of Reproduction*, **22**, 629–33.

Eppig, J.J. (1980b). Regulation of cumulus oophorus expansion by gonadotrophins *in vivo* and *in vitro*. *Biology of Reproduction*, **23**, 545–52.

Eppig, J.J. (1982). The relationship between cumulus cell–oocyte coupling, oocyte meiotic maturation, and cumulus expansion. *Developmental Biology*, **89**, 268–72.

Eppig, J.J. & Downs, S.M. (1988). The role of purines in the maintenance of meiotic arrest in mammalian oocytes. In *Meiotic Inhibition: Molecular Control of Meiosis*, ed. F.P. Haseltine & N.L. First, pp. 103–13. New York: Alan L. Liss Inc.

Eppig, J.J., Peters, A.H.F.M., Telfer, E.E. & Wigglesworth, K. (1993a). Production of cumulus expansion enabling factor by mouse oocytes grown *in vitro*: preliminary characterisation of the factor. *Molecular Reproduction and Development*, **34**, 450–6.

Eppig, J.J., Wigglesworth, K. & Chesnel, F. (1993b). Secretion of cumulus expansion-enabling factor by mouse oocytes: relationship to oocyte growth and competence to resume meiosis. *Developmental Biology*, **158**, 400–9.

Evans, N.P., Dahl, G.E., Glover, B.H. & Karsch, F.J. (1994). Central regulation of pulsatile gonadotropin releasing hormone (GnRH) secretion by estradiol during the period leading up to the preovulatory GnRH surge in the ewe. *Endocrinology*, **134**, 1806–11.

Farstad, W., Mondain-Monval, M., Hyttel, P., Smith, A.J. & Markeng, D. (1989). Periovulatory endocrinology and oocyte maturation in unmated mature blue fox vixens (*Alopex lagopus*). *Acta Veterinaria Scandinavica*, **30**, 313–19.

Fléchon, J.E. (1970). Nature glycoprotéique des granules corticaux de l'oeuf de lapine. *Journal de Microscopie*, **9**, 221–42.

Fléchon, J.E., Kopecny, V., Motlik, J. & Pavlok, A. (1986a). Origin, structure and texture of the zona pellucida of mammalian eggs. *Histochemical Journal*, **18**, 138 (Abstract).

Fléchon, J.E., Motlik, J., Hunter, R.H.F., Fléchon, B., Pivko, J. & Fulka, J. (1986b). Cumulus oophorus mucification during resumption of meiosis in the pig. A scanning electron microscope study. *Reproduction, Nutrition et Développement*, **26**, 989–98.

Fowler, P.A., Messinis, I.E. & Templeton, A.A. (1990). Inhibition of LHRH-induced LH and FSH release by gonadotrophin surge-attenuating factor (GnSAF) from human follicular fluid. *Journal of Reproduction and Fertility*, **90**, 587–94.

Fowler, P.A. & Price, C.A. (1997). Follicle-stimulating hormone stimulates circulating gonadotropin surge-attenuating/inhibiting factor bioactivity in cows. *Biology of Reproduction*, **57**, 278–85.

Fowler, P.A. & Templeton, A. (1996). The nature and function of putative gonadotropin surge-attenuating/inhibiting factor (GnSAF/IF). *Endocrine Reviews*, **17**, 103–20.

Frank-Vaillant, M., Haccard, O., Ozon, R. & Jessus, C. (2001). Interplay between Cdc2 kinase and the c-Mos/MAPK pathway between metaphase I and metaphase II in *Xenopus* oocytes. *Developmental Biology*, **231**, 279–88.

Galli, C. & Moor, R.M. (1991). Somatic cells and the G2 to M-phase transition in sheep oocytes. *Reproduction, Nutrition et Développement*, **31**, 127–34.

Gosden, R.G., Hunter, R.H.F., Telfer, E., Torrance, C. & Brown, N. (1988). Physiological factors underlying the formation of ovarian follicular fluid. *Journal of Reproduction and Fertility*, **82**, 813–25.

Gulyas, B. (1980). Cortical granules in mammalian oocytes. *International Review of Cytology*, **63**, 357–92.

Harper, M.J.K. (1994). Gamete and zygote transport. In *The Physiology of Reproduction*, vol. 1, ed. E. Knobil & J.D. Neill, pp. 123–87. New York: Raven Press.

Hunter, R.H.F. (1967). Polyspermic fertilisation in pigs during the luteal phase of the estrous cycle. *Journal of Experimental Zoology*, **165**, 451–60.

Hunter, R.H.F. (1976). Sperm–egg interactions in the pig: monospermy, extensive polyspermy and the formation of chromatin aggregates. *Journal of Anatomy*, **122**, 43–59.

Hunter, R.H.F. (1979). Ovarian follicular responsiveness and oocyte quality after gonadotrophic stimulation of mature pigs. *Annales de Biologie animale, Biochimie, Biophysique*, **19**, 1511–20.

Hunter, R.H.F. (1980). *Physiology and Technology of Reproduction in Female Domestic Animals*. London & New York: Academic Press.

Hunter, R.H.F., Cook, B. & Baker, T.G. (1976). Dissociation of response to injected gonadotrophin between the Graafian follicle and oocyte in pigs. *Nature (London)*, **260**, 156–8.

Hunter, R.H.F., Lawson, R.A.S. & Rowson, L.E.A. (1972). Maturation, transplantation and fertilisation of ovarian oocytes in cattle. *Journal of Reproduction and Fertility*, **30**, 325–8.

Hunter, R.H.F. & Polge, C. (1966). Maturation of follicular oocytes in the pig after the injection of human chorionic gonadotrophin. *Journal of Reproduction and Fertility*, **12**, 525–31.

Hyttel, P., Fair, T., Callesen, H. & Greve, T. (1997). Oocyte growth, capacitation and final maturation in cattle. *Theriogenology*, **47**, 23–32.

Hyttel, P., Viuff, D., Avery, B., Laurincik, J. & Greve, T. (1996). Transcription and cell-cycle dependent development of intranuclear bodies and granules in two-cell bovine embryos. *Journal of Reproduction and Fertility*, **108**, 263–70.

Ingman, W.V., Owens, P.C. & Armstrong, D.T. (2000). Differential regulation by FSH and IGF-1 of extracellular matrix IGFBP-5 in bovine granulosa cells: effect of association with the oocyte. *Molecular and Cellular Endocrinology*, **164**, 53–8.

Kammerman, S., Canfield, R.E., Kolena, J. & Channing, C.P. (1972). The binding of iodinated hCG to porcine granulosa cells. *Endocrinology*, **91**, 65–74.

Knight, P.G., Lacey, M., Peter, J.L.T. & Whitehead, S.A. (1990). Demonstration of a non-steroidal factor in human follicular fluid that attenuates the self-priming action of gonadotrophin-releasing hormone on pituitary gonadotropes. *Biology of Reproduction*, **42**, 613–18.

Knudson, C.B. & Knudson, W. (1993). Hyaluronan-binding proteins in development, tissue homeostasis and disease. *FASEB Journal*, **7**, 1233–41.

Koppenaal, D.W., Tijssen, A.M.I. & de Koning, J. (1991). The self-priming action of LHRH is under negative FSH control through a factor released by the ovary: observations in female rats *in vivo*. *Journal of Endocrinology*, **129**, 205–11.

Lee, C.Y. (1976). The porcine ovarian follicle. III. Development of chorionic gonadotrophin receptors associated with increase in adenylate cyclase activity during follicle maturation. *Endocrinology*, **99**, 42–8.

Lee, T.H., Klampfer, L., Shows, T.B. & Vilcek, J. (1993). Transcriptional regulation of TSG6, a tumor necrosis factor and interleukin-1-inducible primary response gene coding for a secreted hyaluronan-binding protein. *Journal of Biological Chemistry*, **268**, 6154–60.

Leonardsen, L., Strömstedt, M., Jacobsen, D., *et al.* (2000). Effect of inhibition of sterol Δ-14 reductase on accumulation of meiosis-activating sterol and meiotic resumption in cumulus-enclosed mouse oocytes *in vitro*. *Journal of Reproduction and Fertility*, **118**, 171–9.

Li, R., Norman, R.J., Armstrong, D.T. & Gilchrist, R.B. (2000). Oocyte-secreted factor(s) determine functional differences between bovine mural granulosa cells and cumulus cells. *Biology of Reproduction*, **63**, 839–45.

Littman, B.A. & Hodgen, G.G. (1984). Human menopausal gonadotrophin stimulation in monkeys: blockade of the luteinising hormone surge by a highly transient ovarian factor. *Fertility and Sterility*, **41**, 440–7.

Luck, M.R. (1994). The gonadal extracellular matrix. *Oxford Reviews of Reproductive Biology*, **16**, 33–85.

Masui, Y. (1991). The role of cytostatic factor (CSF) in the control of cell cycles: a summary of twenty years of study. *Development, Growth and Differentiation*, **33**, 543–51.

Masui, Y. & Markert, C.L. (1971). Cytoplasmic control of nuclear behaviour during meiotic maturation of frog oocytes. *Journal of Experimental Zoology*, **177**, 129–46.

McIntush, E.W. & Smith, M.F. (1998). Matrix metalloproteinases and tissue inhibitors of metalloproteinases in ovarian function. *Reviews of Reproduction*, **3**, 23–30.

Messinis, I.E., Lolis, D., Papadopoulos, L., *et al.* (1993). Effect of varying concentrations of follicle stimulating hormone on the production of gonadotrophin surge attenuating factor (GnSAF) in women. *Clinical Endocrinology*, **39**, 45–50.

Messinis, I.E., Millingos, S., Zikopoulos, K., *et al.* (1998). Luteinising hormone response to gonadotrophin-releasing hormone in normal women undergoing ovulation induction with urinary or recombinant follicle stimulating hormone. *Human Reproduction*, **13**, 2415–20.

Messinis, I.E. & Templeton, A.A. (1989). Pituitary response to exogenous LHRH in superovulated women. *Journal of Reproduction and Fertility*, **87**, 633–9.

Messinis, I.E. & Templeton, A. (1990a). Superovulation induction in women suppresses luteinising hormone secretion at the pituitary level. *Clinical Endocrinology*, **32**, 107–14.

Messinis, I.E. & Templeton, A. (1990b). *In vivo* bioactivity of gonadotrophin surge attenuating factor (GnSAF). *Clinical Endocrinology*, **33**, 213–18.

Messinis, I.E. & Templeton, A.A. (1991). Attenuation of gonadotropin release and reserve in superovulated women by gonadotropin surge attenuating factor (GnSAF). *Clinical Endocrinology*, **34**, 259–63.

Moenter, S.M., Caraty, A., Locatelli, A. & Karsch, F.J. (1991). Pattern of GnRH secretion leading up to ovulation in the ewe: existence of a pre-ovulatory GnRH surge. *Endocrinology*, **129**, 1175–82.

Moor, R.M., Osborn, J.C., Cran, D.G. & Walters, D.E. (1981). Selective effects of gonadotrophins on cell coupling, nuclear maturation and protein synthesis in mammalian oocytes. *Journal of Embryology and Experimental Morphology*, **61**, 347–65.

Motlik, J., Fulka, J. & Fléchon, J.E. (1986). Changes in intercellular coupling between pig oocytes and cumulus cells during maturation *in vivo* and *in vitro*. *Journal of Reproduction and Fertility*, **76**, 31–7.

Mroueh, J.M., Argobast, L.K., Fowler, P., *et al.* (1996). Identification of gonadotrophin surge-inhibiting factor (GnSIF)/attenuin in human follicular fluid. *Human Reproduction*, **11**, 490–6.

Nagyová, E., Procházka, R. & Vanderhyden, B.C. (1999). Oocytectomy does not influence synthesis of hyaluronic acid by pig cumulus cells: retention of hyaluronic acid after IGF-1 treatment in serum free medium. *Biology of Reproduction*, **61**, 569–74.

Nagyová, E., Vanderhyden, B.C. & Procházka, R. (2000). Secretion of paracrine factors enabling expansion of cumulus cells is developmentally regulated in pig oocytes. *Biology of Reproduction*, **63**, 1149–56.

Nuttinck, F., Peynot, N., Humblot, P., Massip, A., Dessy, F. & Fléchon, J.E. (2000). Comparative immunohistochemical distribution of connexin 37 and connexin 43 throughout folliculogenesis in the bovine ovary. *Molecular Reproduction and Development*, **57**, 60–6.

Osborn, J.C. & Moor, R.M. (1983). The role of steroid signals in the maturation of mammalian oocytes. *Journal of Steroid Biochemistry*, **19**, 133–7.

Pappa, A., Seferiadis, K., Marselos, M., Tsolas, O. & Messinis, I. (1999). Development and application of competitive ELISA assays for rat LH and FSH. *Theriogenology*, **51**, 911–26.

Pearson, O.P. & Enders, R.K. (1943). Ovulation, maturation and fertilisation in the fox. *Anatomical Record*, **85**, 69–83.

Picton, H.M. & Gosden, R.G. (1999). Oogenesis, in mammals. In *Encyclopedia of Reproduction*, vol. 3, ed. E. Knobil & J.D. Neill, pp. 488–97. San Diego & London: Academic Press.

Pincus, G. & Enzmann, E.V. (1935). The comparative behaviour of mammalian eggs *in vivo* and *in vitro*. 1. The activation of ovarian eggs. *Journal of Experimental Medicine*, **62**, 665–75.

Polge, C. & Dziuk, P. (1965). Recovery of immature eggs penetrated by spermatozoa following induced ovulation in the pig. *Journal of Reproduction and Fertility*, **9**, 357–8.

Procházka, R., Nagyová, E., Brem, G., Schellander, K. & Motlik, J. (1998). Secretion of cumulus expansion-enabling factor (CEEF) in porcine follicles. *Molecular Reproduction and Development*, **49**, 141–9.

Procházka, R., Nagyová, E., Rimkevičova, Z., Nagai, T., Kikuchi, K. & Motlik, J. (1991). Lack of an effect of oocytectomy on expansion of the porcine cumulus. *Journal of Reproduction and Fertility*, **93**, 569–76.

Ralph, J.H., Telfer, E.E. & Wilmut, I. (1995). Bovine cumulus cell expansion does not depend on the presence of an oocyte factor. *Molecular Reproduction and Development*, **42**, 248–53.

Redmer, D.A. & Reynolds, L.P. (1996). Angiogenesis in the ovary. *Reviews of Reproduction*, **1**, 182–92.

Rodgers, R.J., van Wezel, I.L., Rodgers, H.F., Lavranos, T.C., Irvine, C.M. & Krupa, M. (1998). Developmental changes in cells and matrix during follicle growth. In *Gametes: Development and Function*, ed. A. Lauria, F. Gandolfi, G. Enne & L. Gianaroli, pp. 85–99. Rome: Serono Symposia.

Ross, G.T., Cargille, C.M., Lipsett, M.B., *et al.* (1970). Pituitary and gonadal hormones in women during spontaneous and induced ovulatory cycles. *Recent Progress in Hormone Research*, **26**, 1–48.

Salustri, A., Yanagishita, M. & Hascall, V.C. (1990). Mouse oocytes regulate hyaluronic acid synthesis and mucification by FSH-stimulated cumulus cells. *Developmental Biology*, **138**, 26–32.

Salustri, A., Yanagishita, M., Underhill, C.B., Laurent, T.C. & Hascall, V.C. (1992). Localisation and synthesis of hyaluronic acid in the cumulus cells and mural granulosa cells of the preovulatory follicle. *Developmental Biology*, **151**, 541–51.

Sato, E. (1997). Cumulus differentiation and its possible physiological roles. In *Microscopy of Reproduction and Development: A Dynamic Approach*, ed. P.M. Motta, pp. 129–35. Rome: Antonio Delfino Editore.

Sato, E., Inoue, M. & Toyoda, Y. (1993). Morphological profiles of mouse ovarian follicles: extensive accumulation of a strongly negatively-charged substance at specific foci in follicular tissue during oocyte maturation. *Archives for Histology and Cytology*, **56**, 293–302.

Schatten, G. (1994). The centrosome and its mode of inheritance: the reduction of the centrosome during gametogenesis and its restoration during fertilisation. *Developmental Biology*, **165**, 299–335.

Schenken, R.S. & Hodgen, G.D. (1986). Follicle stimulating hormone blocks estrogen-positive feedback during the early follicular phase in monkeys. *Fertility and Sterility*, **45**, 556–60.

Schomberg, D.W. & Tyrey, L. (1972). Uptake of [131I]human chorionic gonadotrophin (HCG) by porcine granulosa cells under incubation and culture conditions. *Biology of Reproduction*, **7**, 127, Abstract.

Schuetz, A.W. & Schwartz, W.J. (1979). Intra-follicular cumulus cell transformation associated with oocyte maturation following gonadotropin hormone stimulation of adult mice. *Journal of Experimental Zoology*, **207**, 399–406.

Schultz, R. (1987). Molecular aspects of oocyte growth and maturation. In *Experimental Approaches to Mammalian Embryonic Development*, ed. J. Rossant & R.A. Pedersen, pp. 195–237. Cambridge: Cambridge University Press.

Shores, E.M. & Hunter, M.G. (2000). Production of tissue inhibitors of metalloproteinases (TIMPs) by pig ovarian cells *in vivo* and the effect of TIMP-1 on steroidogenesis *in vitro*. *Journal of Reproduction and Fertility*, **120**, 73–81.

Simon, A.M., Goodenough, D.A., Li, E. & Paul, D.L. (1997). Female infertility in mice lacking connexin 37. *Nature (London)*, **385**, 525–9.

Singh, B., Zhang, X. & Armstrong, D.T. (1993). Porcine oocytes release cumulus expansion-enabling activity even though porcine cumulus expansion *in vitro* is independent of the oocyte. *Endocrinology*, **132**, 1860–2.

Sobotta, J. (1895). Die Befruchtung und Furchung des Eies der Maus. *Archiv für Mikroskopische Anatomie und Entwicklungsgeschichte*, **45**, 15–93.

Strömstedt, M. & Byskov, A.G. (1999). Mammalian oocyte. In *Encyclopedia of Reproduction*, vol. 3, ed. E. Knobil & J.D. Neill, pp. 468–80. San Diego & London: Academic Press.

Sutovsky, P. & Fléchon, J.E. (1997). Cytoskeletal reorganisation during cumulus expansion in mammals. In *Microscopy of Reproduction and Development: A Dynamic Approach*, ed. P.M. Motta, pp. 119–28. Rome: Antonio Delfino Editore.

Sutovský, P., Fléchon, J.E. & Pavlok, A. (1994). Microfilaments, microtubules and intermediate filaments fulfil differential roles during gonadotrophin-induced expansion of bovine cumulus oophorus. *Reproduction, Nutrition and Development*, **34**, 415–25.

Szöllösi, D. (1967). Development of cortical granules and the cortical reaction in rat and hamster eggs. *Anatomical Record*, **159**, 431–46.

Szöllösi, D. (1978). On the role of gap junctions between follicle cells and the oocyte in the mammalian ovary. *Research in Reproduction*, **10**, 3–4.

Szöllösi, D. (1993). Oocyte maturation. In *Reproduction in Mammals and Man*, ed. C. Thibault, M.-C. Levasseur & R.H.F. Hunter, pp. 307–25. Paris: Ellipses.

Szöllösi, D. & Hunter, R.H.F. (1973). Ultrastructural aspects of fertilisation in the domestic pig: sperm penetration and pronucleus formation. *Journal of Anatomy*, **116**, 181–206.

Thibault, C. (1977). Hammond memorial lecture: Are follicular maturation and oocyte maturation independent processes? *Journal of Reproduction and Fertility*, **51**, 1–15.

Thibault, C. & Gérard, M. (1970). Facteur cytoplasmique nécessaire à la formation du pronucleus mâle dans l'ovocyte de lapine. *Compte rendu hebdomadaire des Séances de l'Académie des Sciences (Paris)*, Séries D, **270**, 2025–6.

Thibault, C. & Gérard, M. (1973). Cytoplasmic and nuclear maturation of rabbit oocytes *in vitro*. *Annales de Biologie Animale Biochimie, Biophysique*, **13**, 145–56.

Thibault, C., Szöllösi, D. & Gérard, M. (1987). Mammalian oocyte maturation. *Reproduction, Nutrition et Développement*, **27**, 865–96.

Tsafriri, A. (1978). Inhibition of nuclear maturation of isolated rat oocytes by follicular constituents. *Annales de Biologie Animale Biochimie, Biophysique*, **18**, 523–8.

Tsafriri, A. (1985). The control of meiotic maturation in mammals. In *Biology of Fertilisation*, vol. 1, ed. C.B. Metz & A. Monroy, pp. 221–52. London & New York: Academic Press.

Tsafriri, A. (1988). Local non-steroidal regulators of ovarian function. In *The Physiology of Reproduction*, ed. E. Knobil & J. Neill, pp. 527–65. New York: Raven Press.

Tsafriri, A. (1999). Follicular development: impact on oocyte quality. In *FSH Action and Intraovarian Regulation*, ed. B.C.J.M. Fauser, pp. 83–105. New York & London: Parthenon Publishing.

Tsafriri, A. & Channing, C.P. (1975). An inhibitory influence of granulosa cells and follicular fluid upon porcine oocyte meiosis *in vitro*. *Endocrinology*, **96**, 922–7.

Tsafriri, A. & Pomerantz, S.H. (1986). Oocyte maturation inhibitor. *Clinics in Endocrinology and Metabolism*, **15**, 157–70.

van Wezel, I.L. & Rodgers, R.J. (1996). Morphological characterisation of bovine primordial follicles and their environment *in vivo*. *Biology of Reproduction*, **55**, 1003–11.

Vanderhyden, B.C. (1993). Species differences in the regulation of cumulus expansion by an oocyte-secreted factor(s). *Journal of Reproduction and Fertility*, **98**, 219–27.

Vanderhyden, B.C., Caron, P.J., Buccione, R. & Eppig, J.J. (1990). Developmental pattern of the secretion of cumulus expansion-enabling factor by mouse oocytes and the role of oocytes in promoting granulosa cell differentiation. *Developmental Biology*, **140**, 307–17.

Vanderhyden, B.C., Telfer, E.E. & Eppig, J.J. (1992). Mouse oocytes promote proliferation of granulosa cells from preantral and antral follicles *in vitro*. *Biology of Reproduction*, **46**, 1196–204.

Whitaker, M. (1996). Control of meiotic arrest. *Reviews of Reproduction*, **1**, 127–35.

Wisniewski, H.G., Hua, J.C., Poppers, D.M., Naime, D., Vilcek, J. & Cronstein, B.N. (1996). TNF/IL-1 inducible protein TSG-6 potentiates plasmin inhibition by inter-α-inhibitor and exerts a strong anti-inflammatory effect *in vivo*. *Journal of Immunology*, **156**, 1609–15.

Zuckerman, S. (1960). Origin and development of oocytes in foetal and mature mammals. *Memoirs of the Society for Endocrinology, No. 7*, 63.

VIII

The process of ovulation and shedding of an oocyte

Introduction

In one sense, at least, the overall focus of this book is on the actual process of ovulation and much of what has been written so far has paved the way for description and analysis of this vital event – shedding of one or more viable gametes from the gonad into the reproductive tract. Phrased more specifically, this chapter will be concerned with release of an oocyte from a mature Graafian follicle into the Fallopian tube, in most species as a secondary oocyte. If subsequent sperm–egg contact at the ampullary–isthmic junction of the tube leads to successful fertilisation and establishment of pregnancy, then the process of ovulation will be held in abeyance for the duration of gestation, or until prenatal loss intervenes and leads to termination of the pregnancy. An exception to this statement is in equine species whose ovaries respond to a chorionic gonadotrophin, PMSG, to form accessory corpora lutea in the early part of pregnancy (i.e. days 60–90). In the absence of fertilisation and assuming that the species is not a seasonal breeder entering a phase of anoestrus, ovarian cyclicity as regulated by pituitary gonadotrophins and uterine or luteal secretion of prostaglandins will lead once more to maturation of a Graafian follicle(s), enhanced secretion of oestradiol, a pre-ovulatory surge of gonadotrophic hormones and the process of ovulation. This sequence should offer a further opportunity for fertilisation and the establishment of pregnancy if coitus or mating occurs at an appropriate time.

The manner in which gonadotrophic hormones programme endocrine and enzymatic changes in the pre-ovulatory follicle(s) but not in those of lesser maturity is clearly of great interest and of fundamental importance. However, perhaps even more exciting is the current focus on white blood cell traffic and activity in the wall of a fully mature Graafian follicle and the contribution such cells may make to the events of ovulation. Although the presence of white cells in ovarian tissues is a long-standing observation, the modern emphasis on

262

regarding ovulation as an inflammatory-like process (Espey, 1980, 1994) has led to renewed interest in the cellular and humoral components of an inflammatory cascade. The cytokine products of white cells passing through ovarian tissues are therefore examined in some detail, but not at the expense of older topics such as the involvement of contractile elements in the follicular wall at the time of shedding an oocyte.

Spontaneous versus induced ovulation

The introductory section refers in particular to species with spontaneous ovulation, those in which cyclic endocrine events culminate in ovulation whether or not a mature male is available and coitally active. In terms of information based on only a small proportion of the estimated 4500 species of mammal, a majority would be classified as spontaneous ovulators. However, species such as rabbits, cats and ferrets, and also camelids such as llama and alpaca, belong to a category termed induced or reflex ovulators in which coitus involving full penile intromission and consequent nervous stimulation of the hypothalamo-pituitary axis is required to trigger the pre-ovulatory surge of gonadotrophic hormones. This reflex is referred to as a neuroendocrine loop and it coordinates the time of semen deposition with that of ovulation, although the former is of course in advance of the latter. Spermatozoa and oocyte(s) should thus enter the female tract in a time relationship optimal for normal fertilisation. In this group of species, ovulation can be induced in the absence of coitus by a single intravenous injection of LH or hCG because responsive follicles will be available and are the source of oestradiol that makes such animals receptive to mating.

Separation of species into the two categories of so-called spontaneous ovulators and induced ovulators may not be fully justified; in fact, there may not be a rigid distinction between animals so classified. For example, in our own species in which a mid-cycle threshold of follicular oestradiol secretion leads to a pre-ovulatory surge of gonadotrophins and thereby spontaneous ovulation, there has long been the suggestion that induced ovulation might occur in specific, if exceptional, circumstances. In other words, a suitable degree of coital and emotional stimulation earlier in the follicular phase than mid cycle might modify the pattern of gonadotrophin secretion, either to accelerate the rate of follicular maturation or to provoke ovulation of a follicle that had not yet achieved the anticipated pre-ovulatory diameter. As a concept, this would seem perfectly feasible. So would the notion that sexual and emotional trauma, as in an incident of rape, could prompt extreme physiological excitation of the hypothalamus and pituitary gland, once again advancing the processes of follicular maturation and ovulation by means of an altered pattern of gonadotrophin secretion. Ovarian innervation might also be making a specific contribution in this unusual setting. The problem is to obtain persuasive experimental rather than circumstantial evidence for precocious ovulation. The latter has been derived in part from unexpected instances of conception following sexual assault early in the follicular phase of the menstrual cycle, and frequently overlooks the fact that spermatozoa of at least some individuals may remain viable in an oestradiol-dominated female tract for 4–5 days. The site of such pre-ovulatory sperm storage has been discussed critically, with the caudal or distal isthmus of the human Fallopian tube – the intramural portion containing viscous secretion – being favoured from diverse points of view (Hunter, 1995, 1998).

The maturity and sensitivity of an ovarian follicle are of immediate relevance to this discussion. Follicles significantly below mature diameter may be induced to ovulate under appropriate conditions, one of which might involve a specific pattern of nervous inputs to the wall of the follicle(s) in question. The network of autonomic fibres in the theca interna of a Graafian follicle has been described in Chapter III. In animals that are characteristically spontaneous ovulators, ovulation of an immature follicle can be induced by a single injection of a gonadotrophin preparation or by twin injections of different gonadotrophins at a suitably spaced interval. However, such experimental ovulations seldom result in normal fertilisation and the establishment of pregnancy. As an example, a single injection of 500 i.u. hCG early in the follicular phase (day 17 or 18) of a 21-day pig oestrous cycle can induce ovulation, but the eggs so liberated are invariably primary oocytes (see Table IV.6) and therefore immature and without a block to polyspermy (Hunter, Cook & Baker, 1976). Even during the midst of the 14- or 15-day luteal phase in the same species, a single injection of 1000 i.u. PMSG followed 4 days later by 500 i.u. hCG will predictably cause ovulation or superovulation (Plate 3) but, once again, a high proportion of the eggs will be immature (Hunter, 1967, 1979). They would therefore not be fertilised normally, even if spermatozoa were to be introduced directly into the Fallopian tubes. These observations illustrate a dichotomy in the response to gonadotrophic stimulation between the somatic cells of a follicle and the germ cell or oocyte.

To conclude these remarks on a note of balance, and following on from the above discussion especially concerning women, the distinction between spontaneous and induced ovulation may be largely one of degree rather than absolute, i.e. spontaneous ovulation may describe what is usually the situation but not invariably so. Indeed, writing of domestic farm animals (cow, sheep, pig), Jöchle (1997) has proposed that spontaneous ovulation in these species is a secondary event and that they are, in reality, facultative induced ovulators.

Mating early in oestrus would accelerate the occurrence of the pre-ovulatory LH peak and thereby ovulation. In Jöchle's view, spontaneous ovulation may have evolved from, or been superimposed on, induced ovulation, with the original ovulatory programme retained and accessible in certain circumstances (see also Jöchle, 1973, 1975).

Relevant to this discussion are the studies of Waberski and Weitze at the Hanover School of Veterinary Medicine, again working with domestic pigs (Waberski, 1997; Waberski *et al.*, 1997). Such animals would be viewed as classical spontaneous ovulators at the 'end' of a 21-day oestrous cycle. Using non-invasive, ultrasonic monitoring of ovarian morphology, Waberski and colleagues observed that mating could 'hasten' or bring forward the time of ovulation by a small number of hours in animals with an extended period of oestrus. This in itself was not a novel contribution, for such a claim was a repeated theme from various Russian laboratories and elsewhere during the 1950s (e.g. Pitkjanen, 1955). However, the Hanover team found that it was primarily the introduction of seminal plasma into the female tract rather than intromission that could advance ovulation in certain groups of animals. Moreover, in a surgically modified model in which seminal plasma could pass up only one horn of the uterus and thus enter only one of the Fallopian tubes, a significant within-animal unilateral advancement of ovulation on the patent side could be distinguished, although, it should be emphasised, not invariably so (Waberski, 1997). Once again, the underlying concept was not completely new for seminal components, especially smooth muscle stimulants, had previously been considered to play a rôle in the events of ovulation in domestic animals, including various species of camel (e.g. the Bactrian camel: Chen, Yuen & Pan, 1985).

What was new in the Hanover studies was fractionation of a pool of boar seminal plasma, with isolation of a putative ovulation-advancing component in a low molecular weight peptide fraction (Waberski *et al.*, 1995; Waberski, Töpfer-Petersen & Weitze, 2000). Further characterisation is awaited, and likewise the mode of signal transduction between the reproductive tract and Graafian follicles, with the thought that specific populations of white blood cells and/or their cytokine products may be involved as sensitive vectors. Nonetheless, there remain the critical points that: (1) these seminal plasma experiments did not advance ovulation in all animals examined; and (2) 'pools' of seminal plasma, even pools derived from the same boar, did not always demonstrate an ability to advance the time of ovulation.

Although the topic will be returned to below, one is left with the suspicion that any specific ovulation-advancing properties of seminal plasma are interacting with other components (i.e. immune, endocrine or autonomic nervous) of a finely tuned programme of follicular modifications at ovulation. The above mention of a possible involvement of white blood cells is strengthened by re-calling that one of the earliest uterine responses to mating and accumulation of seminal plasma is a massive infiltration of polymorphonuclear leucocytes and other white cell populations into the uterine lumen (Lovell & Getty, 1968). Overall, coital advancement of the time of ovulation could be interpreted as a stage or condition situated between spontaneous and induced ovulation. There is certainly no intention here to suggest that whole seminal plasma passes up the female duct system to reach and influence a pre-ovulatory follicle in a direct manner. In fact, in the one instance in which the present author introduced an aliquot of boar seminal plasma directly into a pre-ovulatory follicle at surgery – whilst examining the possibility of pre-ovulatory fertilisation of pig oocytes – respiration of the anaesthetised animal was arrested almost immediately. Combined with profound anaphylactic shock, the animal exhibited a form of *rigor mortis* in less than 3 minutes whilst still on the operating table. She had not previously been exposed to boar seminal plasma or seminal plasma proteins. The experiment was not repeated. It is to be hoped that medical colleagues will not be tempted by this approach to pre-ovulatory fertilisation, even though intraperitoneal insemination is currently practised in fertility (i.e. infertility) clinics and has a long history in domestic animals (see Hunter, 1978).

Timing of ovulation and dimensions of follicle

Despite the contents of the previous section concerning the influence of coitus or mating, ovulation has been noted to occur at a rather precise interval after the systemic surge of gonadotrophic hormones, an interval that is characteristic for individual species. Examples of such times are presented in Table VII.1 and Fig. VIII.1. They can be ascertained by frequent blood sampling for LH assay together with observation of the ovaries; the latter is nowadays made straight-forward by a non-invasive means of ultrasonic scanning. A separate means of judging the gonadotrophin surge to ovulation interval is to pre-empt or mimic the endogenous surge with a single intravenous injection of LH or an LH-rich preparation such as hCG. In extensive studies in laboratory and domestic farm animals, the interval from LH injection to ovulation in animals with mature follicles has been found to parallel the timing noted under spontaneous conditions. It perhaps goes without saying that the pre-ovulatory resumption of meiosis in oocytes contained within suitable mature follicles also responds to the injected gonadotrophin, and proceeds along a time-course that coordinates meiotic stages with the time of induced ovulation and shedding of the oocyte (see Chapter VII).

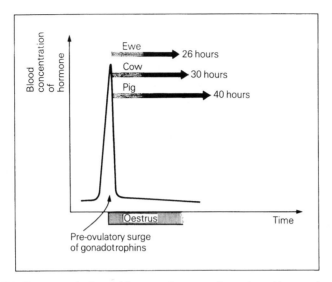

Fig. VIII.1. To portray the interval between the surge of gonadotrophin secretion usually found close to the onset of oestrus in domestic farm animals and ovulation of the follicle 26–40 hours later, according to species. (After Hunter, 1982.)

If the time of ovulation is precise and characteristic for a given species after the endogenous gonadotrophin surge, and encompasses well-timed nuclear and cytoplasmic responses in the oocyte, then it seems reasonable to anticipate corresponding chronological changes in the somatic cells as well, culminating in the rupture or collapse of the follicle at ovulation. With the exception of mucification of the cumulus oophorus and modifications in the corona radiata (see Chapter VII), the time-course of somatic cell changes has yet to be presented in detail for any species. Even so, there can be little doubt that the activation of proteolytic enzymes, for example, must be in an orderly sequence to achieve an accurately programmed ovulation. Likewise, enzymatic transformations in the biosynthetic pathway for steroid hormones are well known, but still have to be defined on a precise chronological basis. As more and more genes involved in the process of ovulation are recognised and cloned, so it should become possible to chronicle the molecular steps that regulate the time of ovulation. This may lead to an understanding of why the final pre-ovulatory modifications in the wall of a human or pig Graafian follicle and in the antral oocyte require approximately three to four times as long as those in a rabbit or hamster. Such molecular details might also help to explain within-species variation, i.e. breed differences in the time interval between the gonadotrophin peak and ovulation (Hunter, 1972).

As already noted, various influences on follicular maturation and the ensuing process of ovulation come from the oocyte, an impressively large cell perhaps best viewed here as a centrally situated computer. However, the extent to which an oocyte acts to influence or regulate the chronological response to the gonadotrophin surge and more specifically the timing of ovulation has yet to be clarified. Because a follicle fails to ovulate when the oocyte dies or is destroyed (Westman, 1934) or is removed microsurgically (El-Fouly *et al.*, 1970), frequently becoming cystic (Hunter & Baker, 1975), then there is a strong suggestion of a critical involvement, although not necessarily in a chronologically precise sense. Similarly, the fact that the oocyte acts to prevent luteinisation of granulosa cells (Westman, 1934; Foote & Thibault, 1969) again indicates a specific involvement in pre-ovulatory events, although not necessarily one of a chronological nature.

As to dimensions of pre-ovulatory follicles, these range enormously despite a relatively consistent size of secondary oocyte (e.g. 80–130 μm in diameter; see Austin, 1961). As examples, fully mature follicles vary from diameters of 0.6 mm in rats, 9–11 mm in pigs, 18–23 mm or even more in women, to as large as 45–55 mm in horses (Table VI.1). There is an approximate proportionality with body mass, and scaling influences have been examined in this connection in a masterly and now classical review by Brambell (1956) and more recently by Gosden & Telfer (1987), but exceptions are not difficult to find. Even within a species, differences may be distinguishable in the diameter of a mature Graafian follicle at puberty compared with dimensions in a fully grown female of large breed or race.

Process of ovulation – general features

The actual process of ovulation remains a fascinating topic of research, in part because successful release of a potentially fertilisable egg from the ovary into the Fallopian tube is such a vital event underlying fertility. The topic is also attractive because of the dramatic modifications to the surface structure of a Graafian follicle that enable escape of the oocyte together with large numbers of somatic cells and much of the follicular fluid. Indeed, ovulation is frequently described as unique because it requires physical disruption of healthy tissue both within the follicular antrum and at the ovarian surface to be followed by transformation of the evacuated follicle into a cyclic corpus luteum. In terms of secretion of progesterone, a fully mature Graafian follicle is already functioning as a corpus luteum in the last hours before its collapse, and synthesis of this particular steroid may be essential for completion of the process of ovulation itself. Its

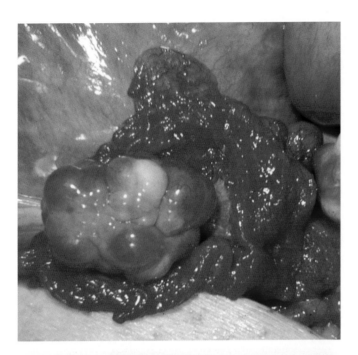

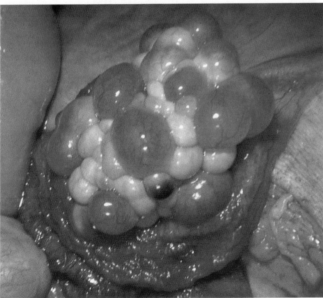

PLATE 05

Portions of the reproductive tract and ovaries of oestrous pigs photographed at mid-ventral laparotomy approximately 24 hours before ovulation. Note the Graafian follicles of 9–10mm diameter, together with corpora albicantia from the previous oestrous cycle, and the intense vascular engorgement of the fimbriated infundibulum (upper photo).

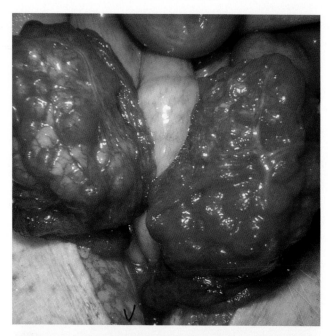

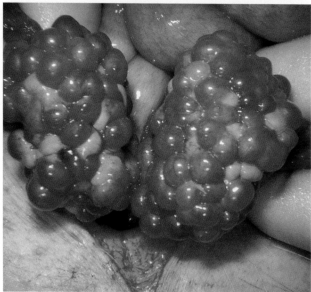

PLATE 06

Ovaries demonstrating an instance of massive spontaneous superovulation presumed to be a response to a tumour of the adenohypophysis and excessive secretion of gonadotrophic hormones.

Upper Ovaries fully embraced by the fimbriated extremities of the Fallopian tubes.

Lower The same ovaries with each fimbriated infundibulum displaced to reveal a total of 97 recent ovulations. (From Hunter, R.H.F, *European Journal of Obstetrics, Gynecology and Reproductive Biology* (1985) 19, 261–6.)

secretion has come about primarily because the pre-ovulatory peak of pituitary gonadotrophins acts to attenuate aromatase availability in the granulosa cells in particular (see Chapters V and VII).

Many individual observations have contributed to the detailed picture of ovulation that can be assembled today. Each has seemed of major importance at the time of discovery, but has then gradually fallen into a lesser perspective as but one small link in a complex sequence of events. The observations come from universities and research institutes around the world, and it may not be fully justifiable to highlight individual laboratories or their members of staff. Even so, in the opinion of the present author, the dedicated contributions during more than 30 years of Professor L. Espey at Trinity University, San Antonio, Texas, USA, and Professor A. Tsafriri at the Weizmann Institute in Israel deserve special mention. Whereas Espey has focused principally on morphological and ultrastructural events in the mature follicle, Tsafriri and colleagues have offered a strong biochemical and endocrine emphasis, doubtless inspired in part by the studies of Lindner at the Weizmann Institute (e.g. Lindner *et al.*, 1977). Both Espey and Tsafriri have sought to integrate their observations into a comprehensive description of pre-ovulatory events and both have repeatedly published major reviews (e.g. Espey, 1974, 1978, 1980, 1999; Espey & Lipner, 1994; Tsafriri *et al.*, 1972, 1976; Tsafriri, Abisogun & Reich, 1987; Tsafriri, Chun & Reich, 1993; Tsafriri & Dekel, 1994). At the time of preparing this monograph, the group demonstrating the greatest activity in the field of ovulation research is probably that of Professor M. Brännström at the University of Goteborg, Sweden.

Other notable contributions in the sphere of ovulation, i.e. changes in the ovulatory follicle, would include those of Hisaw (1947), Blandau (1955, 1969), Zachariae (1958), Nalbandov (1961), Everett (1961), Asdell (1962), Harris (1964, 1969), Rondell (1970), Lipner (1973, 1988), Moor, Hay & Seamark (1975), Thibault (1977), Thibault & Levasseur (1979, 1988), Driancourt & Cahill (1984), Schroeder & Talbot (1985), Driancourt *et al.* (1993), Gougeon (1996, 1997) and Brännström (1997). In a contrary sense, seeking to inhibit the process of ovulation as a means of contraception, mention must be made of Pincus & Chang (1953), Pincus (1965) and Harper (1968). Contributions with a strong biochemical or endocrine slant have been provided, amongst others by Adashi (1990, 2000), Baird (1983, 1990, 1999), Campbell *et al.* (1999), Fortune & Voss (1993), Hillier (1994), Richards (1988, 1994), Scaramuzzi *et al.* (1993), Sirois (1994), Sirois & Doré (1997), Zeleznik & Benyo (1994) and Zeleznik & Hillier (1996).

In one interpretation, the process of ovulation can be viewed as commencing in the somatic cell layers of the mature follicle. This is because, under the influence of gonadotrophins, the steroid synthetic activities of the theca interna and granulosa cells together generate increasing levels of oestradiol-17β secretion. When this secretion achieves a threshold in the systemic circulation, itself a sensitive indicator of follicular maturity, it acts on the hypothalamus in a positive feedback loop to provoke a surge of GnRH release (see Chapter VII), with a consequent surge of pre-ovulatory gonadotrophin secretion from the anterior pituitary gland. This reaches and enters the pre-ovulatory follicles preferentially. By means of tracer studies with radio-labelled preparations of LH, the gonadotrophin was demonstrated to bind to the inner cell layers of such follicles (Rajaniemi & Vanha-Perttula, 1972); it could also be demonstrated in the follicular fluid, observations that have been built on repeatedly during the past 30 years. This binding of gonadotrophins to receptors in suitable follicles sets in train a cascade of endocrine, morphological and biochemical events. As already noted, such events are frequently likened to an inflammatory reaction (Espey, 1980, 1994), culminating in a well-timed release of an egg, usually at the stage of secondary oocyte, in a sexually receptive animal. The descriptions that follow will attempt to highlight key morphological and biochemical events in the Graafian follicle, a majority of which appear common to nearly all mammalian species so far studied. A useful starting point would be the paper of Schochet (1915).

Process of ovulation – morphological highlights

In response to the gonadotrophin surge, a pre-ovulatory follicle embarks on a final phase of growth and expansion, with enhanced accumulation of antral fluid. As noted in Chapter VII, the cumulus cells are stimulated by the gonadotrophin surge into active secretion of hyaluronic acid and this major component of the extracellular matrix attracts water and increases the volume of the antral cavity. As a consequence particularly of water uptake and proteoglycan expansion, a follicle achieves its maximum diameter, which falls within a range characteristic of the species (Table VI.1). Growth and remodelling of the basement membrane must be exceptionally rapid to accommodate these final stages, and could in part explain why growth soon gives way to breakdown and eventual dissolution of the membrane.

Within the follicle, there is a detectable loosening of granulosa cells resting on the basement membrane, and their columnar organisation is gradually modified (Bjersing & Cajander, 1974). The classical gap junctions between these cells disappear, enabling such cells to become partially dissociated. As noted, the cumulus component of the granulosa cells commences to expand due to production of hyaluronic acid and other proteoglycans, and yet the gap junctions between cells of the corona radiata and oocyte are not markedly altered. Shortly before ovulation and as suggested above, there is complete dissolution of the basement membrane that has separated granulosa cells from those of the theca interna and hitherto prevented inward access of the capillary and lymphatic beds (for a review, see Motta *et al.*, 1995).

There are substantial modifications to the vascular network in the follicle wall (see Chapter III, and Macchiarelli *et al.*, 1995), with initial dilatation of capillaries and hyperaemia of the thecal layer seen as a progressive change in colour of the pre-ovulatory follicles from clear to pinkish and even to red, owing to increased capillary permeability (Plate 4). Indeed, in rabbits for example, a species with approximately a 10-hour pre-ovulatory interval, there is a measurable increase in ovarian blood flow within 4–6 hours of the gonadotrophin surge, and ovarian blood content may increase as much as five-fold (Espey, 1997, 1999) or even up to seven-fold (Driancourt *et al.*, 1993). Similar observations have been made in rats (Tanaka, Espey & Okamura, 1989). The hyperaemia is thought to be mediated by local histamine and PGE_2 (see Cavender & Murdoch, 1988). This is followed by an increasing degree of capillary (petechial) haemorrhage observed around much of the follicle wall (Plate 4), but local ischaemia and a locally defined avascular region are usually noted towards the apex of the follicle.

Increased vascular permeability with seepage of plasma from distended capillaries causes the theca externa to become oedematous. Such increased vascular permeability may have been influenced by platelet activating factor (PAF), a phospholipid involved in inflammatory reactions (Espey *et al.*, 1989). Specific angiogenic factors may also have contributed (Koos, 1995). Overall, there are good parallels with an inflammatory-like response. Such changes in the vascular bed are closely coincident with capillary invasion of the granulosa cell layer and then extensive capillary proliferation after breakdown of the basement membrane, again promoted by the pre-ovulatory surge of gonadotrophins. Proliferation of the capillary network within the granulosa cell layers is associated with the onset of luteinisation and provision of adequate substrate for steroidogenesis, and reflects an active local influence of potent angiogenic factors.

In a majority of species so far examined, there is a reduction in intra-follicular pressure, the follicle(s) becoming flaccid and pendulous (Plate 5) and losing tensile strength in the wall. Ovulation is certainly not explosive in nature, at least not in the large domestic species (sheep, pigs, cows and horses) in which a final phase of softening and deformation is characteristic – a change that can be detected by simple rectal palpation in cattle and horses. The widely used

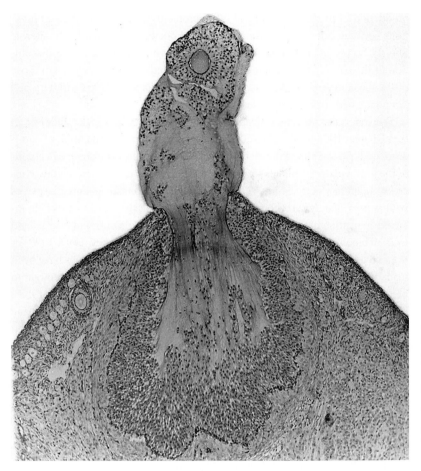

Fig. VIII.2. Section through the ovulating follicle of a rabbit to show the liberated oocyte surrounded by its corona radiata being displaced in the escaping follicular fluid. Note the groups of granulosa cells carried in the 'streaming' follicular fluid, and the undulating appearance of the follicle wall. (Courtesy of Professor R.J. Blandau. After Hunter, 1988.)

term 'rupture' therefore appears inappropriate. Collapse of the follicle would seem more descriptive of ovulation in these species. On the other hand, in rats, rabbits and cats, a spurt of follicular fluid containing the so-called egg mass has frequently been noted (e.g. Blandau, 1955, 1969), suggesting that 'rupture' may be correct in these species and perhaps suggesting an involvement of smooth muscle activity. Rhythmic contractions of the ovary and apparent contraction of theca externa cells (see Chapter III) would facilitate displacement of the oocyte–cumulus complex from the follicular antrum (see Thomson, 1919; Guttmacher & Guttmacher, 1921; Schroeder & Talbot, 1985).

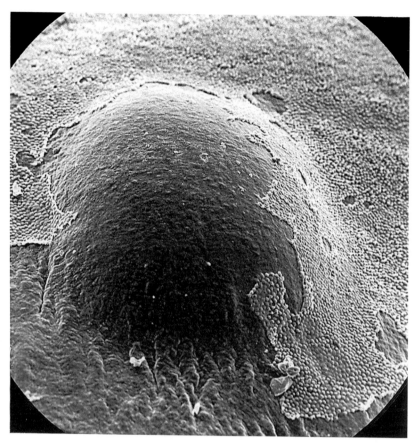

Fig. VIII.3. Scanning electron micrograph of a porcine Graafian follicle to illustrate some of the pre-ovulatory changes at or towards the apex. In particular, note the desquamation of the follicle wall. (Courtesy of Dr J.E. Fléchon. After Hunter, 1980.)

The ovulatory follicle(s) protrudes progressively further from the ovarian surface as shedding of the oocyte becomes imminent. A thinning followed by gradual degradation of the apical region of the follicle wall may give a translucent region and leads to formation of an aperture (Fig. VIII.2), with or without prior formation of a stigma (see Motta, Cherney & Didio, 1971; Motta & Van Blerkom, 1975). If such a stigma can be detected, it is usually devoid of blood vessels in its central portion. Cells are sloughed off from the region of the stigma, eventually permitting seepage of follicular fluid (Figs. VIII.2 and VIII. 3). This stage is followed by rapid release of the non-viscous fluid and then an oozing of the viscous portion of the follicular fluid together with the cumulus mass enclosing the egg (see Harper, 1963; Blandau, 1973). After an aperture has formed in the follicle wall, an oocyte and its investment of loosened cells are usually extruded within no more than 1–2 minutes (Espey, 1999).

Process of ovulation – biochemical events

As already mentioned, local ischaemia causes death of epithelial cells at the apex of the follicle, and the dying cells release hydrolases that seemingly contribute to dissolution of the underlying cell layers. A final breakdown of the follicle wall at the eventual site of aperture formation depends on appropriate degradation of collagenous connective tissue and other constituents of the extracellular matrix. Indeed, the increase in volume of a pre-ovulatory follicle is facilitated by a progressive dissociation of collagen fibres, a process that commences in the tunica albuginea in the hour preceding follicular rupture (Espey, 1999). The connective tissue of the underlying theca shows similar changes. Fibroblasts become active and proliferate, perhaps triggered by elements of the inflammatory cascade and/or by seepage of plasma from distended thecal capillaries. Such fibroblasts are thought to secrete proteolytic enzymes, especially collagenase, that will degrade the extracellular matrix, facilitate their own movement, and lead to a thinning of the follicle wall at the apex.

Plasmin is also prominent in these events, and plasminogen activator increases in response to elevated gonadotrophic hormone secretion (Reich *et al.*, 1991; Espey & Lipner, 1994). Mural granulosa and cumulus cell production of plasminogen activator enables generation of plasmin (Beers, Strickland & Reich, 1975), with maximum activity noted at the apex of the follicle. Plasmin, the active enzyme of the plasminogen system, is a trypsin-like proteinase that can cleave and degrade a wide spectrum of substrates. One function of plasmin would be to dissociate the connective tissue matrix of collagen fibres and activate the precursor of collagenase (a procollagenase). Collagenolytic activity is said to be maximal at the time of ovulation (Driancourt *et al.*, 1993). Despite this logical sequence of events, a more precise involvement of plasminogen activator in the terminal events within a pre-ovulatory follicle remains unclear. Somewhat surprisingly, a mouse model bearing a targeted deletion of either of the genes for tissue plasminogen activator (tPA) or urokinase plasminogen activator (uPA) does not yield an anovulatory phenotype (Cameliet *et al.*, 1994).

Also implicated in inflammatory reactions are various members of the prostaglandin family. Significant increases in these low molecular weight substances in follicular fluid have been reported during the pre-ovulatory interval since the first observations of LeMaire *et al.* (1973, 1975), Ainsworth, Baker & Armstrong (1975) and Bauminger & Lindner (1975). Increases in several types of prostaglandin in rat follicular fluid are especially impressive, with concentrations of PGE_2 and $PGF_{2\alpha}$ in the tissues and follicular fluid of ovulatory follicles increasing 70-fold and 20-fold, respectively. Elevated titres of prostaglandins would facilitate changes in both biochemical pathways and contractile activity of ovarian tissues. Moreover, treatment with inhibitors of prostaglandins such as the anti-inflammatory drug indomethacin during late pro-oestrus blocks ovulation in rats (Armstrong & Grinwich, 1972; Tsafriri *et al.*, 1972) and rabbits (O'Grady *et al.*, 1972), thereby suggesting a critical rôle for prostaglandin activity in the events of follicular breakdown. In our own species, high doses of non-steroidal anti-inflammatory drugs will prevent ovulation and can provoke luteinisation of the unruptured follicles (Killick & Elstein, 1987).

A detailed examination has been made of the prostaglandin content of pig follicular fluid (Hunter & Poyser, 1985), a species with a pre-ovulatory interval of 40 hours that closely parallels the human LH peak to ovulation interval of approximately 38 hours. The concentration of PGE_2, $PGF_{2\alpha}$ and PGI_2 measured as 6-oxo-$PGF_{1\alpha}$ were low 24–32 hours before the expected time of ovulation and the concentration of PGE_2 increased only slowly during the next 22 hours (Table IV.8). However, in a period of 2 hours before anticipated ovulation, the mean concentrations of PGE_2, $PGF_{2\alpha}$ and 6-oxo-$PGF_{1\alpha}$ increased by 119-, 11- and 5-fold, respectively. A single sample of follicular fluid contained one of the highest concentrations of 6-oxo-$PGF_{1\alpha}$ (54.4 ng/ml) reported for any species, but this value was still lower than those of PGE_2 (529 ng/ml) and $PGF_{2\alpha}$ (94.6 ng/ml). These findings emphasise the increase in prostaglandin concentration in follicular fluid shortly before ovulation whereas concentrations remained at baseline in follicles that failed to ovulate, again suggesting an essential involvement of prostaglandins in the process of ovulation. They may have a direct bearing on modifications to the follicle wall and also on the physiological maturity of the oocyte about to be released (see Szöllösi *et al.*, 1978). However, the precise mechanism(s) of action of prostaglandins in facilitating ovulation remains to be revealed, although a contribution of these molecules in the regulation of collagenase and stimulation of plasminogen activator is certainly appreciated, as is an influence on the vascular bed and in chemotaxis of leucocytes.

One other molecule of specific interest should be mentioned here – ovarian kallikrein. As reflected in its mRNA status, kallikrein activity increases as the time of ovulation approaches (for a summary, see Espey, 1997), and kinin formation might contribute particularly to the ovarian vascular changes during ovulation: kinins promote vasodilation, as specifically demonstrable for bradykinin and also histamine. Other factors considered to have a potential involvement in a follicular inflammatory response include cytokines (see below), growth factors and metalloproteases (see Chapter VII). An involvement of PAF has been referred to above. Nerve growth factor and its receptor increase on theca cells in response to the gonadotrophin surge (Dissen *et al.*, 1996), and apparently uniquely so in ovulatory follicles, at least in rats (Espey, 1997). It would be exciting to reveal a specific contribution of the autonomic nervous system to the events of ovulation rather than more generally in a paracrine context.

To place matters in one form of perspective: there can be little doubt that more and more molecules will be identified in the context of final maturation of a Graafian follicle and the process of ovulation. The challenge will be to distinguish those that are essential to these events in contrast to those that facilitate the process, and then to present an integrated description of their chronological and topographical involvement (see Zeleznik & Hillier, 1996; Espey, 1997, 1999; Driancourt *et al.*, 2001). This clearly indicates a need to understand their numerous interactions and specific synergies. Nonetheless, there remains the fact that ovulation – absolutely vital in a strict reproductive sense – has a number of 'fail-safe' mechanisms. In other words, certain components of the ovulatory programme downstream from the gonadotrophin surge can be compromised or blocked experimentally and yet the terminal event will occur, albeit less efficiently than would be anticipated under spontaneous circumstances (e.g. lower number of ovulations, perhaps not all oocytes functional, poorer overall fertility).

Contribution of leucocytes and cytokines

Because a quite detailed account has already been presented in Chapter V concerning a contribution of cytokines to ovarian follicular physiology, the paragraphs that follow will be kept reasonably precise. The relevance here concerns the presence of enhanced populations of leucocytes in the tissues of pre-ovulatory follicles and the ability of such white blood cells to synthesise and secrete diverse cytokine molecules. The observation of changing populations of leucocytes in ovarian tissues is one of long-standing (e.g. Brambell, 1956; Zachariae, Asboe-Hansen & Boseila, 1958; Parkes & Deanesly, 1966; Parr, 1974), but greater attention has been paid to these cells in recent years because of a suspicion that their increased numbers might be involved in the terminal events of follicular growth and maturation (for a review, see Bukulmez & Arici, 2000).

Before becoming immersed in detail, it is well worth reflecting on whether cells of the immune system are attracted to – and active in – a follicle primarily in order to contribute to rupture or collapse of the structure with release of its oocyte. Alternatively, but not exclusively so, are they attracted by major changes in the vascular bed, transudation of increasing quantities of plasma, specific breakdown of capillaries and reorganisation of populations of follicular cells as luteal cells that are already secreting progesterone? In other words, are white cells in the wall of a mature Graafian follicle more involved in a weakening and degrading action of proteolytic enzymes on the collagenous connective tissue, or are they principally involved in modulating the events of steroidogenesis in a newly organised population of luteal cells? Organisation and support of incipient luteal cells includes a conspicuous phase of angiogenesis, and the molecular products of white cells could be involved here in regulating the expression of appropriate genes.

A large body of work, especially from the laboratory of Brännström, has argued strongly that products emanating from infiltrating leucocytes do indeed make an important contribution to the process of ovulation. Others with a long-standing interest in the field have been more circumspect and advised caution. According to Espey (1999), macrophages or other derivatives of leucocytes are seldom observed in non-vascular thecal tissue of a mature follicle before actual ovulation. Whereas Brännström and colleagues believe that leucocytes or their derivatives contribute to the primary events of ovulation, for example by releasing proteolytic enzymes to degrade follicular connective tissue, Espey's view is that leucocytes accumulate in ovulatory follicles as a response to an already-established acute inflammatory reaction. It is the latter reaction that activates thecal fibroblasts and initiates breakdown of the follicle wall prior to any infiltration of leucocytes (Espey, 1999). The present author finds no serious incompatibility here: leucocyte-derived enzymes could still be contributing to the process of ovulation. Even so, a critical question would be the extent to which recruitment of leucocytes is programmed by the gonadotrophin surge. Alternatively, is this a response to downstream events at some stage removed from binding of gonadotrophins to follicular tissues?

An involvement of cytokines with the oocyte and its cellular investment as a preparation for release into the lumen of the Fallopian tube should also be considered. In this interpretation, the ultimate influence of follicular leucocyte activity might be more on post-ovulatory events in the genital tract than on pre-ovulatory events in the gonad itself.

As a brief reminder concerning cytokines (see Chapter V), these are soluble proteins or glycoproteins that perform specific functions in a paracrine or autocrine manner. They act by binding to specific receptors, which themselves are coupled to systems of intracellular signalling. Cytokines tend to be produced not individually but rather in combination with other cytokines, and in patterns characteristic of a particular cell and stimulus. Thus has arisen the concept of specific cytokine networks. In such networks, there are stimulatory cytokines, inhibitory cytokines, cytokine inhibitors (not the same thing!) and cytokine inducers, all able to function in a particular physiological context (see Brännström, 1997).

White cells in the blood system that become prominent in the peri-ovulatory ovary include neutrophils (granulocytes) and monocytes (non-granular). Such cells are essentially spherical in the circulation but are transformed to an amoeboid shape upon entering extra-vascular sites. Attraction to such sites may involve the processes of adhesion, transmigration and chemotaxis, and the transformed white cells may be prompted to perform diverse functions. Polymorphonuclear leucocytes are certainly implicated in inflammatory responses and release secretory granules containing (1) enzymes such as collagenase, elastase, myeloperoxidase, cathepsins and lysozyme, and (2) cytokines such as IL-1 and TNF-α. These white cells may also secrete eicosanoids (see Chapter V), plasminogen activator and PAF (Brännström, 1997).

Diverse studies on the presence of polymorphonuclear leucocytes in the thecal tissue of Graafian follicles have noted a pre-ovulatory increase in the number of these cells. In rats, neutrophils are located mainly in the medulla before ovulation, but there is a pronounced increase in their numbers in both the medullary region and in the theca of ovulatory follicles after the LH peak or hCG administration (for a summary, see Brännström, 1997). Similarly, in rabbits and sheep, there is a marked increase in the numbers of polymorphonuclear leucocytes in the follicular wall close to the time of ovulation (Cavender & Murdoch, 1988). As to our own species, neutrophils were reported in the thecal layer of the follicle wall, in the cortical stroma and in the tunica albuginea (Brännström et al., 1994). Decreasing neutrophil counts in rats after administration of a neutrophil-depleting monoclonal antibody resulted in a reduced ovulation rate (Brännström et al., 1995).

In contrast to the finding in humans, eosinophilic granulocytes are the most abundant leucocyte sub-type in pig ovaries, and especially so in the thecal layer of pre-ovulatory follicles (Standaert, Zamor & Chew, 1991). Considerable numbers of eosinophils are also found in the theca externa of ovulating sheep follicles (Cavender & Murdoch, 1988). As to macrophages, large numbers are seen in the hilus region of the rat ovary but, during ovulation, a five-fold increase in the density of macrophages occurs in the thecal layer of such mature follicles. This is also the finding in the thecal layer and tunica albuginea of human follicles shortly before ovulation (Brännström et al., 1994). Large numbers of macrophages accumulate in the apical region of follicles undergoing ovulation. The potential rôle of macrophages in the process of ovulation could be related to secretion of plasminogen activator, collagenases, eicosanoids or nitric oxide.

Building on the observation of a marked increase in the number of leucocytes in the ovary close to the time of ovulation, various experiments have
been performed in laboratory rodents, especially in rats, which of course have
multiple ovulations. Using an *in vitro* model, LH-induced ovulation could be
stimulated by perfusion with additional or supplementary leucocytes (Hellberg
et al., 1991), suggesting some contribution from these cells. The leucocytes
appeared to accumulate on the theca–granulosa border, precisely where the
capillary network would be arrested by the basement membrane. Nonetheless,
if leucocytes are attracted towards follicular fluid (Brännström, 1997), and the
basement membrane of the follicle commences to undergo dissolution as ovulation approaches, doubtless with some seepage of follicular fluid, then migration
of these leucocytes towards the theca and theca–granulosa interface would be
anticipated. This could certainly be of physiological significance.

Mast cells resemble basophilic granulocytes circulating in the blood, but in
fact are an independent cell type. The mast cell of connective tissue is widely
distributed, including in the ovary, where mast cells are most abundant in the
central portion, clustering in the vicinity of the larger blood vessels, at least
in the rat ovary (Brännström, 1997). These cells contain histamine, heparin,
kininogenase, serotonin (5-hydroxytryptamine) and TNF-α, all of which may
be released upon activation. An increase in the concentration of histamine has
been noted in rabbit follicles before ovulation. Histamine will provoke ovulation
in vitro in the perfused rat ovary model (Schmidt, Owman & Sjöberg, 1986),
seemingly by causing vasodilation and perhaps also by increasing contractile
activity. Even so, a critical question remains concerning the extent to which
mast cell products that could influence ovulation are released as a consequence
of the pre-ovulatory gonadotrophin surge.

Cytokines of potential importance in the ovary during ovulation may be
secreted by migrating leucocytes and/or the follicular cells themselves. They
would include IL-1, TNF-α and colony stimulating factors (CSFs). IL-1 expression is induced by LH, has diverse pro-inflammatory influences, and is
extensively secreted by activated macrophages and neutrophils. Relatively high
concentrations of IL-1 have been reported in human follicular fluid (Khan *et al.*,
1988), and rat ovaries secrete IL-1 bioactivity during ovulation (Brännström
et al., 1994). The amount of IL-1β mRNA increased in rat ovaries at the time
of ovulation, principally in the thecal tissues (Hurwitz *et al.*, 1991), and human
theca and granulosa cells together contain a complete IL-1β system (Hurwitz
et al., 1992). This is also true for mouse pre-ovulatory follicles when examined
by immunohistochemistry (Simón *et al.*, 1994).

Stimulation of the ovulatory process with recombinant human IL-1β could
be demonstrated in the perfused rat ovary (Brännström, Wang & Norman,

1993b). IL-1 may be influencing ovulation principally by stimulating synthesis of progesterone and prostaglandins, and likewise by stimulating production of nitric oxide. Somewhat paradoxically in the context of ovulation, IL-1 appears to inhibit peri-ovulatory increases in plasminogen activator activity (Bonello, Norman & Brännström, 1993), and the expression of plasminogen activator inhibitors is increased in human granulosa cells by IL-1 (Piquette *et al.*, 1994).

TNF-α is also a potent paracrine mediator of several inflammatory and immune functions. It is secreted by activated monocytes and neutrophilic granulocytes, and also from endothelial cells and fibroblasts. Pre-ovulatory follicles reveal an extensive distribution of TNF-α mRNA (Roby & Terranova, 1989; Sancho-Tello *et al.*, 1992), rat ovaries close to ovulation secrete TNF-α bioactivity (Brännström *et al.*, 1994), and TNF-α immunoactivity is found in human follicular fluid (Wang *et al.*, 1992). Once again using the perfused rat ovary *in vitro* model, TNF-α will potentiate the extent of LH-induced ovulation. An increase in the secretion of both progesterone and prostaglandins may be the primary route of involvement of TNF-α in the events of ovulation (Fukumatsu *et al.*, 1992; Wang *et al.*, 1992).

CSFs are a family of glycosylated cytokines. Granulocyte-CSF (GCSF) and granulocyte–macrophage CSF (GMCSF) are known to recruit, regulate and activate granulocytes and macrophages in peripheral tissues. These cell types are found within the stromal and thecal regions of the ovary before and after ovulation (Brännström, Mayrhofer & Robertson, 1993a; Jasper *et al.*, 2000). The perfused rat ovary model suggested an involvement of GMCSF in ovulation, with increased amounts of GMCSF produced just before so-called rupture of the follicles (Brännström *et al.*, 1994).

Integration of the diverse observations reported above is not straightforward, but Brännström (1997) has proposed the following, doubtless oversimplified, model. The pre-ovulatory gonadotrophin surge triggers an increased expression of IL-1 and TNF-α. These cytokines will enhance the secretion of progesterone by granulosa cells. They will also induce expression of adhesion molecules and chemokines to recruit neutrophilic granulocytes (NG) and monocytes (M) to the extracellular matrix of the theca cells. Activation of these white cell types by GMCSF and other CSFs would lead to release of proteases and vasoactive substances (Fig. VIII.4). Brännström (1997) concluded provocatively that IL-1 may limit the proteolysis of a pre-ovulatory follicle to the apical region by decreasing the influence of plasminogen activator, a proposal that awaits experimental support.

Despite these remarks focusing on cytokines and related molecules, one must not lose sight of the contribution of histamine, various kinins, prostaglandins and perhaps lipoxygenase products of arachidonate metabolism. Indeed, the

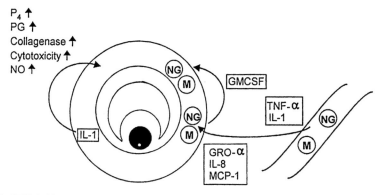

Fig. VIII.4. Simplified model depicting a mature Graafian follicle and the potential involvement of leucocytes and various cytokines in the terminal events that culminate in ovulation. P_4, progesterone; PG, prostaglandin; IL, interleukin; NG, neutrophilic granulocytes; M, monocytes; GMCSF, granulocyte–macrophage colony stimulating factor; TNF, tumour necrosis factor; GRO, growth regulated oncogene; MCP, monocyte chemoattractant protein. (Adapted from Brännström, 1997.)

interconnected lipoxygenase and cyclo-oxygenase pathways may optimise ovulation and facilitate steroidogenesis (Mikuni *et al.*, 1998). Similarly, an involvement of serine proteases, kallikrein and collagenase activity should enter into the picture (see Espey & Lipner, 1994). The critical problem remains the manner in which such multiple molecular contributions – and doubtless many others – are coordinated and integrated. As proposed elsewhere in this monograph, a sensitive influence of the oocyte in these vital events cannot be neglected.

Rôle of fimbriated extremity of Fallopian tube

One physical aspect that tends to be overlooked in descriptions of the process of ovulation is the massaging influence of the fimbriated infundibulum on the follicular surface in species without an encompassing ovarian bursa. In seeking a comprehensive explanation for formation of an aperture at the apex of a Graafian follicle, with or without the appearance of a stigma, insufficient consideration has been given to the local effects of this contact between gonad and genital tract. The inner epithelial surface of the fimbriated infundibulum will irritate the apex of protruding follicles as it sweeps back and forth, a stimulus that will be accentuated by rapid beating of the dense tracts of cilia. At the very least, such intimacy could prompt the release of histamine and other vasoactive substances. It should be recalled that the beat of Fallopian tube cilia is most active at the time of ovulation (Blandau, 1969, 1973), being programmed by a particular ratio of oestradiol to progesterone (Borell, Nilsson & Westman, 1957).

In an attempt to determine whether the fimbrial movement so described has a specific influence on the events of ovulation, the fimbriated extremity was surgically removed from one Fallopian tube, leaving the contralateral tube intact by way of control. In a small series of domestic pigs examined close to the anticipated time of hCG-induced ovulation, no chronological advantage could be found in the timing of ovulation on the ovary embraced by an intact fimbria, nor was the morphology of ovulation detectably different between follicles on the two ovaries (Fléchon & Hunter, 1980). Further studies might have yielded a more intriguing result, for the experiment was by no means perfect in the sense that even after fimbrectomy, intestinal movements and other forms of tissue contact would have massaged the ovarian surface, and especially that of protruding follicles. Accordingly, it seems worth keeping an open mind on a potential involvement of tracts of cilia on the inner surface of the fimbriated infundibulum in modifications to the apical region of a pre-ovulatory follicle (see Hunter, 1988).

Ischaemic model for studying ovulation

This chapter has been concerned with examining the process of ovulation under physiological conditions. It is doubtless banal to say that much more remains to be learnt and that one way forward could depend upon developing suitable surgical models. An experimental preparation that may prove valuable has been noted in the domestic pig, a species that ovulates approximately 40 hours after the gonadotrophin surge (Table VII.1). This interval is mentioned once again, since it has reasonably close parallels with the interval in women between the gonadotrophin surge and ovulation, and also because it is of sufficient length to permit a useful stepwise analysis of intermediate events.

In essence, the model is an acute ischaemic ovarian preparation. If the ovarian blood supply is arrested at some stage during the last hour before hCG-induced ovulation by placing ligatures around the ovarian ligaments or clamping with appropriate forceps, then pre-ovulatory follicles of 9–10 mm diameter will rapidly become first haemorrhagic and then flaccid. Such changes resemble the condition seen during spontaneous ovulation and occur within 2–3 minutes of impeding ovarian blood flow. This observation was made repeatedly during an earlier series of experiments in which animals were ovariectomised shortly before ovulation to furnish follicular tissue for organ culture (Baker, Hunter & Neal, 1975; Hunter & Baker, 1975). Dramatic changes were clearly occurring within follicles with a reduced or arrested blood supply and, although a grossly unphysiological approach, a local degree of follicular ischaemia may initiate part of the final ovulation cascade.

One other model involving pig ovaries might usefully be exploited to reveal more of the events preceding ovulation. In experiments in mated animals in which the objective was to recover recently ovulated eggs and spermatozoa from the same Fallopian tube without any possibility of direct contact or interaction between the male and female gametes (for use in subsequent *in vitro* fertilisation studies), double ligatures were positioned around the Fallopian tube tissues and adjacent vascular and lymphatic arcades in the region of the ampullary–isthmic junction. Such ligatures were introduced at mid-ventral laparotomy during the pre-ovulatory interval in oestrous animals and, for absolute confidence in the separated suspensions of gametes, the tissues were transected between the ligatures. Mature follicles in the ovary adjacent to the ligated Fallopian tube failed to ovulate and became cystic, whereas follicles in the contralateral ovary adjoining an intact Fallopian tube ovulated as anticipated. This model suggests a number of points bearing on the process of ovulation, not least the vital contribution of an appropriate blood supply and perhaps a critical traffic in white blood cells involving lymphatic pathways. It offers a surgical preparation open to exploitation in a number of ways, concerning both the pre-ovulatory Graafian follicles and the oocytes within.

Genes involved in the ovulatory cascade

At a molecular level, characterisation of genes involved in the ovulatory cascade has so far added detail rather than specific new insights into the actual process of ovulation, here meaning shedding of an oocyte from a collapsing or ruptured follicle. For example, genes that programme enzyme activity in the biosynthetic pathway for steroid hormones have now been highlighted and their pattern of expression corresponds with the current understanding of steroid synthesis and secretion in the various cell layers of a Graafian follicle. Such information has been extended to enzymes involving members of the prostaglandin family. Concerning genes for prostaglandin synthase (PGS), there are known to be at least two isoforms: PGS-1, a 69 kDa isoform said to be constitutively expressed in the ovary and localised to theca (and luteal) cells. By contrast, PGS-2, a 72 kDa isoform, is transiently induced by the gonadotrophin surge and said to be localised to granulosa cells of those follicles destined to ovulate (Wong & Richards, 1991; Sirois, Simmons & Richards, 1992). Comparable information could be offered in respect of diverse hormones, enzymes and growth factors, but it would be repetitive of material elsewhere in the text.

Notable amongst the earlier papers in this field were contributions by Fortune & Voss (1993) and Einspanier *et al.* (1994). The reviews of Richards (1994) and Richards *et al.* (1995) remain highlights. In more recent publications, the application of differential display techniques for identification of genes involved in the process of ovulation has been emphasised (e.g. Espey *et al.*, 1999). Even so, in a multi-factorial, multi-compartmental event such as mammalian ovulation, drawing a distinction between genes that are essential and gene activity that is permissive will seldom be straightforward, no matter which experimental model is employed.

Concluding remarks

This chapter has ranged widely over topics related to ovulation, with an initial consideration of the distinction between so-called spontaneous and induced ovulation. The range in size of Graafian follicles close to ovulation was under-lined with reference to laboratory rodents, domestic farm animals and humans, and the significance of mature follicular diameter interpreted from several angles. The analysis then examined the process of ovulation itself, with bio-chemical detail focusing on ultimate breakdown of the follicular apex and re-lease of the oocyte in its investment of cumulus cells and escaping follicular fluid (Fig. VIII.5). The sequence of events in the follicular wall is far from under-stood, although there is persuasive evidence that the classical gonadal hor-mone progesterone has an important influence. Because impending ovulation shows many features of an inflammatory process, there is a bewildering array of molecules to consider, even if various lines of evidence suggest that not each one of these is absolutely essential in its own right.

The vascular changes in the follicle wall leading to dramatic changes of colour reflect an increased blood flow and permeability of the capillary bed; thecal tissues become oedematous and there is an associated proteolytic cas-cade that underlies breakdown of the wall. The dogma remains that this cascade involves plasminogen activator, plasmin and collagenases, with enhanced pro-teolytic activity and the extent of tissue degradation being regulated locally by specific inhibitors. These include members of the inhibitors of metallopro-teases type 1 family, such as TIMP-1 which is prominent in the regulation of collagenase activity (see Chapter VII) and remodelling of follicular tissues. Prominent also in current interpretations of the ovulatory cascade are contribu-tions from prostaglandins, cytokines, PAF, components of the renin–angiotensin system (Yoshimura *et al.*, 1992, 1993), and the ever-more fashionable nitric oxide.

Relevant to secretion of many of the above molecules, an involvement of migrating populations of leucocytes within the follicular wall and antrum was

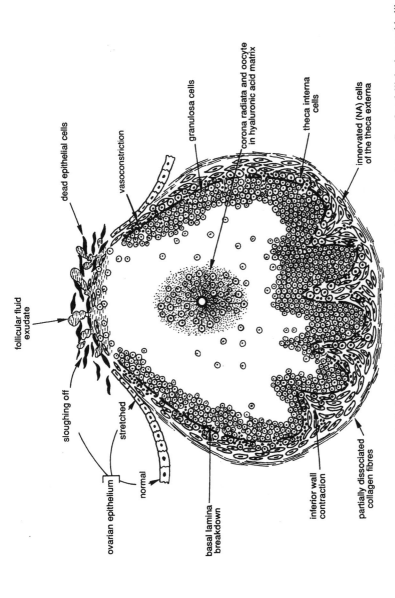

Fig. VIII.5. An attractive drawing of the morphological and histological changes occurring in a pre-ovulatory Graafian follicle that enable liberation of the oocyte within a loosened investment of cumulus cells. NA, noradrenergic. (Adapted from Thibault & Levasseur, 1988.)

assessed. Activated white blood cells could be exerting potent influences in a highly localised manner. Even so, caution was advised in this interpretation of potential leucocytic contributions, not least when the remarkable precision of ovulation time after the surge of gonadotrophic hormones was recalled. Could a migrating population of cells of varying size really bestow such precision? Whether there was any rôle in the events of ovulation for a form of irritation by the fimbriated extremity on the follicular apex was questioned from limited experimental data. Thinning and weakening in this portion of the follicle wall may or may not be reflected in a protruding stigma. This seems to vary both between species and between follicles in the same ovary. The relevance of autonomic nervous inputs within the thecal layers of pre-ovulatory follicles has been touched on in previous chapters, not least in terms of prompting contractile movements and expulsion of antral contents, but no new experimental evidence was brought to light. It remains an exciting and fruitful theme for future studies.

A final thought bears on the gonad in general and its twin functions of liberating gametes and secreting hormones in particular. One of the complications in clarifying the events of ovulation is to disentangle the two. Whilst the endocrine activity of the ovary after puberty and before the menopause is always prominent, at least a proportion of the hormones synthesised and secreted at around the time of ovulation may not be making an essential contribution to actual release of an oocyte. For those who lose sight of this dichotomy, the scope for confusion is considerable.

References

Adashi, E.Y. (1990). The potential relevance of cytokines to ovarian physiology: the emerging role of resident ovarian cells of the white blood cell series. *Endocrine Reviews*, **11**, 454–64.

Adashi, E.Y. (ed.) (2000). *Ovulation: Evolving Scientific and Clinical Concepts.* Serono Symposia USA MA. Norwell, New York: Springer-Verlag New York Inc.

Ainsworth, L., Baker, R.D. & Armstrong, D.T. (1975). Pre-ovulatory changes in follicular fluid prostaglandin F levels in swine. *Prostaglandins*, **9**, 915–25.

Armstrong, D.T. & Grinwich, D.L. (1972). Blockade of spontaneous and LH-induced ovulation in rats by indomethacin, an inhibitor of prostaglandin biosynthesis. *Prostaglandins*, **1**, 21–8.

Asdell, S.A. (1962). Mechanism of ovulation. In *The Ovary*, ed. S. Zuckerman, pp. 435–49. London & New York: Pergamon Press.

Austin, C.R. (1961). *The Mammalian Egg.* Oxford: Blackwell Scientific Publications.

Baird, D.T. (1983). The ovary. In *Reproduction in Mammals*, Book 3, *Hormonal Control of Reproduction*, ed. C.R. Austin & R.V. Short, pp. 91–114. Cambridge: Cambridge University Press.

Baird, D.T. (1990). The selection of the follicle of the month. In *From Ovulation to Implantation*, ed. J.L.H. Evers & M.J. Heineman, pp. 3–19. International Congress Series 917. Amsterdam: Excerpta Medica.

Baird, D.T. (1999). Folliculogenesis and gonadotrophins. In *Gonadotrophins and Fertility in Women*, vol. 3, ed. E.Y. Adashi, D.T. Baird & P.G. Crosignani, pp. 1–10. Rome: Serono Fertility Series.

Baker, T.G., Hunter, R.H.F. & Neal, P. (1975). Studies on the maintenance of porcine Graafian follicles in organ culture. *Experientia*, **31**, 133–5.

Bauminger, S. & Lindner, H.R. (1975). Periovulatory changes in ovarian prostaglandin formation and their hormonal control in the rat. *Prostaglandins*, **9**, 737–51.

Beers, W.H., Strickland, S. & Reich, E. (1975). Ovarian plasminogen activator: relationship to ovulation and hormonal regulation. *Cell*, **6**, 387–94.

Bjersing, L. & Cajander, S. (1974). Ovulation and the mechanism of follicle rupture. I. Light microscopic changes in rabbit ovarian follicles prior to induced ovulation. *Cell and Tissue Research*, **149**, 287–300.

Blandau, R.J. (1955). Ovulation in the living albino rat. *Fertility and Sterility*, **6**, 391–404.

Blandau, R.J. (1969). Gamete transport – comparative aspects. In *The Mammalian Oviduct*, ed. E.S.E. Hafez & R.J. Blandau, pp. 129–62. Chicago & London: University of Chicago Press.

Blandau, R.J. (1973). Gamete transport in the female mammal. In *Handbook of Physiology*, Section 7, *Endocrinology*, vol. II, ed. R.O. Greep & E.B. Astwood, pp. 153–63. Washington, DC: American Physiological Society.

Bonello, N., Norman, R.J. & Brännström, M. (1993). Interleukin-1β inhibits luteinising hormone-induced plasminogen activatory activity in rat preovulatory follicles *in vitro*. *Endocrine*, **3**, 49–54.

Borell, U., Nilsson, O. & Westman, A. (1957). Ciliary activity in the rabbit Fallopian tube during oestrus and after copulation. *Acta Obstetricia et Gynecologica Scandinavica*, **36**, 22–8.

Brambell, F.W.R. (1956). Ovarian changes. In *Marshall's Physiology of Reproduction*, 3rd edn, vol. 1, part 1, ed. A.S. Parkes, pp. 397–542. London: Longmans, Green & Co.

Brännström, M. (1997). Intra-ovarian immune mechanisms in ovulation. In *Microscopy of Reproduction and Development: A Dynamic Approach*, ed. P.M. Motta, pp. 163–8. Rome: Antonio Delfino Editore.

Brännström, M., Bonello, N., Norman, R.J. & Robertson, S.A. (1995). Reduction of ovulation rate in the rat by administration of a neutrophil-depleting monoclonal antibody. *Journal of Reproductive Immunology*, **29**, 265–70.

Brännström, M., Mayrhofer, G. & Robertson, S.A. (1993a). Localisation of leukocyte subsets in the rat ovary during the periovulatory period. *Biology of Reproduction*, **48**, 277–86.

Brännström, M., Norman, R.J., Seamark, R.F. & Robertson, S.A. (1994). The rat ovary produces cytokines during ovulation. *Biology of Reproduction*, **50**, 88–94.

Brännström, M., Wang, L. & Norman, R.J. (1993b). Ovulatory effect of interleukin-1β on the perfused rat ovary. *Endocrinology*, **132**, 399–404.

Bukulmez, O. & Arici, A. (2000). Leukocytes in ovarian function. *Human Reproduction Update*, **6**, 1–15.

Cameliet, P., Schoonjans, L., Kieckens, L., *et al.* (1994). Physiological consequences of loss of plasminogen activator gene function in mice. *Nature (London)*, **368**, 419–24.

Campbell, B.K., Dobson, H., Baird, D.T. & Scaramuzzi, R.J. (1999). Examination of the relative role of FSH and LH in the mechanism of ovulatory follicle selection in sheep. *Journal of Reproduction and Fertility*, **117**, 355–67.

Cavender, J.L. & Murdoch, W.J. (1988). Morphological studies of the microcirculatory system of periovulatory ovine follicles. *Biology of Reproduction*, **39**, 989–97.

Chen, B.X., Yuen, Z.X. & Pan, G.W. (1985). Semen-induced ovulation in the Bactrian camel (*Camelus bactrianus*). *Journal of Reproduction and Fertility*, **74**, 335–9.

Dissen, G.A., Hill, D.F., Costa, M.E., Dees, W.L., Lara, H.E. & Ojeda, S.R. (1996). A role for TrkA nerve growth factor receptors in mammalian ovulation. *Endocrinology*, **137**, 198–209.

Driancourt, M.A. & Cahill, L.P. (1984). Pre-ovulatory follicular events in sheep. *Journal of Reproduction and Fertility*, **71**, 205–11.

Driancourt, M.A., Gougeon, A., Monniaux, D., Royère, D. & Thibault, C. (2001). Folliculogenèse et ovulation. In *La Reproduction chez les Mammifères et l'Homme*, ed. C. Thibault & M.-C. Levasseur, pp. 316–47. Paris: Ellipses.

Driancourt, M.A., Gougeon, A., Royère, D. & Thibault, C. (1993). Ovarian function. In *Reproduction in Mammals and Man*, ed. C. Thibault, M.C. Levasseur & R.H.F. Hunter, pp. 281–305. Paris: Ellipses.

Einspanier, A., Ivell, R., Rune, G. & Hodges, J.K. (1994). Oxytocin gene expression and oxytocin immunoactivity in the ovary of the common marmoset monkey (*Callithrix jacchus*). *Biology of Reproduction*, **50**, 1216–22.

El-Fouly, M.A., Cook, B., Nekola, M. & Nalbandov, A.V. (1970). Role of the ovum in follicular luteinisation. *Endocrinology*, **87**, 288–93.

Espey, L.L. (1974). Ovarian proteolytic enzymes and ovulation. *Biology of Reproduction*, **10**, 216–35.

Espey, L.L. (1978). Ovarian contractility and its relationship to ovulation: a review. *Biology of Reproduction*, **19**, 540–51.

Espey, L.L. (1980). Ovulation as an inflammatory reaction – a hypothesis. *Biology of Reproduction*, **22**, 73–106.

Espey, L.L. (1994). Current status of the hypothesis that mammalian ovulation is comparable to an inflammatory reaction. *Biology of Reproduction*, **50**, 233–8.

Espey, L.L. (1997). Mammalian ovulation: morpho-physiological aspects. In *Microscopy of Reproduction and Development: A Dynamic Approach*, ed. P.M. Motta, pp. 143–50. Rome: Antonio Delfino Editore.

Espey, L.L. (1999). Ovulation. In *Encyclopaedia of Reproduction*, vol. 3, ed. E. Knobil & J.D. Neill, pp. 605–14. San Diego & London: Academic Press.

Espey, L.L. & Lipner, H. (1994). Ovulation. In *The Physiology of Reproduction*, 2nd edn, ed. E. Knobil & J.D. Neill, pp. 725–80. New York: Raven Press.

Espey, L.L., Tanaka, N., Woodard, D.S., Harper, M.J.K. & Okamura, H. (1989). Decrease in ovarian platelet-activating factor during ovulation in the gonadotropin-primed immature rat. *Biology of Reproduction*, **41**, 104–10.

Espey, L.L., Vladu, B., Skelsey, M. & Richards, J.S. (1999). Application of differential display to the isolation of genes unique to ovulation in the rat. Unpublished manuscript, available from the author.

Everett, J.W. (1961). The mammalian female reproductive cycle and its controlling mechanisms. In *Sex and Internal Secretions*, 3rd edn, vol. 1, ed. W.C. Young, pp. 497–555. Baltimore, MD: Williams & Wilkins.

Fléchon, J.E. & Hunter, R.H.F. (1980). Ovulation in the pig: a light and scanning electron microscope study. *Proceedings of the Society for the Study of Fertility*, Abstract 5, 9.

Foote, W.D. & Thibault, C. (1969). Recherches expérimentales sur la maturation *in vitro* des ovocytes de truie et de veau. *Annales de Biologie Animale, Biochimie, Biophysique*, **9**, 329–49.

Fortune, J.E. & Voss, A.K. (1993). Oxytocin gene expression and action in bovine preovulatory follicles. *Regulatory Peptides*, **45**, 257–61.

Fukumatsu, Y., Katabuchi, H., Naito, M., Takeya, M., Takahashi, K. & Okamura, H. (1992). Effect of macrophages on proliferation of granulosa cells in the ovary in rats. *Journal of Reproduction and Fertility*, **96**, 241–9.

Gosden, R.G. & Telfer, E. (1987). Scaling of follicular sizes in mammalian ovaries. *Journal of Zoology*, **211**, 157–68.

Gougeon, A. (1996). Regulation of ovarian follicular development in primates: facts and hypotheses. *Endocrine Reviews*, **17**, 121–55.

Gougeon, A. (1997). Kinetics of human ovarian follicular development. In *Microscopy of Reproduction and Development: A Dynamic Approach*, ed. P.M. Motta, pp. 67–77. Rome: Antonio Delfino Editore.

Guttmacher, M.S. & Guttmacher, A.F. (1921). Morphological and physiological studies on the musculature of the mature Graafian follicle of the sow. *Johns Hopkins Hospital Bulletin*, **32**, 394–9.

Harper, M.J.K. (1963). Ovulation in the rabbit: the time of follicular rupture and expulsion of the eggs, in relation to injection of luteinising hormone. *Journal of Endocrinology*, **26**, 307–16.

Harper, M.J.K. (1968). Pharmacological control of reproduction in women. In *Progress in Drug Research*, vol. 12, ed. E. Jucker, pp. 47–136. Basel & Stuttgart: Birkhäuser Verlag.

Harris, G.W. (1964). Sex hormones, brain development and brain function. *Endocrinology*, **75**, 627–48.

Harris, G.W. (1969). Ovulation. *American Journal of Obstetrics and Gynecology*, **105**, 659–69.

Hellberg, P., Thomsen, P., Janson, P.O. & Brännström, M. (1991). Leukocyte supplementation increases the luteinising hormone-induced ovulation rate in the *in vitro*-perfused rat ovary. *Biology of Reproduction*, **44**, 791–7.

Hillier, S.G. (1994). Hormonal control of folliculogenesis and luteinisation. In *Molecular Biology of the Female Reproductive System*, ed. J.K. Findlay, pp. 1–37. London: Academic Press.

Hisaw, F.L. (1947). Development of the Graafian follicle and ovulation. *Physiological Reviews*, **27**, 95–119.

Hunter, R.H.F. (1967). Polyspermic fertilisation in pigs during the luteal phase of the oestrous cycle. *Journal of Experimental Zoology*, **165**, 451–60.

Hunter, R.H.F. (1972). Ovulation in the pig: timing of the response to injection of human chorionic gonadotrophin. *Research in Veterinary Science*, **13**, 356–61.

Hunter, R.H.F. (1978). Intraperitoneal insemination, sperm transport and capacitation in the pig. *Animal Reproduction Science*, **1**, 167–79.

Hunter, R.H.F. (1979). Ovarian follicular responsiveness and oocyte quality after go-nadotrophic stimulation of mature pigs. *Annales de Biologie Animale, Biochimie, Biophysique*, **19**, 1511–20.

Hunter, R.H.F. (1980). *Physiology and Technology of Reproduction in Female Domestic Animals*. London & New York: Academic Press.

Hunter, R.H.F. (1982). *Reproduction of Farm Animals*. London & New York: Longman.

Hunter, R.H.F. (1988). *The Fallopian Tubes: Their Rôle in Fertility and Infertility*. Berlin, Heidelberg & New York: Springer-Verlag.

Hunter, R.H.F. (1995). Human sperm reservoirs and Fallopian tube function: a rôle for the intra-mural portion? *Acta Obstetricia et Gynecologica Scandinavica*, **74**, 677–81.

Hunter, R.H.F. (1998). Have the Fallopian tubes a vital rôle in promoting fertility? *Acta Obstetricia et Gynecologica Scandinavica*, **77**, 475–86.

Hunter, R.H.F. & Baker, T.G. (1975). Development and fate of porcine Graafian follicles identified at different stages of the oestrous cycle. *Journal of Reproduction and Fertility*, **43**, 193–6.

Hunter, R.H.F., Cook, B. & Baker, T.G. (1976). Dissociation of response to injected gonadotropin between the Graafian follicle and oocyte in pigs. *Nature (London)*, **260**, 156–8.

Hunter, R.H.F. & Poyser, N.L. (1985). Ovarian follicular fluid concentrations of prostaglandins E_2, $F_{2\alpha}$ and I_2 during the pre-ovulatory period in pigs. *Reproduction, Nutrition et Développement*, **25**, 909–17.

Hurwitz, A., Loukides, J., Ricciarelli, E., *et al.* (1992). Human intraovarian interleukin-1 (IL-1) system: highly compartmentalized and hormonally dependent regulation of the genes encoding IL-1, its receptor, and its receptor antagonist. *Journal of Clinical Investigation*, **89**, 1746–54.

Hurwitz, A., Ricciarelli, E., Botero, L., Rohan, R.M., Hernandez, E.R. & Adashi, E.Y. (1991). Endocrine and autocrine mediated regulation of rat ovarian (theca-interstitial) interleukin-1β gene expression: gonadotropin-dependent preovulatory acquisition. *Endocrinology*, **129**, 3427–9.

Jasper, M.J., Robertson, S.A., Van der Hoek, K.H., Bonello, N., Brännström, M. & Norman, R.J. (2000). Characterisation of ovarian function in granulocyte–macrophage colony-stimulating factor deficient mice. *Biology of Reproduction*, **62**, 704–13.

Jöchle, W. (1973). Coitus-induced ovulation. *Contraception*, **7**, 523–64.

Jöchle, W. (1975). Current research in coitus-induced ovulation: a review. *Journal of Reproduction and Fertility*, Supplement, **22**, 165–207.

Jöchle, W. (1997). Letter to the Editor. *Reproduction in Domestic Animals*, **32**, 325.

Khan, S.A., Schmid, K., Hallin, P., Paul, R.D., Geyter, C.D. & Nieschlag, E. (1988). Human testis cytosol and ovarian follicular fluid contains high amounts of interleukin-1-like factor(s). *Molecular and Cell Endocrinology*, **58**, 221–30.

Killick, S. & Elstein, M. (1987). Pharmacologic production of luteinised unrup-tured follicles by prostaglandin synthetase inhibitors. *Fertility and Sterility*, **47**, 773–7.

Koos, R.D. (1995). Increased expression of vascular endothelial growth/permeability factor in the rat ovary following an ovulatory gonadotropin stimulus: potential roles in follicle rupture. *Biology of Reproduction*, **52**, 1426–35.

LeMaire, W.J., Leidner, R. & Marsh, J.M. (1975). Pre- and post-ovulatory changes in the concentration of prostaglandins in rat Graafian follicles. *Prostaglandins*, **9**, 221–9.

LeMaire, W.J., Yang, N.S.T., Behrman, H.R. & Marsh, J.M. (1973). Preovulatory changes in the concentration of prostaglandins in rabbit Graafian follicles. *Prostaglandins*, **3**, 367–76.

Lindner, H.R., Amsterdam, A., Salomon, Y., *et al.* (1977). Intraovarian factors in ovulation: determinants of follicular response to gonadotrophins. *Journal of Reproduction and Fertility*, **51**, 215–35.

Lipner, H. (1973). Mechanism of mammalian ovulation. In *American Handbook of Physiology*, Section 7, *Endocrinology*, vol. II, part 1, ed. R.O. Greep & E.B. Astwood, pp. 409–37. Washington, DC: American Physiological Society.

Lipner, H. (1988). Mechanism of mammalian ovulation. In *The Physiology of Reproduction*, vol. 1, ed. E. Knobil & J.D. Neill, pp. 447–88. New York: Raven Press.

Lovell, J.E. & Getty, R. (1968). Fate of semen in the uterus of the sow: histologic study of endometrium during the 27 hours after natural service. *American Journal of Veterinary Research*, **29**, 609–25.

Macchiarelli, G., Nottola, S.A., Vizza, E., Correr, S. & Motta, P.M. (1995). Changes of ovarian microvasculature in hCG stimulated rabbits. A scanning electron microscopic study of corrosion casts. *Italian Journal of Anatomy and Embryology*, **100**, Suppl. 1, 469–77.

Mikuni, M., Yoshida, M., Hellberg, P., *et al.* (1998). The lipoxygenase inhibitor, nordihydroguaiaretic acid, inhibits ovulation and reduces leukotriene and prostaglandin levels in the rat ovary. *Biology of Reproduction*, **58**, 1211–16.

Moor, R.M., Hay, M.F. & Seamark, R.F. (1975). The sheep ovary: regulation of steroidogenic, haemodynamic and structural changes in the largest follicle and adjacent tissue before ovulation. *Journal of Reproduction and Fertility*, **45**, 595–604.

Motta, P., Cherney, D.D. & Didio, L.J.A. (1971). Scanning and transmission electron microscopy of the ovarian surface in mammals with special reference to ovulation. *Journal of Submicroscopy and Cytology*, **3**, 85–100.

Motta, P.M., Nottola, S.A., Familiari, G., Macchiarelli, G., Vizza, E. & Correr, S. (1995). Ultrastructure of human reproduction from folliculogenesis to early embryo development. A review. *Italian Journal of Anatomy and Embryology*, **100**, 9–72.

Motta, P.M. & Van Blerkom, J. (1975). A scanning electron microscopic study of the luteo-follicular complex. II. Events leading to ovulation. *American Journal of Anatomy*, **143**, 241–64.

Nalbandov, A.V. (1961). Mechanisms controlling ovulation of avian and mammalian follicles. In *Control of Ovulation*, ed. C.A. Villee, pp. 122–31. Oxford, New York & London: Pergamon Press.

O'Grady, J.P., Caldwell, B.V., Auletta, F.J. & Speroff, L. (1972). The effects of an inhibitor of prostaglandin synthesis (indomethacin) on ovulation, pregnancy and pseudopregnancy in the rabbit. *Prostaglandins*, **1**, 97–106.

Parkes, A.S. & Deanesly, R. (1966). The ovarian hormones. In *Marshall's Physiology of Reproduction*, 3rd edn, vol. 3, ed. A.S. Parkes, pp. 570–828. London & New York: Longmans, Green & Co.

Parr, E.L. (1974). Histological examination of the rat ovarian follicle wall prior to ovulation. *Biology of Reproduction*, **11**, 483–503.

Pincus, G. (1965). *The Control of Fertility*. New York & London: Academic Press.

Pincus, G. & Chang, M.C. (1953). Effects of progesterone and related compounds on ovulation and early development in rabbit. *Acta Physiologica Latino-Americana*, **3**, 177–83.

Piquette, G.N., Simon, C., Danasouri, I.E., Frances, A. & Polan, M.L. (1994). Gene regulation of interleukin-1β, interleukin-1 receptor type 1, and plasminogen activator inhibitor-1 and -2 in human granulosa-luteal cells. *Fertility and Sterility*, **62**, 760–70.

Pitkjanen, I.G. (1955). Ovulation, fertilisation and early embryonic development in the pig. *Izvestia Akademii Nauk SSSR, Series Biologicheskaya*, **3**, 120–31.

Rajaniemi, H. & Vanha-Perttula, T. (1972). Specific receptor for LH in the ovary: evidence by autoradiography and tissue fractionation. *Endocrinology*, **90**, 1–9.

Reich, R., Daphina-Iken, D., Chun, S.Y., *et al.* (1991). Preovulatory changes in ovarian expression of collagenases and tissue metalloproteinase inhibitor messenger ribonucleic acid: role of eicosanoids. *Endocrinology*, **129**, 1869–75.

Richards, J.S. (1988). Molecular aspects of hormone action in ovarian follicular development, ovulation and luteinisation. *Annual Review of Physiology*, **50**, 441–63.

Richards, J.S. (1994). Hormonal control of gene expression in the ovary. *Endocrine Reviews*, **15**, 725–51.

Richards, J.S., Fitzpatrick, S.L., Clemens, J.W., Morris, J.K., Alliston, T. & Sirois, J. (1995). Ovarian cell differentiation: a cascade of multiple hormones, cellular signals and regulated genes. *Recent Progress in Hormone Research*, **50**, 223–54.

Roby, K.F. & Terranova, P.F. (1989). Localisation of tumor necrosis factor (TNF) in rat and bovine ovary using immunocytochemistry and cell blot: evidence for granulosal production. In *Growth Factors and the Ovary*, ed. A.N. Hirshfield, pp. 273–8. New York: Plenum Press.

Rondell, P. (1970). Biophysical aspects of ovulation. *Biology of Reproduction*, Supplement, **2**, 64–89.

Sancho-Tello, M., Perez-Roger, I., Imakawa, K., Tilzer, L. & Terranova, P.F. (1992). Expression of tumor necrosis factor-α in the rat ovary. *Endocrinology*, **130**, 1359–64.

Scaramuzzi, R.J., Adams, N.R., Baird, D.T., *et al.* (1993). A model for follicle selection and the determination of ovulation rate in the ewe. *Reproduction, Fertility and Development*, **5**, 459–78.

Schmidt, G., Owman, C. & Sjöberg, N.O. (1986). Histamine induces ovulation in the isolated perfused rat ovary. *Journal of Reproduction and Fertility*, **78**, 159–66.

Schochet, S.S. (1915). A suggestion as to the process of ovulation and ovarian cyst formation. *Anatomical Record*, **10**, 447–57.

Schroeder, P.C. & Talbot, P. (1985). Ovulation in the animal kingdom: a review with an emphasis on the role of contractile processes. *Gamete Research*, **11**, 191–221.

Simón, C., Tsafriri, A., Chun, S.Y., Piquette, G.N., Dang, W. & Polan, M.L. (1994). Interleukin-1 receptor antagonist suppresses human chorionic gonadotropin-induced ovulation in the rat. *Biology of Reproduction*, **51**, 662–7.

Sirois, J. (1994). Induction of prostaglandin endoperoxide synthase-2 by human chorionic gonadotrophin in bovine preovulatory follicles *in vivo*. *Endocrinology*, **135**, 841–8.

Sirois, J. & Doré, M. (1997). The late induction of prostaglandin G/H synthase-2 in equine preovulatory follicles support its role as a determinant of the ovulatory process. *Endocrinology*, **138**, 4427–34.

Sirois, J., Simmons, D.L. & Richards, J.S. (1992). Hormonal regulation of messenger ribonucleic acid encoding a novel isoform of prostaglandin endoperoxide H synthase in rat preovulatory follicles. *Journal of Biological Chemistry*, **267**, 11586–92.

Standaert, F.S., Zamor, C.S. & Chew, B.P. (1991). Quantitative and qualitative changes in blood leukocytes in the porcine ovary. *American Journal of Reproductive Immunology*, **215**, 163–8.

Szöllösi, D., Gérard, M., Ménézo, Y. & Thibault, C. (1978). Permeability of ovarian follicle: corona cell–oocyte relationship in mammals. *Annales de Biologie Animale, Biochimie, Biophysique*, **18**, 511–21.

Tanaka, N., Espey, L.L. & Okamura, H. (1989). Increase in ovarian blood volume during ovulation in the gonadotropin-primed immature rat. *Biology of Reproduction*, **40**, 762–8.

Thibault, C. (1977). Are follicular maturation and oocyte maturation independent processes? *Journal of Reproduction and Fertility*, **51**, 1–15.

Thibault, C. & Levasseur, M.C. (1979). *La Fonction Ovarienne chez les Mammifères*. Paris & New York: Masson.

Thibault, C. & Levasseur, M.C. (1988). Ovulation. *Human Reproduction*, **3**, 513–23.

Thomson, A. (1919). The ripe human Graafian follicle, together with some suggestions as to its mode of rupture. *Journal of Anatomy*, **54**, 1–40.

Tsafriri, A. & Dekel, N. (1994). Molecular mechanisms in ovulation. In *Molecular Biology of the Female Reproductive System*, ed. J.K. Findlay, pp. 207–58. London: Academic Press.

Tsafriri, A., Abisogun, A.O. & Reich, R. (1987). Steroids and follicular rupture at ovulation. *Journal of Steroid Biochemistry*, **27**, 359–63.

Tsafriri, A., Chun, S.Y. & Reich, R. (1993). Follicular rupture and ovulation. In *The Ovary*, ed. E.Y. Adashi & P.P.K. Leung, pp. 227–44. New York: Raven Press.

Tsafriri, A., Lieberman, M.E., Koch, S., *et al.* (1976). Capacity of immunologically purified FSH to stimulate cyclic AMP accumulation and steroidogenesis in Graafian follicles and to induce ovum maturation and ovulation in the rat. *Endocrinology*, **98**, 655–61.

Tsafriri, A., Lindner, H.R., Zor, U. & Lamprecht, S.A. (1972). Physiological role of prostaglandins in the induction of ovulation. *Prostaglandins*, **2**, 1–10.

Waberski, D. (1997). Effects of semen components on ovulation and fertilisation. *Journal of Reproduction and Fertility*, Supplement, **52**, 105–9.

Waberski, D., Claassen, R., Hahn, T., *et al.* (1997). LH profile and advancement of ovulation after transcervical infusion of seminal plasma at different oestrous stages in gilts. *Journal of Reproduction and Fertility*, **109**, 29–34.

Waberski, D., Südhoff, H., Hahn, T., *et al.* (1995). Advanced ovulation in gilts by the intra-uterine application of a low molecular mass pronase-sensitive fraction of boar seminal plasma. *Journal of Reproduction and Fertility*, **105**, 247–52.

Waberski, D., Töpfer-Petersen, E. & Weitze, K.F. (2000). Does seminal plasma contribute to gamete interaction in the porcine female tract? In *Boar Semen*

Preservation IV, ed. L.A. Johnson & H.D. Guthrie, pp. 165–72. Lawrence, KA: Allen Press Inc.

Wang, L.J., Brännström, M., Robertson, S.A. & Norman, R.J. (1992). Tumor necrosis factor α in the human ovary: presence in follicular fluid and effects on cell proliferation and prostaglandin production. *Fertility and Sterility*, **58**, 934–40.

Westman, A. (1934). Über die hormonale Funktion des unbefruchteten Eies. *Archiv für Gynäekologie*, **156**, 550–65.

Wong, W.Y.L. & Richards, J.S. (1991). Evidence for two antigenically distinct molecular weight variants of prostaglandin H synthase in the rat ovary. *Molecular Endocrinology*, **5**, 1269–79.

Yoshimura, Y., Karube, M., Koyama, N., Shiokawa, S., Nanno, T. & Nakamura, Y. (1992). Angiotensin II directly induces follicle rupture and oocyte maturation in the rabbit. *FEBS Letters* **307**, 305–8.

Yoshimura, Y., Karube, M., Oda, T., *et al.* (1993). Locally produced angiotensin II induces ovulation by stimulating prostaglandin production in *in vitro* perfused rabbit ovaries. *Endocrinology*, **133**, 1609–16.

Zachariae, F. (1958). Studies on the mechanism of ovulation. Permeability of the blood–liquor barrier. *Acta Endocrinologica*, **27**, 339–42.

Zachariae, F., Asboe-Hansen, G. & Boseila, A.W.A. (1958). Studies on the mechanism of ovulation. Migration of basophilic leucocytes from blood to genital organs at ovulation in the rabbit. *Acta Endocrinologica*, **28**, 547–52.

Zeleznik, A.J. & Benyo, D.F. (1994). Control of follicular development, corpus luteum function, and the recognition of pregnancy in higher primates. In *The Physiology of Reproduction*, 2nd edn, ed. E. Knobil & J.D. Neill, pp. 751–82. New York: Raven Press.

Zeleznik, A.J. & Hillier, S.G. (1996). The ovary: endocrine function. In *Scientific Essentials of Reproductive Medicine*, ed. S.G. Hillier, H.C. Kitchener & J.P. Neilson, pp. 133–46. London & Philadelphia: W.B. Saunders.

IX

Post-ovulatory fate of follicle and oocyte: contributions of somatic cells and follicular fluid

Introduction

A substantial monograph discussing development and maturation of a Graafian follicle and the ensuing process of ovulation cannot end adequately at the stage of rupture and/or collapse of an individual follicle. Although there is perhaps no simple means of achieving a satisfactory conclusion, a final perspective on physiological events might consider either development of a corpus luteum or the fate of the so-called 'products of ovulation' (Harper, 1973). To the present author at least, formation and development of a functional corpus luteum appear to have received abundant attention ever since the classical work of Corner (1919). This is not to suggest that new and fascinating findings on the physiology of luteal cells will cease to be published, especially concerning expression of their gene programmes. However, apart from a brief section on key aspects of luteinisation, an alternative approach to a conclusion seemed open and attractive.

Rather than maintaining a focus on the ovary itself, the destiny of the follicular fluid, somatic cells and oocyte after release from a Graafian follicle at the time of ovulation appeared important topics. As one much interested in the events of fertilisation and early embryonic development, and also in the physiology of the Fallopian tubes, this is the course that has been adopted. Apart from a long-standing enthusiasm, there is the hope of offering a balance not readily available in other contemporary treatments. There is an associated feeling that observations made in veterinary clinical departments, predominantly on large domestic species rather than laboratory rodents, might be of value to medical colleagues. Within this declaration of interest, the following pages offer a concise post-ovulatory view of events shortly before and after the time of fertilisation.

Collapsed follicle evolves into corpus luteum

The destiny of a Graafian follicle after ovulation is to become a functional corpus luteum, synthesising and secreting increasing concentrations of progesterone up to a threshold or plateau sufficient to support a potential pregnancy. Putting the relatively minor adreno-cortical synthesis of progesterone to one side, a further source of this steroid involved in the maintenance of pregnancy may be found in the placenta, as for example in sheep and primates, or be derived from accessory corpora lutea, as in equids. Although secretion of progesterone is essential for the establishment of pregnancy, formation of a fully functional corpus luteum cannot be taken for granted. Seasonally breeding animals may form an inadequate luteal structure towards the end of a breeding season (see Haresign *et al.*, 1975; Webb *et al.*, 1977; Shelton *et al.*, 1990; Hunter, 1991), a time when initiation of pregnancy may not be in the best interest of an individual or species. Inadequate corpora lutea may also be generated in response to hormone treatments for the induction of ovulation, for example in anoestrous sheep (Shareha, Ward & Birchall, 1976) and post-partum suckled cows (Webb *et al.*, 1977). Luteal insufficiency may be found after induction of ovulation in women undergoing *in vitro* fertilisation treatment (see Chapter XI). In such instances, the physiological maturity of follicles induced to ovulate would need to be questioned, as expressed not least in the number of LH receptors on the granulosa cells (see Chapter VI).

Turning briefly to some essential details, a corpus luteum is formed principally by differentiation of mural granulosa cells remaining in the collapsed follicle. Hypertrophy of such granulosa cells is a predominant feature rather than hyperplasia: mitotic divisions diminish markedly after the pre-ovulatory LH surge and there is no evidence that actual luteal cells undergo replication (Hirshfield, 1984). Despite the apparent confidence of this remark, repeated in a number of modern reviews, Corner (1956) demonstrated an increase in the number of 'small' luteal cells (see below) commencing on about day 10 of the human luteal phase. Structural and functional remodelling are classical cellular features of luteinisation, with cell migration and expansion inwards towards the central blood clot that remains after ovulation (Clark, 1900). Fibroblasts and endothelial cells contribute to the structural organisation. As already noted, granulosa cells commence to luteinise before ovulation in response to the surge of LH. As a consequence, loss of aromatase activity occurs in the hours before ovulation, so granulosa cells cease to synthesise oestradiol. Instead, they become granulosa lutein cells that secrete increasing quantities of progesterone which enter the rapidly proliferating capillary bed and thereby the systemic circulation (see Baird, 1984). Extremely active angiogenesis is a major feature of luteinisation, generating a rich blood supply within a mature corpus luteum.

Locally produced angiogenic factors play a key rôle, with VEGF being strongly implicated in the process (Zeleznik, 1999). The rate of blood flow to a mature corpus luteum is exceptional, usually exceeding that to other tissues, and is considered essential for the provision of substrates and nutrition. It follows that mature corpora lutea have a high metabolic rate.

Cells of the theca interna may also contribute to the developing corpus luteum (Corner, 1919) and, in species in which this is characteristic (e.g. primates), they may enable the corpus luteum to synthesise and secrete significant quantities of oestradiol simultaneously with progesterone (see Baird *et al.*, 1975). The two sources of luteal cells remain in reasonably distinct clusters in primates whereas, in other species examined, there is an intermingling of the two cell types (Fig. IX.1). Large and small luteal cells may be distinguished, seemingly representing the contributions of granulosa (20–40 μm diameter) and theca (<20 μm diameter) cells, respectively. However, if granulosa cells are changing in size, then clearly caution will be required in such an interpretation. Moreover, the ratio between the two cell types may change with the stage of the cycle or pregnancy as small cells become larger, and other subsets of luteal cells may also be distinguished (Zeleznik, 1992). It is frequently stated that small luteal cells are less responsive to gonadotrophic stimulation, producing far less progesterone per cell than do large luteal cells. But this is surely what would be anticipated if small luteal cells possess fewer LH receptors and less cytoplasm and cytoplasmic organelles (e.g. smooth endoplasmic reticulum and mitochondria) than large (i.e. larger) cells.

A corpus luteum develops into a roughly spherical body, its mature size generally corresponding with the diameter of a pre-ovulatory follicle. An outstanding difference is, of course, that a fully formed corpus luteum is a solid mass of cells whereas a mature Graafian follicle is predominantly a fluid-filled sac in most mammals so far examined (but see mention of tenrecs in Chapter IV (p. 107)). After an initial phase of downregulation or desensitisation, cells of a developing corpus luteum display an extensive distribution of LH receptors. These enable gonadotrophic support of steroidogenesis, but other pituitary or placental (chorionic) hormones may also contribute (e.g. prolactin), thereby forming a luteotrophic complex. As a consequence of LH binding to luteal cells, there is stimulation of adenylyl cyclase and intracellular concentrations of cAMP, linking events to the PKA system of signalling.

Rôle of fimbriated infundibulum, cilia and myosalpinx

A possible involvement of the fimbriated extremity of the Fallopian tube in provoking surface changes in the ovulatory follicle was discussed in Chapter VIII.

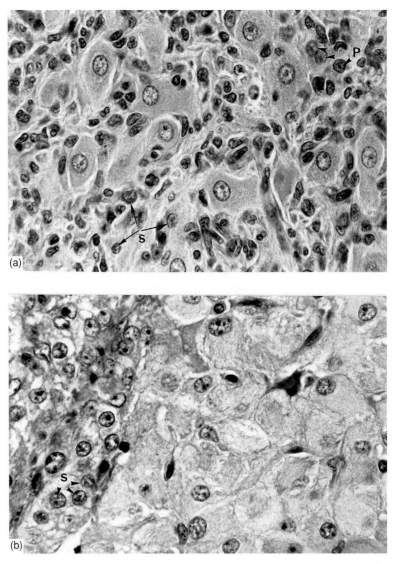

Fig. IX.1. Sections to indicate luteal cells of different size, in part due to their origin as theca or granulosa cells. (After Thibault, Levasseur & Hunter, 1993.)

Whether or not the dense tracts of cilia that line the fimbriated epithelium contribute significantly to irritation and degradation of follicular tissues (see Chapter VIII), the cilia are certainly involved in displacing the oocyte and associated somatic cells from the surface of the follicle into the ostium of the

Fallopian tube as a cumulus–oocyte complex. Under the prevailing ratio of circulating concentrations of oestradiol and progesterone, the beat of the cilia is strongly coordinated and direction orientated, enabling tracts of these organelles to grip and sweep the cumulus cell–oocyte complex as it emerges from the aperture at the follicular apex. This activity ensures disruption of any remaining anchorage to the mural granulosa and moves the complex across the engorged folds of the fimbriated infundibulum and so into the ampulla of the Fallopian tube (Hafez & Blandau, 1969; Hunter, 1977). The conspicuously curved tips of the cilia are critical in these events (Fig. IX.2), their specific shape enabling the synchronised beating of the organelles to obtain a purchase on the somatic cells. The interaction between cilia and cumulus investment is the vital one rather than contact between cilia and an exposed portion of the zona pellucida surface. Indeed, if the cumulus cells are removed from the egg by enzymatic treatment and/or microdissection and the denuded egg(s) or resin spheres that simulate eggs are then transplanted on to the inner surface of the fimbriated folds close to the time of ovulation, they are not transported but simply rotate *in situ* on the fimbrial folds (Blandau, 1969, 1973).

Tracts of cilia line the ampullary portion of the tube and contribute in a major way to transport of the cumulus–oocyte complex from the ostium to the site of fertilisation at the ampullary–isthmic junction. Their beat is similarly programmed by the prevailing ratio of ovarian steroid hormone concentrations close to the time of ovulation and is strongly ab-ovarian. In conjunction with the influence of waves of myosalpingeal contraction and perhaps contractile activity in the adjoining mesenteries too, displacement of eggs to the site of fertilisation occurs rapidly in most species so far examined, requiring 6–15 minutes in rabbits (Harper, 1961a,b; Blandau, 1969) and no more than 30–45 minutes in large domestic species (Andersen, 1927; Oxenreider & Day, 1965; Hunter, 1974). Although such periods of time before commencement of fertilisation may be viewed as short, especially when compared with the subsequent sojourn in the Fallopian tube measured in days (e.g. 2–5 days or more, depending on species), they are of sufficient duration for major changes to take place in the egg surface. This is not least since progression of eggs to the ampullary–isthmic junction is not smooth and direct, but rather takes place in intermittent rushes in phase with waves of contraction in the ampulla (Harper, 1961a,b; Wintenberger-Torrès, 1961). Changes would certainly have time to occur in the human egg, which may take 30 hours or more to reach the ampullary–isthmic junction (Croxatto & Ortiz, 1975; Pauerstein, 1975; Eddy *et al.*, 1976).

The suggested modifications would include: (1) loosening and shedding of many of the follicular cells; (2) changing the environment of the egg from

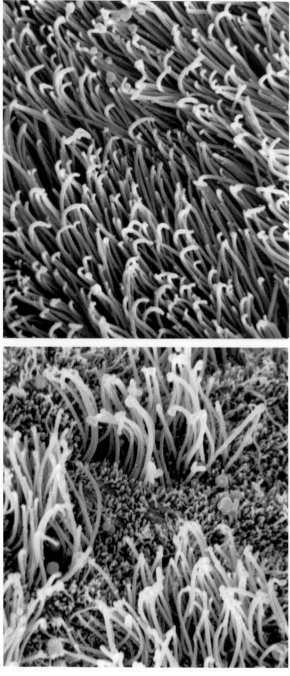

Fig. IX.2. Scanning electron micrographs to show portions of the ciliated epithelium in the Fallopian tube and, in particular, the curved extremities of these organelles. (Adapted from Hunter, Fléchon & Fléchon, 1991.)

one dominated by ovarian follicular fluid to one dominated by constituents of Fallopian tube fluid, including its secretion of unique glycoproteins (Fazleabas & Verhage, 1986; Robitaille *et al.*, 1988; Gandolfi *et al.*, 1991; Wagh & Lippes, 1993; Hunter, 1994); and (3) seemingly exposing the oocyte to gradients in temperature (David, Vilensky & Nathan, 1972; Hunter & Nichol, 1986). Subtle alterations in oocyte metabolism would be anticipated. Overall, it is reasonable to conclude that the egg surface is fully and finally prepared during this phase of transport for interaction with a competent spermatozoon. The preceding word has been written intentionally in the singular, for initial sperm : egg ratios at the site of fertilisation are close to unity in many mammalian species (Moricard & Bossu, 1951; Stefanini, Oura & Zamboni, 1969; Zamboni, 1972; Yanagimachi & Mahi, 1976; Shalgi & Kraicer, 1978; Overstreet *et al.*, 1978; Hunter, 1993, 1996; Yanagimachi, 1994).

Although dense tracts of cilia within the fimbriated infundibulum have been given prominence in the above description of egg capture and the initial phase of transport, the author has frequently observed an alternative scenario. During surgical intervention close to the time of ovulation in mature pigs, rather than the fimbria alone embracing the ovary tightly in anticipation of oocyte release, a substantial portion of the thin-walled and highly distended ampulla surrounded the ovarian surface. The ampulla had presumably assumed this position primarily as a consequence of myosalpingeal activity. In such circumstances, tracts of ampullary cilia might also be brought into play in stripping the cumulus–oocyte complex from the surface of an ovulating follicle. It is perhaps appropriate to mention that if the prevailing ratio of ovarian steroid hormones is modified experimentally, then the nature and direction of the cilial beat will change as a consequence, as will the nature of myosalpingeal activity.

Liberated follicular cells as paracrine tissue

Substantial numbers of somatic cells – very largely granulosa cells – are displaced from the Graafian follicle into the Fallopian tube at the time of ovulation. In this regard, there are in essence two populations: those still attached to or associated with the oocyte(s) as a cumulus and corona cell investment and those seen in the vicinity as groups of cells in a fluid suspension. As indicated above, cumulus cells associated with the oocyte are progressively liberated with time, an influence doubtless brought about in part by the nature of the cilial beat and contractile activity within the ampullary portion of the Fallopian tube and also by constituents of the tubal luminal fluid (Swyer, 1947): bicarbonate ion has long been proposed to be one such factor in rabbits, at least as far as corona radiata cells are concerned (Stambaugh, Noriega & Mastroianni, 1969). In addition, in animals mated before the time of ovulation, there is an influence of sperm head enzymes, especially acrosomal hyaluronidase, on loosening the mucified cumulus mass and liberation of cells from the zona pellucida. Eggs are denuded more rapidly in mated animals than in those not mated or inseminated. This remark is made despite the low initial sperm : egg ratios referred to above. The somewhat paradoxical situation is perhaps best explained by release of hyaluronidase and other proteolytic enzymes from dead or dying spermatozoa arrested in the isthmus portion of the Fallopian tube. Such hyaluronidase would be displaced or diffuse through regional fluid compartments of the lumen (see below) towards the site of fertilisation.

Even though dissociated, follicular cells remain in fluid suspension in the vicinity of the oocyte or newly fertilised egg (embryo). Furthermore, many of these cells are demonstrably viable and active in a secretory sense. In a monograph on Fallopian tube physiology (Hunter, 1988), the argument was presented that since follicular cells can be successfully cultured *in vitro* as granulosa cells, and for example actively synthesise steroid hormones when furnished with suitable substrates (Channing, 1966), a similar rôle *in vivo* should not be overlooked. Not only may these cells synthesise steroids, but they may also retain an ability to secrete diverse peptides including growth factors, thus influencing both the process of fertilisation and the first stages of embryonic development (Hunter, 1995a). There is evidence, moreover, that isolated cumulus cells readily form pyruvate in the presence of glucose and lactate (Leese & Barton, 1985), suggesting a specific rôle in embryonic nutrition.

In fact, it is now widely appreciated that the suspension of follicular cells in the vicinity of the egg can act as a paracrine tissue and may perform vital functions to support development of the newly fertilised egg through the first few cleavage divisions (Motta, Nottola & Familiari, 1997). So, in addition to the contribution of endosalpingeal secretions and components of blood transudate in tubal fluid, physiologists should also take account of specific products from the suspension of granulosa cells in the vicinity of a developing embryo (Fig. IX.3). These cells may be numerous in species that shed many eggs at ovulation, such as mice, rats, hamsters, rabbits and pigs. An influence of this cell suspension on the secretory activity and molecular spectrum of the endosalpinx needs to be considered, and also on angiogenesis in the mucosa and musculature. Such programming might explain in part the regional differences in the composition of Fallopian tube fluid (see below).

Whilst detached granulosa cells may be acting to influence the embryo, a dynamic relationship may also be occurring in the other direction. From the

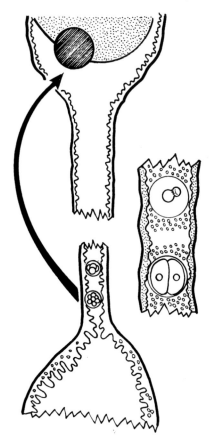

Fig. IX.3. Diagrammatic representation of a Fallopian tube containing developing embryos. The enlarged portion of the tube shows a pronucleate egg and cleaving embryo with a suspension of cumulus (granulosa) cells in close proximity. (Adapted in part from Hunter, 1988.)

earliest stages of zygote formation, it seems probable that the somatic cell suspension acts to amplify embryonic signals that are then transmitted locally to the ovulatory ovary (Fig. IX.3). The activity of the developing corpus luteum, in particular its increasing secretion of progesterone, could thus be regulated in a highly sensitive manner. Indeed, one could argue that the larger the population of granulosa cells liberated at ovulation, the greater the scope for amplification of embryonic signals and the greater the need for a developing corpus luteum to be so stimulated. In such a scheme, so-called early pregnancy factors would therefore represent not only molecules derived from the embryo but also those from somatic cells of granulosa origin.

Contribution and fate of follicular fluid

As noted in Chapter IV, the volume of follicular fluid can be substantial by the time of ovulation, especially in women (5.0–7.5 ml or more) or large domestic species such as cows (2.0–3.2 ml) and horses (5–15 ml or more). Accordingly, a major question arises concerning the fate of such fluid at the time of follicular collapse and the extent to which it contributes to the composition of tubal fluid. Because the ovary is intimately embraced by the fimbriated extremity of the Fallopian tube at ovulation, or in some species (e.g. golden hamster) there may be an ovarian bursa, there can be little doubt that follicular fluid released from the collapsing follicle makes initial contact with the inner surface of the fimbrial folds and at least the cranial portion of the ampulla. The same remarks almost certainly apply to seepage of fluid from a pre-ovulatory follicle whilst structural modifications occur at the apex before rupture or collapse. Nonetheless, a number of lines of evidence indicate that the bulk of the follicular fluid refluxes rather promptly from the Fallopian tube into the peritoneal cavity. In fact, follicular fluid is a significant constituent of peritoneal fluid. Even putting this consideration to one side, the volume of follicular fluid released at ovulation could not be accommodated within the Fallopian tube lumen unless a conspicuous or exceptional degree of dilatation were to occur. Such dilatation is seldom observed during surgical intervention at the time of ovulation nor would it be possible, for example, in the remarkably small Fallopian tube of the mare when set against the volume of follicular fluid released.

There is a further consideration of size, one imposed by the dimensions and distensibility of the isthmus. This portion of the Fallopian tube has a prominent and powerful smooth muscle coat composed of both circular and longitudinal layers (Fig. IX.4) and, at the time of ovulation, the myosalpinx is tightly constricted (Brundin, 1969; Hunter, 1977, 1988). This renders the isthmus scarcely patent and is one of the factors restricting ad-ovarian progression of viable spermatozoa before ovulation, contributing to the extremely low initial sperm : egg ratios. Accordingly, passage of ovarian follicular fluid through the isthmus appears improbable shortly after ovulation. Moreover, the caudal portion of the isthmus contains an extremely viscous glycoprotein secretion at this time in a number of species (e.g. rabbit, pig, human; for a review, see Hunter, 1994, 1998), which would also impede movement of fluid, even if it were able to reach the caudal isthmus.

Whilst a fairly prompt reflux of the bulk of follicular fluid into the peritoneal cavity represents its principal fate, a contribution to the composition of tubal fluid should certainly be accepted. This would be minor on a volumetric basis, but could be significant in a qualitative sense. Furthermore, bearing in mind the

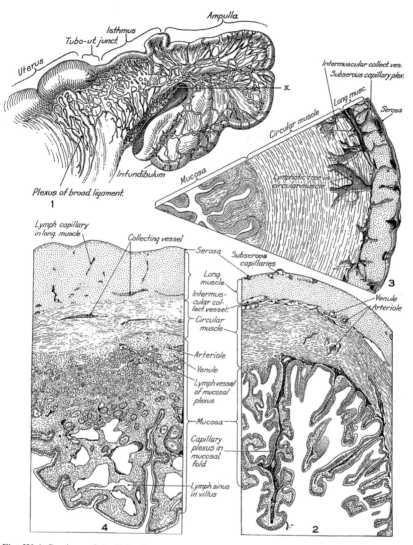

Fig. IX.4. Sections of the Fallopian tube to indicate the extensive layers of the myosalpinx, especially that of the circular muscle, in this instance in pigs. Tubo-ut., tubo-uterine; collect., collection; musc., muscle; long., longitudinal. (Adapted from Andersen, 1927, and Hunter, 1988.)

origin of follicular fluid and the involvement of granulosa cells in its formation, then the post-ovulatory suspension of cumulus cells could be elaborating a form of follicular fluid in the vicinity of the oocyte(s) and/or embryo(s). Despite this interpretation, a set of observations using the progesterone content of ovarian follicular fluid as a tracer molecule indicated that less than 1% of the fluid was retained in the Fallopian tube shortly after ovulation (Hansen, Srikandakumar & Downey, 1991).

There remains one other aspect to consider. Even though bulk contact of follicular fluid with the tubal lumen is largely transient, it could be exerting a potent influence via the endosalpinx and mucosa to modify the composition of tubal luminal fluid. Bearing in mind the constituents of follicular fluid, including high concentrations of steroid hormones, prostaglandins, cytokines and growth factors, this seems highly probable. Not only might follicular fluid be able to influence contractile activity of the myosalpinx and the nature of cilial beat, but there could also be influences on blood flow in the proliferating capillary bed of the ampulla and on the extent of transudation. In addition, the secretory activity of the endosalpinx might be modified, not least in its generation of unique glycoproteins, which are conspicuous components of Fallopian tube fluid at this time (Oliphant & Ross, 1982; Sutton *et al.*, 1984; Sutton, Nancarrow & Wallace, 1986; Gerena & Killian, 1990; Wagh & Lippes, 1993). As an overall interpretation, the contribution of follicular fluid might be viewed as modifying the tubal environment for the benefit of gametes (both oocytes and spermatozoa), and also in terms of metabolic support for a newly formed embryo (for a review, see Hunter, 1988, 1994).

Regional fluid environments within the Fallopian tube(s)

Because this chapter has focused quite largely on components of fluid in the lumen of the Fallopian tubes, it is pertinent to add some complementary observations concerning composition of the fluid. Despite the heightened activity of Fallopian tubes around the time of ovulation, seen as both enhanced smooth muscle activity and enhanced beat of cilia that would be expected to promote movement and mixing of luminal fluid, regional differences can be detected in its composition. This was first noted simply as differences between fluid in the ampullary portion versus fluid in the isthmus (Roblero, Biggers & Lechene, 1976) and was expressed as elemental composition (Table IX.1). However, subsequent studies have examined the fluids more closely on the basis of acute sampling by micropuncture and highlighted regional differences – in effect a series of fluid compartments (Leese, 1988; Hunter, 1990; Nichol *et al.*, 1992). In other words, regional specialisations in the Fallopian tube fluids need to be considered, since they might influence male and female gametes, and subsequently bear on development of the embryo. Changing substrate requirements during early cleavage have long been appreciated (Brinster, 1963, 1965, 1969).

Table IX.1. *The weighted mean concentrations (mM/l) of various elements in serum and the microenvironments of one- and two-celled mouse embryos in the ampulla and isthmus, respectively, of the oviduct*

Element	Ampulla ($N = 25$)	Isthmus ($N = 17$)	Serum ($N = 12$)
Na	137 ± 4.78	142 ± 3.42	148 ± 1.72
Cl	147 ± 4.98	145 ± 6.20	140 ± 2.10
Ca	1.62 ± 0.13	1.48 ± 0.14	3.82 ± 0.16
K	17.8 ± 1.38	29.7 ± 3.63^a	5.26 ± 0.15
Mg	0.63 ± 0.08	0.81 ± 0.07	1.38 ± 0.20
S	5.13 ± 0.53	9.13 ± 0.92^a	21.8 ± 0.32
P	3.90 ± 0.64	8.46 ± 1.79^a	6.97 ± 0.27

^{a}Significantly different ($P < 0.01$) from the concentration in ampullary fluid (Fisher–Behren's test). All other comparisons are not statistically significant at the $P = 0.05$ level.
Adapted from Roblero *et al.*, 1976.

The extent to which the 'products of ovulation' exert an influence on the composition of regional fluids remains uncertain. As suggested above, ovarian follicular fluid would be expected to modify the composition of ampullary fluid and perhaps also that of fluid at the ampullary–isthmic junction, but not of fluids within the isthmus. An involvement of follicular cells in modifying fluid composition has been discussed above, both directly in terms of their secretory products and indirectly via a programming influence on the epithelium. Degenerating and moribund follicular cells would also influence the composition of tubal fluid. In a more sensitive manner, the oocyte itself may act to modify tubal fluid composition both in terms of (1) its secretions and (2) their influence on the endosalpinx. And, finally, despite reasoning previously that the bulk of follicular fluid released at ovulation will reflux from the ampulla into the abdominal cavity, in a contrary sense there may be passage of minor quantities of peritoneal fluid into the Fallopian tube. For example, some of the peritoneal fluid bathing or making contact with the ovary close to the time of ovulation would almost certainly be displaced into the lumen of the tube. Paradoxically, the one time during the peri-ovulatory phase when this might not occur would be during bulk movement of follicular fluid at the actual moment of ovulation.

Local ovarian influences on tubal physiology

A widely accepted dogma amongst mammalian physiologists is that events in the reproductive tract are powerfully influenced by the secretory activity of the gonads. The cranial portion of the female reproductive tract has evolved as the Fallopian tubes, and the primary endocrine influence of the ovaries on these structures is by means of steroid hormones, especially oestradiol-17β and progesterone. The ever-changing ratio of the concentrations of these two hormones in the circulating blood is specifically important and becomes critical around the time of ovulation. Other ovarian hormones of significance in the peri-ovulatory period include prostaglandins, peptide hormones such as oxytocin and relaxin, and diverse growth factors; these have been discussed earlier in this monograph. Follicular fluid concentrations of some of these hormones peak shortly before ovulation, and this is particularly true of prostaglandins (see Chapter VIII). Incisive programming of the neighbouring Fallopian tube would be anticipated.

Ovarian hormones are able to influence the genital tract by diverse routes. At the time of ovulation, potent concentrations of hormones will be released into the Fallopian tube with the follicular fluid. This has been considered above, with the caveat that most of the follicular fluid would be refluxed into the peritoneal cavity. Ovarian hormones would also reach the reproductive tract via the systemic circulation and, in this regard, comparable concentrations of hormones would be expected in the arterial supply to each side of the reproductive tract and therefore to each Fallopian tube. A third means of delivering ovarian hormones, and one of specific significance close to the time of ovulation and shortly thereafter, involves a local counter-current transfer of appropriately sized molecules (e.g. steroid hormones and prostaglandins) between the ovarian vein and the tubal branch of the ovarian artery (Fig. IX.5; Hunter, Cook & Poyser, 1983). And not to be overlooked is the generation of ovarian hormones by cumulus cells in suspension in tubal fluid acting as a paracrine tissue. Delivery of ovarian hormones to the Fallopian tubes via the lymphatic system has not been mentioned, but should be considered as a corresponding vascular route.

Concerning a counter-current exchange mechanism, the proximity of the ovarian vein to the highly convoluted ovarian artery in the region of the gonadal pedicle would facilitate a short-circuit routing of appropriate ovarian molecules to the Fallopian tubes; in reality, there is a region of intimate contact between the two blood vessels (Del Campo & Ginther, 1972, 1973). An overall significance of this arrangement in a context of fertilisation is that the structure that will shed the female gamete, the Graafian follicle, is able to coordinate the final phases of sperm maturation and release from the sperm reservoir in the caudal portion of the Fallopian tube (see Hunter, 1995a, 1997). This is achieved in a highly sensitive manner, since local transfer of hormones to the vascular arcade of the Fallopian tube prevents dilution in the systemic circulation and, in

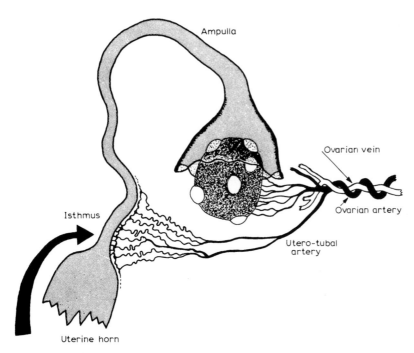

Fig. IX.5. A semi-diagrammatic representation of the arterial supply to the ovary and isthmus of the pig Fallopian tube. A portion of the ovarian vein is also shown. A counter-current transfer of follicular hormones was demonstrated from the ovarian vein to the corresponding artery and thus into the utero-tubal branch. (After Hunter *et al.*, 1983.)

large measure, avoids complexing of hormones with binding proteins. If steroid hormone concentrations in the vascular arcade to the isthmus are perturbed experimentally, then conspicuous influences on the pattern of sperm transport and normality of fertilisation will be found (Hunter, 1972).

As to the transport of oocytes within the Fallopian tube(s), already noted to involve myosalpingeal and ciliary activity and perhaps fluid flow, there is abundant evidence that ovarian steroid hormones exert control and can be modified to provoke anomalous transport of eggs or embryos. This is seen as either accelerated transport through the tubes into the uterus or, in marked contrast, as so-called 'tube locking' (i.e. arrest). The specific influence will depend on the steroid hormone in question, the dose administered, and the timing of such treatment (for reviews, see Harper & Pauerstein, 1976; Hunter, 1977, 1988). Alternative approaches to modifying steroid hormone concentrations can involve superovulation treatment or ovariectomy. However, the extent to which passage of eggs within the Fallopian tubes is influenced locally by ovarian hormones delivered via a counter-current transfer system still requires clarification. The same remark holds for the subsequent transport of the fertilised egg and cleaving embryo through the isthmus into the uterus (see Croxatto, 1996). Although this discussion has focused on local mechanisms as a sensitive means of controlling tubal physiology, there would always be the further influence of hormones delivered in the systemic circulation.

An outstanding point is the ultimate fate of the suspension of follicular cells described in the lumen of the Fallopian tube. A response cannot be offered with certainty but, in domestic farm animals, a suspension of follicular cells enters the uterus at the same time as developing embryos (Hunter, 1978, 1988). In other words, the somatic cell suspension seemingly remains in close proximity to the embryo(s), strengthening its proposed involvement in a paracrine rôle. The ultimate destiny of follicular cells in the uterus is unknown, but they might be incorporated by elements of the trophoblast (Hunter, 1978). This could be especially so in ungulates with a dramatically remodelled and elongating blastocyst (conceptus) and with subsequent epithelio-chorial or syndesmo-chorial placentation.

Increasing progesterone secretion modifies tubal physiology

As indicated in the introductory section, this chapter has opted not to discuss development of the corpus luteum, the spherical body of cells that forms after release of the oocyte and follicular fluid from a Graafian follicle at ovulation. Even so, it is important to note that progesterone, the major steroid hormone synthesised and secreted by the corpus luteum (the so-called hormone of pregnancy) increases in concentration in the systemic circulation with increase in number and size of luteal cells. Under the influence of increasing secretion of progesterone and an ever-widening ratio of progesterone to oestradiol concentration, the physiology of the Fallopian tubes is modified in anticipation of successful fertilisation and early embryonic development. Progesterone has a major influence on release of viable spermatozoa from the reservoir in the caudal isthmus of the tube in diverse species (for reviews, see Hunter, 1995a,b); this is probably also true of the human Fallopian tube. Increasing secretion of progesterone leads to enhanced numbers of spermatozoa progressing to the site of fertilisation and beyond, and thus to increasing sperm : egg ratios. However, this situation arises only after fertilisation is complete and an irreversible block to polyspermy has been established in the zona pellucida and/or plasma membrane of the zygote.

Increasing titres of circulating progesterone also act to increase the patency of the Fallopian tube, further facilitating progression of spermatozoa. This is achieved by reducing the extent of oedema in the mucosa and reducing the tone of the myosalpinx, especially in the circular muscle layers of the isthmus. The latter is brought about since progesterone promotes activation of β-adrenergic receptors in the myosalpinx, whereas contraction is initiated under oestradiol dominance and activation of α-adrenergic receptors (Brundin, 1969; Hunter, 1977). Progressive relaxation of the isthmus is associated with a controlled passage of embryos towards the utero-tubal junction and so into the uterus some 2–5 days or more after ovulation, but with a qualification according to species as discussed below. Although transport of embryos is passive, there is a contribution from ab-ovarian waves of myosalpingeal contraction. However, their direction of propagation, amplitude and frequency have all been modified in the interval since the time of ovulation (see Guttmacher & Guttmacher, 1921; Wislocki & Snyder, 1933; Hafez & Blandau, 1969; Harper & Pauerstein, 1976). The rôle of a diminished volume of tubal fluid in displacement of embryos into the uterus has not been examined specifically, although it has been shown that the direction of fluid flow in the isthmus has changed from ad-ovarian to ab-ovarian at this time (Bellve & McDonald, 1968).

Progression of fertilised versus unfertilised eggs

In animals or human patients not subjected to superovulation or steroid hormone therapy, fertilised eggs progress to the uterus as cleaving embryos on a timescale that has been reasonably well characterised according to species. One of the most rapid passages through the Fallopian tubes after spontaneous ovulation occurs in pigs, in which species embryos enter the uterus at the four-cell stage some 46–48 hours after ovulation or the onset of fertilisation (Assheton, 1898; Pomeroy, 1955). In laboratory species such as rats, mice, guinea-pigs and rabbits, embryos remain for approximately 3 days in the Fallopian tubes and enter the uterus as young morulae of 8–16 cells. Similarly, the embryos of domestic ruminants such as sheep or cows require about 3 days for tubal transit, entering the uterus 72–78 hours after ovulation as young morulae of 8–16 cells (for a summary, see Hunter, 1980, 1988).

There may be considerably greater variation in the timing of egg progression to the uterus in primates. Available figures suggest that a 3- to 4-day sojourn in the Fallopian tubes is common in both rhesus monkeys and women, with morulae of 8–16 cells passing to the uterus 72–100 hours after ovulation (Harper & Pauerstein, 1976; Edwards, 1980; Hearn, 1986); such estimates agree with the earlier figures of Hertig, Rock & Adams (1956). In marked contrast to all the laboratory and large domestic species so far examined, there is the possibility of tubal ectopic pregnancy in primates if the normal mechanisms of embryo transport are compromised. This is not to suggest advanced development of

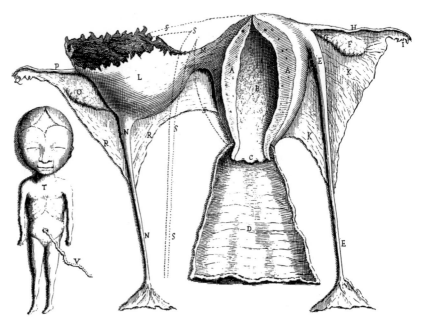

Fig. IX.6. A classical depiction of an ectopic tubal pregnancy that appears in Mauriceau's (1675) *Traité des Maladies des Femmes Grosses*. (Courtesy of the Bibliothèque Nationale, Paris.)

an embryo arrested in the tube, for such an occurrence is pathological, highly dangerous, and can lead to massive haemorrhage and rapid death upon rupture. Early diagnosis of the condition traditionally depended upon alertness and clinical skills, but nowadays when suspected can be revealed in a straightforward manner by means of ultrasonic scanning. Early intervention is usually also straightforward by means of laparoscopy and, with appropriate microsurgical techniques, much of the damaged Fallopian tube may be conserved by reconstruction. The present author has a special interest in tubal ectopic pregnancy (Fig. IX.6), and has proposed an aetiology involving fragments of endometrium displaced to proliferate in a Fallopian tube (Hunter, 1988, 1998, 2002).

Equids are also exceptional, but in quite another sense. Whereas embryos progress to the uterus in 5–6$\frac{1}{2}$ days or so as advanced morulae or young blastocysts (Oguri & Tsutsumi, 1972; Betteridge, Eaglesome & Flood, 1979), unfertilised eggs may have a somewhat different fate. Most unfertilised eggs are not transported to the uterus but rather remain in the Fallopian tubes – frequently for months – to degenerate *in situ* (van Niekerk & Gerneke, 1966; Betteridge & Mitchell, 1974). Unfertilised eggs from successive cycles and from accessory ovulations (induced spontaneously by equine chorionic gonadotrophin during early pregnancy) may all accumulate in the tube. A key factor in this differential transport of fertilised and unfertilised eggs is the timescale, i.e. the 5–7-day residence of embryos in the Fallopian tube. On the one hand, this enables a differentiating embryo to become capable of secreting factors that will influence contractile activity of the myosalpinx, thereby regulating its own progression to the uterus. In this regard, key embryonic factors consist principally of prostaglandin molecules, especially prostaglandin E_2 (Weber *et al.*, 1991a, b). On the other hand, unfertilised eggs will have degenerated conspicuously during the 5–7-day interval, with advanced wastage of the cytoplasm and obvious distortion of the zona pellucida. Many such degenerate eggs appear flat or dish shaped, are soft and malleable, and would not be effectively displaced by contractile activity along a Fallopian tube in the mid-luteal phase (Hunter, 1989; Hunter & Nichol, 1993). Nonetheless, a small proportion of degenerate equine eggs may reach the uterus. These are thought to have been displaced either due to their appropriate orientation at the time of contractile activity or to have been located in front of (distal to) a healthy embryo arising from a subsequent ovulation and fertilisation (Hunter, 1989, 1991).

There is also evidence in hamsters (Ortiz *et al.*, 1986; Croxatto, 1996; Villalón, Velasquez & Croxatto, 1999) and in certain species of bat (Rasweiler, 1972, 1979) that unfertilised eggs are transported to the uterus on significantly different timescales from those of embryos. Such observations lead to consideration of locally mediated signals such as diverse early pregnancy factors, unfortunately beyond the scope of the present volume. This is except in terms of the amplification of early embryonic signals by the tubal suspension of ovarian follicular (granulosa) cells acting as a paracrine tissue, as referred to above (see Fig. IX.3).

Post-ovulatory ageing of oocytes

In the light of the preceding paragraphs, a few brief comments on the post-ovulatory ageing of mammalian eggs are appropriate. Allowing for the fact that introduction of a sperm suspension into the female at coitus may not be suitably timed before ovulation for a normal sequence of events at fertilisation, an essential question arises as to the viable lifespan of a mammalian egg. By this is meant the period of time in the Fallopian tube during which it remains capable not only of being penetrated and activated by a fertilising spermatozoon, but also competent to develop into a viable embryo. It is widely accepted that the ovulated egg remains fertilisable for longer than it retains the capacity to give rise to a normal embryo (Austin, 1970). Ageing and indeed visibly degenerating eggs in some species may still be penetrable by spermatozoa

(Blandau & Jordan, 1941; Chang, 1952; Braden, 1959). Unfortunately, the distinction between the functional lifespan of an ovulated egg and the penetrable lifespan has not always been clear in published papers.

There is accumulating evidence that the viable lifespan of an oocyte in the female tract is not a constant figure for a given species, but varies according to the maturity of the oocyte at ovulation and the composition of fluids in the Fallopian tubes; the latter are influenced especially by gonadal steroid hormones. In fact, the quality of oocytes shed in a series of spontaneous ovarian cycles in the same animal may be demonstrably different. Even so, viable lifespans of little more than 6–8 hours would be prudent estimates for a majority of eutherian mammals so far examined. Figures of 10–12 hours or even 12–18 hours have been published for the 'lifespan' of the ovulated oocyte in certain species (e.g. cow and sheep: Dauzier & Wintenberger-Torrès, 1952; Thibault, 1967; Killeen & Moore, 1970), but a rigorous distinction has not been offered between penetrable lifespan and viable lifespan in such instances. Apart from oocyte maturity at the time of ovulation, other factors that might influence the lifespan of an oocyte would be environmental (e.g. season of the year, ambient temperature, nutritional considerations, diverse stress factors), breed or strain of an animal within a species, and modifications to the endocrine background culminating in ovulation, such as regulation of menstrual cycles or synchronisation of oestrous cycles.

As to the nature of post-ovulatory degeneration of an oocyte, consideration needs to be given to the egg surface, to cytoplasmic organelles and finally to nuclear structures. The surface of the zona pellucida contains receptors or binding sites for spermatozoa, referred to as part of the ZP_3/ZP_2 system of glycoproteins (see p. 41), as classified in the mouse model of Wassarman (1988, 1991) and Wasserman, Liu & Litscher (1997). These receptors are gradually lost with time, although whether primarily by leaching or by molecular coating may depend on species. As suggested above, secretions are added to the surface of the zona pellucida during passage of an egg down the Fallopian tube whilst it is progressively denuded of cumulus and corona cells. Such ampullary secretions may gradually mask sperm receptors, and lead eventually to a complete loss of fertilisability.

Concerning cytoplasmic organelles, a principal focus is usually on the so-called cortical granules, which are, in reality, very small vesicles, each approximately 1 μm in diameter, containing granular enzymatic material (Austin, 1956; Szöllösi, 1962; Fléchon, 1970). Shortly before ovulation, these have migrated to just beneath the plasma membrane, being formed in the Golgi apparatus deeper in the cytoplasm (Guraya, 1983). Close proximity to the egg surface is essential if, upon activation by a fertilising spermatozoon with the resultant cytoplasmic Ca^{2+} oscillations, the cortical granules are to undergo fusion with the overlying plasma membrane in order to release their enzymatic contents

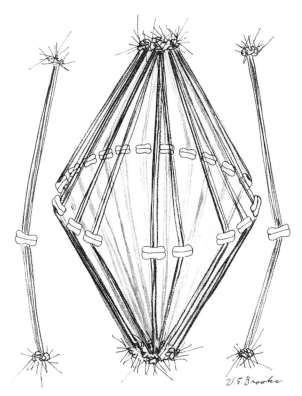

Fig. IX.7. Diagrammatic representation to show early stages in degeneration of the second meiotic spindle, with lateral loss of microtubules and associated chromosomes from the spindle apparatus. (After Szöllösi, 1975.)

into the fluid-filled perivitelline space. This last step leads to a change in the nature of the zona pellucida and/or establishment of a block to polyspermy. An essential point with post-ovulatory ageing is that the cortical granules commence to swell and lose their peripheral location – and thus their ability to fuse with the plasma membrane. They drift in a disorganised manner in the egg cortex, thereby compromising or preventing formation of an effective block to polyspermy. Seemingly at random, such a centripetal movement is doubtless an expression of modifications to the cytoskeleton. It leaves an egg open to the risk of multiple sperm penetration.

Turning to the nucleus, a characteristic feature of post-ovulatory ageing is progressive disorganisation of the microtubules that constitute the meiotic spindle. Pairs of microtubules become gradually dissociated laterally from the spindle with ageing by losing their polar attachments to organelles termed kinetochores (Fig. IX.7). The structural and genetic integrity of the metaphase plate is thereby disrupted, for loss of spindle elements necessarily entails a loss of their attached chromosomes from the division apparatus. Even if activation by a fertilising spermatozoon were still possible and the second meiotic division resumed, the overall chromosome complement would be depleted and any resultant zygote would be aneuploid. Early embryonic death is the most frequent sequel. Other structural features of degeneration could be described both in the nucleus and in the cytoplasm. Taken together, they would prevent normal fertilisation and establishment of a viable embryo.

The extent to which post-ovulatory ageing of the oocyte represents a spontaneous degenerative process as distinct from programmed cell death remains to be resolved. Apoptotic mechanisms almost certainly come into play when the process of ageing is advanced. However, surface changes in the zona pellucida are seemingly not triggered in such a manner, nor perhaps would this be anticipated, since the zona is an (accessory) investment, enveloping the egg proper. There could be reservations, too, as to whether the seemingly random scattering of the cortical granules into deeper regions of the cytoplasm with post-ovulatory age is primarily prompted by an apoptotic cascade, and the same comment might apply to certain other structural modifications.

Despite the apparent emphasis throughout the above section on deleterious influences of post-ovulatory ageing of the mammalian egg, it is nonetheless appropriate to recall that there may be a brief but critical phase of post-ovulatory maturation during progression of the oocyte(s) to the site of fertilisation at the ampullary–isthmic junction of the Fallopian tube.

Concluding remarks

The above pages focus on matters for which there is some understanding, however inadequate and incomplete. Now to move into unknown territory. Follicular fluid is not evacuated completely from a Graafian follicle at ovulation, and the beneficial rôle of residual fluid might be a fruitful topic of enquiry. Whilst its more viscous portion becomes at least partially associated with blood products and undergoes a degree of clotting in the cavity of the formative corpus luteum, its potent content of hormones and growth factors could still exert a trophic influence on the development and function of surrounding luteal cells. There is certainly scope here for experimental manipulation *in vivo* by introducing different samples of follicular fluid taken from follicles at different stages of maturity into the cavity of Graafian follicles shortly after ovulation. Such samples of follicular fluid could be supplemented with known concentrations of appropriate hormones and growth factors.

In addition to the actual fluid not being fully expelled from the follicle at ovulation, there are rare instances of the oocyte remaining *in situ* (Brambell, 1956; Rasweiler, 1984). One could speculate on the reasons for such retention, not least from an endocrine point of view involving maturation of the oocyte and insufficient mucification and expansion of the cumulus oophorus (see Chapter VII). However, an over-riding point here is that some degree of anchorage by follicular cells would probably be involved. If coitus were to be appropriately timed and transport of spermatozoa to the upper reaches of the Fallopian tubes and thence to the ovarian surface occurred, then there is a possibility of fertilisation of the oocyte in or on the surface of the collapsed follicle. Such an event would pave the way to an ovarian ectopic pregnancy.

Looking back over the present chapter, one is conscious that the emphasis has moved from the gonad to the genital tract, but ovarian developments can only ultimately bear fruit upon shedding of an oocyte into the Fallopian tube of a mated animal. Moreover, whilst one level of conversation between a Graafian follicle and tissues of the Fallopian tube necessarily takes place via the vascular system, the full nature of more intimate conversations certainly awaits exploration. It is in this context that contributions from the follicular fluid, the somatic cells and of course the oocyte itself have been examined within the lumen of the Fallopian tube. The suspension of follicular (granulosa) cells that acts as a paracrine tissue in the immediate vicinity of the egg or embryo during its tubal transit has received particular emphasis, for in one very real sense such suspended cells provide a continuity with the antral environment of a mature Graafian follicle and a powerful means of amplifying molecular signals from the oocyte and very young embryo (zygote) to the mother.

References

Andersen, D.H. (1927). The rate of passage of the mammalian ovum through various portions of the Fallopian tube. *American Journal of Physiology*, **82**, 557–69.

Assheton, R. (1898). The development of the pig during the first ten days. *Quarterly Journal of Microscopical Science*, **41**, 329–59.

Austin, C.R. (1956). Cortical granules in hamster eggs. *Experimental Cell Research*, **10**, 533–40.

Austin, C.R. (1970). Ageing and reproduction: post-ovulatory deterioration of the egg. *Journal of Reproduction and Fertility, Supplement*, **12**, 39–53.

Baird, D.T. (1984). The ovary. In *Reproduction in Mammals*, 2nd edn, vol. 3, ed. C.R. Austin & R.V. Short, pp. 91–114. Cambridge: Cambridge University Press.

Baird, D.T., Baker, T.G., McNatty, K.P. & Neal, P. (1975). Relationship between the secretion of the corpus luteum and the length of the follicular phase of the ovarian cycle. *Journal of Reproduction and Fertility*, **45**, 611–19.

Bellve, A.R. & McDonald, M.F. (1968). Directional flow of Fallopian tube secretion in the Romney ewe. *Journal of Reproduction and Fertility*, **15**, 357–64.

Betteridge, K.J., Eaglesome, M.D. & Flood, P.F. (1979). Embryo transport through the mare's oviduct depends upon cleavage and is independent of the ipsilateral corpus luteum. *Journal of Reproduction and Fertility, Supplement*, **27**, 387–94.

Betteridge, K.J. & Mitchell, D. (1974). Direct evidence of retention of unfertilised ova in the oviduct of the mare. *Journal of Reproduction and Fertility*, **39**, 145–8.

Blandau, R.J. (1969). Gamete transport – comparative aspects. In *The Mammalian Oviduct*, ed. E.S.E. Hafez & R.J. Blandau, pp. 129–62. Chicago: University of Chicago Press.

Blandau, R.J. (1973). Gamete transport in the female mammal. In *Handbook of Physiology*, Section 7, *Endocrinology*, vol. II, ed. R.O. Greep & E.B. Astwood, pp. 153–63. Washington, DC: American Physiological Society.

Blandau, R.J. & Jordan, E.J. (1941). The effect of delayed fertilisation on the development of the rat ovum. *American Journal of Anatomy*, **68**, 275–87.

Braden, A.W.H. (1959). Are nongenetic defects of the gametes important in the etiology of prenatal mortality? *Fertility and Sterility*, **10**, 285–98.

Brambell, F.W.R. (1956). Ovarian changes. I. Development of the ovary and oogenesis. In *Marshall's Physiology of Reproduction*, 3rd edn, vol. 1, part 1, ed. A.S. Parkes, pp. 397–542. London & New York: Longmans, Green & Co.

Brinster, R.L. (1963). A method for *in vitro* cultivation of mouse ova from two-cell to blastocyst. *Experimental Cell Research*, **32**, 205–8.

Brinster, R.L. (1965). Studies of the development of mouse embryos *in vitro*: energy metabolism. In *Ciba Foundation Symposium on Preimplantation Stages of Pregnancy*, ed. G.E.W. Wolstenholme & M. O'Connor, pp. 60–74. London: Churchill.

Brinster, R.L. (1969). Mammalian embryo culture. In *The Mammalian Oviduct*, ed. E.S.E. Hafez & R.J. Blandau, pp. 419–44. Chicago: University of Chicago Press.

Brundin, J. (1969). Pharmacology of the oviduct. In *The Mammalian Oviduct*, ed. E.S.E. Hafez & R.J. Blandau, pp. 251–69. Chicago: University of Chicago Press.

Chang, M.C. (1952). Effects of delayed fertilisation on segmenting ova, blastocysts and foetuses in rabbits. *Federation Proceedings, American Societies for Experimental Biology*, **11**, 24.

Channing, C.P. (1966). Progesterone biosynthesis by equine granulosa cells growing in tissue culture. *Nature (London)*, **210**, 1266.

Clark, J.G. (1900). The origin, development and degeneration of the blood vessels of the human ovary. *Johns Hopkins Hospital Reports*, **9**, 593–676.

Corner, G.W. (1919). On the origin of the corpus luteum of the sow from both granulosa and theca interna. *American Journal of Anatomy*, **26**, 117–83.

Corner, G.W. Jr (1956). The histological dating of the human corpus luteum of menstruation. *American Journal of Anatomy*, **98**, 37–65.

Croxatto, H.B. (1996). Gamete transport. In *Reproductive Endocrinology, Surgery and Technology*, ed. E.Y. Adashi, J.A. Rock & Z. Rozenwaks, pp. 385–402. Philadelphia: Lippincott-Raven.

Croxatto, H.B. & Ortiz, M.E.S. (1975). Egg transport in the Fallopian tube. *Gynaecological Investigation*, **6**, 215–25.

Dauzier, L. & Wintenberger-Torrès, S. (1952). Recherches sur la fécondation chez les mammifères: la remontée des spermatozoïdes dans le tractus génital de la brebis. *Comptes Rendus de la Société de Biologie*, **146**, 67–70.

David, A., Vilensky, A. & Nathan, H. (1972). Temperature changes in the different parts of the rabbit's oviduct. *International Journal of Gynecology and Obstetrics*, **10**, 52–6.

Del Campo, C.H. & Ginther, O.J. (1972). Vascular anatomy of the uterus and ovaries and the unilateral luteolytic effect of the uterus: guinea pigs, rats, hamsters and rabbits. *American Journal of Veterinary Research*, **33**, 2561–78.

Del Campo, C.H. & Ginther, O.J. (1973). Vascular anatomy of the uterus and ovaries and the unilateral luteolytic effect of the uterus: horses, sheep and swine. *American Journal of Veterinary Research*, **34**, 305–16.

Eddy, C.A., Garcia, R.G., Kraemer, D.C. & Pauerstein, C.J. (1976). Ovum transport in non-human primates. In *Symposium on Ovum Transport and Fertility Regulation*, ed. M.J.K. Harper & C.J. Pauerstein, pp. 390–403. Copenhagen: Scriptor.

Edwards, R.G. (1980). *Conception in the Human Female*. London: Academic Press.

Fazleabas, A.T. & Verhage, H.G. (1986). The detection of oviduct-specific proteins in the baboon (*Papio anubis*). *Biology of Reproduction*, **35**, 455–62.

Fléchon, J.E. (1970). Nature glycoprotéique des granules corticaux de l'oeuf de lapine. *Journal de Microscopie*, **9**, 221–42.

Gandolfi, F., Modina, S., Brevini, T.A.L., Galli, C., Moor, R.M. & Lauria, A. (1991). Oviduct ampullary epithelium contributes a glycoprotein to the zona pellucida, perivitelline space and blastomeres membrane of sheep embryos. *European Journal of Basic and Applied Histochemistry*, **35**, 383–92.

Gerena, R.L. & Killian, G.J. (1990). Electrophoretic characterisation of proteins in oviduct fluid of cows during the estrous cycle. *Journal of Experimental Zoology*, **256**, 113–20.

Guraya, S.S. (1983). Recent progress in the structure, origin, composition and function of cortical granules in animal eggs. *International Review of Cytology*, **78**, 257–360.

Guttmacher, M.S. & Guttmacher, A.F. (1921). Morphological and physiological studies on the musculature of the mature Graafian follicle of the sow. *Johns Hopkins Hospital Bulletin*, **32**, 394–9.

Hafez, E.S.E. & Blandau, R.J. (eds.) (1969). *The Mammalian Oviduct*. Chicago: University of Chicago Press.

Hansen, C., Srikandakumar, A. & Downey, B.R. (1991). Presence of follicular fluid in the porcine oviduct and its contribution to the acrosome reaction. *Molecular Reproduction and Development*, **30**, 148–53.

Haresign, W., Foster, J.P., Haynes, N.B., Crichton, D.B. & Lamming, G.E. (1975). Progesterone levels following treatment of seasonally anoestrous ewes with synthetic LH-releasing hormone. *Journal of Reproduction and Fertility*, **43**, 269–79.

Harper, M.J.K. (1961a). The mechanisms involved in the movement of newly ovulated eggs through the ampulla of the rabbit Fallopian tube. *Journal of Reproduction and Fertility*, **2**, 522–4.

Harper, M.J.K. (1961b). Egg movement through the ampullar region of the Fallopian tube of the rabbit. In *Proceedings of the 4th International Congress on Animal Reproduction*, The Hague, p. 375.

Harper, M.J.K. (1973). Stimulation of sperm movement from the isthmus to the site of fertilisation in the rabbit oviduct. *Biology of Reproduction*, **8**, 369–77.

Harper, M.J.K. & Pauerstein, C.J. (1976). *Symposium on Ovum Transport and Fertility Regulation*. Copenhagen: Scriptor.

Hearn, J.P. (1986). The embryo-maternal dialogue during early pregnancy in primates. *Journal of Reproduction and Fertility*, **76**, 809–19.

Hertig, A.T., Rock, J. & Adams, E.C. (1956). A description of 34 human ova within the first 17 days of development. *American Journal of Anatomy*, **98**, 435–94.

Hirshfield, A.N. (1984). Continuous [^{3}H]thymidine infusion: a method for the study of follicular dynamics. *Biology of Reproduction*, **30**, 485–91.

Hunter, M.G. (1991). Characteristics and causes of the inadequate corpus luteum. *Journal of Reproduction and Fertility, Supplement*, **43**, 91–9.

Hunter, R.H.F. (1972). Local action of progesterone leading to polyspermic fertilisation in pigs. *Journal of Reproduction and Fertility*, **31**, 433–44.

Hunter, R.H.F. (1974). Chronological and cytological details of fertilisation and early embryonic development in the domestic pig, *Sus scrofa*. *Anatomical Record*, **178**, 169–86.

Hunter, R.H.F. (1977). Function and malfunction of the Fallopian tubes in relation to gametes, embryos and hormones. *European Journal of Obstetrics, Gynaecology and Reproductive Biology*, **7**, 267–83.

Hunter, R.H.F. (1978). Intraperitoneal insemination, sperm transport and capacitation in the pig. *Animal Reproduction Science*, **1**, 167–79.

Hunter, R.H.F. (1980). *Physiology and Technology of Reproduction in Female Domestic Animals*. London & New York: Academic Press.

Hunter, R.H.F. (1988). *The Fallopian Tubes: Their Role in Fertility and Infertility*. Berlin, Heidelberg, New York: Springer-Verlag.

Hunter, R.H.F. (1989). Differential transport of fertilised and unfertilised eggs in equine Fallopian tubes: a straightforward explanation. *Veterinary Record*, **125**, 304–5.

Hunter, R.H.F. (1990). Physiology of the Fallopian tubes, with special reference to gametes, embryos and microenvironments. In *From Ovulation to Implantation*, ed. J.L.H. Evers & M.J. Heineman, International Congress Series, No. 917, pp. 101–19. Amsterdam: Excerpta Medica.

Hunter, R.H.F. (1991). Fertilisation in the pig and horse. In *A Comparative Overview of Mammalian Fertilisation*, ed. B.S. Dunbar & M.G. O'Rand, pp. 329–49. New York: Plenum Press.

Hunter, R.H.F. (1993). Sperm : egg ratios and putative molecular signals to modulate gamete interactions in polytocous mammals. *Molecular Reproduction and Development*, **35**, 324–7.

Hunter, R.H.F. (1994). Modulation of gamete and embryonic microenvironments by oviduct glycoproteins. *Molecular Reproduction and Development*, **39**, 176–81.

Hunter, R.H.F. (1995a). Ovarian endocrine control of sperm progression in the Fallopian tubes. *Oxford Reviews of Reproductive Biology*, **17**, 85–124.

Hunter, R.H.F. (1995b). Human sperm reservoirs and Fallopian tube function: a rôle for the intra-mural portion? *Acta Obstetricia et Gynecologica Scandinavica*, **74**, 677–81.

Hunter, R.H.F. (1996). Ovarian control of very low sperm : egg ratios at the commencement of mammalian fertilisation to avoid polyspermy. *Molecular Reproduction and Development*, **44**, 417–22.

Hunter, R.H.F. (1997). Sperm dynamics in the female genital tract: interactions with Fallopian tube microenvironments. In *Microscopy of Reproduction and Development. A Dynamic Approach*, Proceedings of Second International Malpighi Symposium, Rome, pp. 35–45. Rome: Antonio Delfino Editore.

Hunter, R.H.F. (1998). Have the Fallopian tubes a vital rôle in promoting fertility? *Acta Obstetricia et Gynecologica Scandinavica*, **77**, 475–86.

Hunter, R.H.F. (2002). Tubal ectopic pregnancy: a patho-physiological explanation involving endometriosis. *Human Reproduction*, **17**, 1688–91.

Hunter, R.H.F., Cook, B. & Poyser, N.L. (1983). Regulation of oviduct function in pigs by local transfer of ovarian steroids and prostaglandins: a mechanism to influence sperm transport. *European Journal of Obstetrics, Gynaecology and Reproductive Biology*, **14**, 225–32.

Hunter, R.H.F., Fléchon, B. & Fléchon, J.E. (1991). Distribution, morphology and epithelial interactions of bovine spermatozoa in the oviduct before and after ovulation: a scanning electron microscope study. *Tissue and Cell*, **23**, 641–56.

Hunter, R.H.F. & Nichol, R. (1986). A pre-ovulatory temperature gradient between the isthmus and ampulla of pig oviducts during the phase of sperm storage. *Journal of Reproduction and Fertility*, **77**, 599–606.

Hunter, R.H.F. & Nichol, R. (1993). Passage of unfertilised horse eggs transplanted to pig oviducts. *Equine Veterinary Journal*, **25**, 544–6.

Killeen, I.D. & Moore, N.W. (1970). Transport of spermatozoa and fertilisation in the ewe following cervical and uterine insemination early and late in oestrus. *Australian Journal of Biological Sciences*, **23**, 1271–7.

Leese, H.J. (1988). The formation and function of oviduct fluid. *Journal of Reproduction and Fertility*, **82**, 843–56.

Leese, H.J. & Barton, A.M. (1985). Production of pyruvate by isolated mouse cumulus cells. *Journal of Experimental Zoology*, **234**, 231–6.

Moricard, R. & Bossu, J. (1951). Arrival of the fertilising sperm at the follicular cell of the secondary oocyte: a study of the rat. *Fertility and Sterility*, **2**, 260–6.

Motta, P.M., Nottola, S.A. & Familiari, G. (1997). Cumulus corona cells at fertilisation and segmentation. A paracrine organ. In *Microscopy of Reproduction and Development: A Dynamic Approach*, ed. P.M. Motta, pp. 177–86. Rome: Antonio Delfino Editore.

Nichol, R., Hunter, R.H.F., Gardner, D.K., Leese, H.J. & Cooke, G.M. (1992). Concentrations of energy substrates in oviductal fluid and blood plasma of pigs during the peri-ovulatory period. *Journal of Reproduction and Fertility*, **96**, 699–707.

Oguri, N. & Tsutsumi, Y. (1972). Non-surgical recovery of equine eggs, and an attempt at non-surgical egg transfer in horses. *Journal of Reproduction and Fertility*, **31**, 187–95.

Oliphant, G. & Ross, P.R. (1982). Demonstration of production and isolation of three sulfated glycoproteins from the rabbit oviduct. *Biology of Reproduction*, **26**, 537–44.

Ortiz, M.E., Bedregal, P., Carvajal, M.I. & Croxatto, H.B. (1986). Fertilised and unfertilised ova are transported at different rates by the hamster oviduct. *Biology of Reproduction*, **34**, 777–81.

Overstreet, J.W., Cooper, G.W. & Katz, D.F. (1978). Sperm transport in the reproductive tract of the female rabbit. II. The sustained phase of transport. *Biology of Reproduction*, **19**, 115–32.

Oxenreider, S.L. & Day, B.N. (1965). Transport and cleavage of ova in swine. *Journal of Animal Science*, **24**, 413–17.

Pauerstein, C.J. (1975). Clinical implications of oviductal physiology and biochemistry. *Gynecological Investigation*, **6**, 253–64.

Pomeroy, R.W. (1955). Ovulation and the passage of the ova through the Fallopian tubes in the pig. *Journal of Agricultural Science*, **45**, 327–30.

Rasweiler, J.J. (1972). Reproduction in the long-tongued bat, *Glossophaga soricina*. 1. Preimplantation development and histology of the oviduct. *Journal of Reproduction and Fertility*, **31**, 249–62.

Rasweiler, J.J. (1979). Differential transport of embryos and degenerating ova by the oviducts of the long-tongued bat, *Glossophaga soricina*. *Journal of Reproduction and Fertility*, **55**, 329–34.

Rasweiler, J.J. (1984). Reproductive failure due to the entrapment of oocytes in luteinised follicles of the little bulldog bat (*Noctilio albiventris*). *Journal of Reproduction and Fertility*, **71**, 95–101.

Robitaille, G., St-Jacques, S., Potier, M. & Bleau, G. (1988). Characterisation of an oviductal glycoprotein associated with the ovulated hamster oocyte. *Biology of Reproduction*, **38**, 687–94.

Roblero, L., Biggers, J.D. & Lechene, C.P. (1976). Electron probe analysis of the elemental microenvironment of oviducal mouse embryos. *Journal of Reproduction and Fertility*, **46**, 431–4.

Shalgi, R. & Kraicer, P.F. (1978). Timing of sperm transport, sperm penetration and cleavage in the rat. *Journal of Experimental Zoology*, **204**, 353–60.

Shareha, A.M., Ward, W.R. & Birchall, K. (1976). Effect of continuous infusion of gonadotrophin-releasing hormone in ewes at different times of the year. *Journal of Reproduction and Fertility*, **46**, 331–40.

Shelton, K., Gayerie de Abreu, M.F., Hunter, M.G., Parkinson, T.J. & Lamming, G.E. (1990). Luteal inadequacy during the early luteal phase of subfertile cows. *Journal of Reproduction and Fertility*, **90**, 1–10.

Stambaugh, R., Noriega, C. & Mastroianni, L. (1969). Bicarbonate ion: the corona cell dispersing factor of rabbit tubal fluid. *Journal of Reproduction and Fertility*, **18**, 51–8.

Stefanini, M., Oura, C. & Zamboni, L. (1969). Ultrastructure of fertilisation in the mouse. II. Penetration of sperm into the ovum. *Journal of Submicroscopical Cytology*, **1**, 1–23.

Sutton, R., Nancarrow, C.D. & Wallace, A.L.C. (1986). Estrogen and seasonal effects on the production of an estrus-associated glycoprotein in oviducal fluid of sheep. *Journal of Reproduction and Fertility*, **77**, 645–53.

Sutton, R., Nancarrow, C.D., Wallace, A.L.C. & Rigby, N.W. (1984). Identification of an oestrus-associated glycoprotein in oviducal fluid of the sheep. *Journal of Reproduction and Fertility*, **72**, 415–22.

Swyer, G.I.M. (1947). A tubal factor concerned in the denudation of rabbit ova. *Nature (London)*, **159**, 873–4.

Szöllösi, D.G. (1962). Cortical granules: a general feature of mammalian eggs? *Journal of Reproduction and Fertility*, **4**, 223–4.

Szöllösi, D. (1975). Mammalian eggs ageing in the Fallopian tubes. In *Ageing Gametes*, ed. R.J. Blandau, pp. 98–121. Basel: Karger.

Thibault, C. (1967). Analyse comparée de la fécondation et de ses anomalies chez la brebis, la vache et la lapine. *Annales de Biologie Animale, Biochimie, Biophysique*, **7**, 5–23.

Thibault, C., Levasseur, M.-C. & Hunter, R.H.F. (1993). *Reproduction in Mammals and Man*. Paris: Ellipses.

Van Niekerk, C.H. & Gerneke, W.H. (1966). Persistence and parthenogenetic cleavage of tubal ova in the mare. *Onderstepoort Journal of Veterinary Research*, **33**, 195–231.

Villalón, M., Velasquez, L. & Croxatto, H. (1999). Oocyte and embryo transport. In *Encyclopedia of Reproduction*, vol. 3, ed. E. Knobil & J.D. Neill, pp. 459–68. San Diego & London: Academic Press.

Wagh, P.V. & Lippes, J. (1993). Human oviductal fluid proteins. V. Identification of human oviductin-1 as alpha-fetoprotein. *Fertility and Sterility*, **59**, 148–56.

Wassarman, P.M. (1988). Zona pellucida glycoproteins. *Annual Review of Biochemistry*, **57**, 415–42.

Wassarman, P.M. (1991). Fertilisation in the mouse. In *A Comparative Overview of Mammalian Fertilisation*, ed. B.S. Dunbar & M.G. O'Rand, pp. 151–65. New York & London: Plenum Press.

Wassarman, P.M., Liu, C. & Litscher, E.S. (1997). Building the mammalian egg zona pellucida: an organelle that regulates fertilisation. In *Microscopy of Reproduction and Development: A Dynamic Approach*, ed. P.M. Motta, pp. 169–76. Rome: Antonio Delfino Editore.

Webb, R., Lamming, G.E., Haynes, N.B., Hafs, H.D. & Manns, J.G. (1977). Response of cyclic and post-partum suckled cows to injections of synthetic LH-RH. *Journal of Reproduction and Fertility*, **50**, 203–10.

Weber, J.A., Freeman, D.A., Vanderwall, D.K. & Woods, G.L. (1991a). Prostaglandin E$_2$ secretion by oviductal transport-stage equine embryos. *Biology of Reproduction*, **45**, 540–3.

Weber, J.A., Freeman, D.A., Vanderwall, D.K. & Woods, G.L. (1991b). Prostaglandin E$_2$ hastens oviductal transport of equine embryos. *Biology of Reproduction*, **45**, 544–6.

Wintenberger-Torrès, S. (1961). Mouvements des trompes et progression des oeufs chez la brebis. *Annales de Biologie Animale, Biochimie, Biophysique*, **1**, 121–33.

Wislocki, G.B. & Snyder, F.F. (1933). The experimental acceleration of the rate of transport of ova through the Fallopian tube. *Bulletin of the Johns Hopkins Hospital*, **52**, 379–86.

Yanagimachi, R. (1994). Mammalian fertilisation. In *The Physiology of Reproduction*, 2nd edn, ed. E. Knobil & J.D. Neill, pp. 189–317. New York: Raven Press.

Yanagimachi, R. & Mahi, C.A. (1976). The sperm acrosome reaction and fertilisation in the guinea pig: a study *in vivo*. *Journal of Reproduction and Fertility*, **46**, 49–54.

Zamboni, L. (1972). Fertilisation in the mouse. In *Biology of Mammalian Fertilisation and Implantation*, ed. K.S. Moghissi & E.S.E. Hafez, pp. 213–62. Springfield, IL: Charles C. Thomas.

Zeleznik, A.J. (1992). Luteinisation and luteolysis. In *Gonadal Development and Function*, vol. 94, ed. S.G. Hillier, Serono Symposia Publications, pp. 178–87. New York: Raven Press.

Zeleznik, A.J. (1999). Luteinisation. In *Encyclopedia of Reproduction*, vol. 2, ed. E. Knobil & J.D. Neill, pp. 1076–83. San Diego & London: Academic Press.

X

Failure of ovulation: status of the gonads

Introduction

In a text that has examined the events associated with mammalian ovulation in some depth, there remains an obligation to take note of instances in which physiological mechanisms have been in error and successful ovulation has been compromised or failed. Such instances would, of course, be quite distinct from anovulation during pregnancy (except in equids), in early lactation (with an appropriate frequency of suckling), and during seasonal anoestrus in many species of animal, both wild and domestic. Failure of ovulation may be a major cause of infertility but its actual incidence will depend on both genetic and environmental factors. The genetic backcloth is certainly important in terms of species, race or strain, or even family lineage. As to environmental factors, these include: diet and nutritional status; seasonal influences including temperature and travel across time zones; exposure to drugs including alcohol and nicotine; overt pollution as from traffic or industrial fumes or more subtle olfactory considerations; and, as a prominent feature of modern society, so-called noise pollution.

The manner in which ovulation is compromised may also be diverse, although a straightforward partition would be between systemic endocrine factors and local ovarian factors. As to the latter, they may involve hormones, growth factors, cytokines or be primarily structural but, in reality, systemic and local factors are closely intertwined. There will clearly be molecular influences on structural anomalies, yet to be elucidated or fully clarified.

This chapter makes no attempt to be comprehensive or to treat its selected topics exhaustively. They are offered simply as examples. The principal purpose of the chapter is to indicate that even in the presence of demonstrable follicular growth and development, the culminating event of ovulation – as distinct from atresia – cannot be taken for granted. This remark holds true for our own species and for laboratory and domestic farm animals. Examples are also provided of conditions in which follicular growth is not possible due to an absence of germ cells or of appropriate trophic hormones.

Cystic follicles in animals

Graafian follicles that fail to complete the final phases of maturation and ovulate may enter the pathway to atresia and regress into islands of scar tissue. An alternative destiny is to become cystic, again with a permanent loss of the potential for ovulation, even in response to high doses of injected gonadotrophic hormones. Although unable to ovulate, such structures can continue to enlarge. They may still resemble follicles superficially, but features of degeneration can invariably be distinguished. One of the earliest is dissolution of the basement membrane between granulosa and theca cell layers, with progressive dissociation of cells within the mural granulosa and disruption of the cumulus oophorus supporting the oocyte (see Sturgis, 1961). The oocyte itself soon shows signs of disorganisation, in both the cytoplasm and nucleus. Aggregations of chromatin may appear in the germinal vesicle and, following breakdown of the nuclear membrane, there may be partial formation of a spindle apparatus. As a sequel to changes in the granulosa cells and oocyte, degeneration may be detected in the theca layers, both in the cells themselves and in the vascular bed.

Comments concerning the vascular bed take account of the fact that cysts may appear as clear-walled structures or be haemorrhagic. In the former, the capillary network has shrunk to become vestigial. The extensive network of surface blood vessels characteristic of mature follicles is no longer detectable macroscopically, giving the structure a pale yellow or almost clear appearance. Alternatively, as a major contrast, there may have been a prominent breakdown of blood vessels giving the follicle a strongly haemorrhagic appearance. The colour of individual cysts ranges from bright pinkish-red to a deep bluish-red, often almost purple. Not only has breakdown of vessels occurred within the wall of the follicle but the antral fluid is also filled with blood components. Some degree of clotting may be found and residual fluid may be difficult to aspirate. It is not easy to determine the ultimate destiny of such structures but, in domestic farm animals at least, they may remain detectable for periods of months. In the absence of surgical intervention, there is little prospect of resolving this ovarian condition with resumption of ovulatory cycles, although treatment with analogues of hypothalamic releasing factors has had a limited success (see below).

Formation of follicular cysts in farm animals is a significant cause of infertility, especially in high-yielding dairy cows (Webb *et al.*, 1998). Such cysts may give rise to intense and persistent sexual behaviour (oestrus) expressed as the condition of nymphomania. Although the granulosa cells may have largely degenerated, the follicles can continue to enlarge and secrete oestrogens and usually also potent concentrations of androgens from the theca cells. However, other forms of cyst can arise in domestic animals, such as luteinised follicles in which layers of luteal cells appear macroscopically as luteal tissue surrounding a fluid-filled or haemorrhagic cavity. These different forms of cyst may originate in an endocrine imbalance involving gonadotrophic hormones, frequently during a phase of heavy lactation. They tend to persist and require clinical intervention rather than regressing spontaneously. Cystic ovarian problems may also increase with age in farm animals, and often underlie instances of so-called reproductive failure (Perry & Pomeroy, 1956; Laing, 1970). At a microscopic level, development of cystic follicles is thought to express an imbalance between cell proliferation and cell death (Isobe & Yoshimura, 2000).

Formation of cystic ovarian follicles was a familiar sequel to many of the pioneer hormonal treatments for synchronising oestrous cycles in farm animals. The pharmacological approach was to suppress oestrus and ovulation with a natural or synthetic steroid hormone, such as injections of a solution of progesterone in oil or administration of an oral progestagen, for a period of time sufficient to bring all animals in a group under the influence of the treatment. The steroid blockade was then stopped abruptly, permitting animals to rebound spontaneously into the follicular phase of an oestrous cycle. Whereas daily injections of a solution of progesterone were noted to give an excellent suppression of oestrus and ovulation, this form of pituitary gland and hypothalamic blockade was associated with a high incidence of cystic follicles at the oestrus following treatment (e.g. Ulberg, Christian & Casida, 1951a; Casida, 1976). As a consequence, fertility was generally poor or negligible.

Experimental studies in pigs have also frequently resulted in cystic ovarian follicles when manipulation of the oestrous cycle has been attempted (e.g. Ulberg, Grummer & Casida, 1951b; Nellor, 1960; Dziuk, 1962). Two different approaches have also acted to generate cystic follicles in this species. In a series of unpublished studies, the author noted that this condition could be induced by the following.

1. Unilateral ovariectomy shortly before the anticipated time of ovulation, perhaps associated with a consequent endocrine imbalance influencing the remaining ovary.
2. Again, with intervention shortly before ovulation, the placement of double ligatures around the Fallopian tube and neighbouring vascular and

lymphatic arcades close to the ampullary–isthmic junction would result in failure of ovulation and development of cystic follicles (see Chapter VIII).

Puncture of maturing Graafian follicles on days 19 or 20 of the oestrous cycle was also found to promote the formation of cysts in many instances (Hunter & Baker, 1975).

As to resolution of the cystic condition, there is a body of work from cattle in which this is a major cause of so-called 'repeat breeding' (see Kesler & Garverick, 1982). Early experiments suggested that cystic follicles may respond more predictably to treatment with GnRH than with LH or hCG (Kittok, Britt & Convey, 1973; Britt, 1975). However, if the follicle is already showing extensive morphological degeneration, an ovulatory or luteinisation response even to endogenous gonadotrophins is difficult to envisage and yet this has more recently been reported, at least as far as a luteinisation response is concerned (Osawa *et al.*, 1995; Garverick, 1997). Lysis of luteinised cystic follicles may be obtained using $PGF_{2\alpha}$ injected systemically or, less straightforwardly, on a protocol of GnRH injections in the presence of an intra-vaginal progesterone-releasing device followed by a $PGF_{2\alpha}$ injection (Thatcher *et al.*, 1993). Practical approaches to the problem of cystic follicles are clearly required, for 7–13% of dairy cattle may suffer from the condition during the post-partum period (Borsberry & Dobson, 1989; Garverick, 1997). Novel therapies are currently being sought based upon experimental models for inducing the formation of cysts in cattle (Webb *et al.*, 1998) and in rabbits (López-Béjar *et al.*, 1998).

Polycystic ovarian disease in women

Frequently referred to as the polycystic ovarian syndrome (PCOS), this is a superficially well-recognised disorder in women (Franks, 1989; Ben-Shlomo, Franks & Adashi, 1995; Udoff & Adashi, 1995). As suggested by the terminology, the condition is characterised morphologically by the presence of multiple cystic follicles (2–8 mm in diameter) in the subcapsular region close to the surface of ovaries, themselves enlarged by an increase in the central stroma. In the developed world, the condition may be found in up to 10% of young women (Crosignani & Nicolosi, 2001). It was widely, if loosely, referred to in the 1960s and 1970s as the Stein–Leventhal syndrome and yet, even today, it remains inadequately defined (Legro, 1999).

A proportion of women presenting with the condition may have almost normal reproductive health and, despite a high incidence of isolated polycystic ovarian morphology (22%), no uniformly adverse influence on fertility has

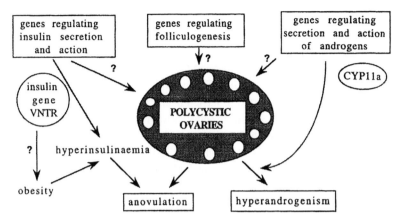

Fig. X.1. Potential aetiological factors in the polycystic ovary syndrome, as proposed by Franks *et al.* (1997). VNTR, variable number tandem repeats. (Courtesy of Professor S. Franks.)

been reported (Clayton *et al.*, 1992). Indeed, in comparisons within an *in vitro* fertilisation programme between apparently normal recipients and those with polycystic ovaries, there was no significant difference in the incidence of pregnancy and live births (MacDougall *et al.*, 1993). Conversely, adverse effects on reproductive performance have been noted, such as disturbed menstrual cycles and chronic anovulation. In one large study, 46% primary and 26% secondary infertility were recorded (Balen *et al.*, 1995). As a consequence of associated endocrine disorders, there may be an elevated incidence of miscarriages (Regan *et al.*, 1990).

Turning specifically to endocrine aspects, the polycystic ovarian condition is frequently associated with enhanced plasma concentrations of androgens (both androstenedione and testosterone), LH and insulin, and may be expressed as a decreased insulin sensitivity (Dunaif *et al.*, 1995). These endocrine perturbations are generally associated with obesity and hirsutism. An excess of adrenal androgen secretion may be noted, but the principal origin of elevated androgens is the cystic ovary. The theca layers of the multiple follicles together with the hypertrophied stromal tissue are the culprits here. Moreover, the elevated androgen titres may offer a partial explanation for the failure of development of follicles to pre-ovulatory size; rather, premature induction of atresia may be a consequence. Adipose tissue conversion of precursor molecules into androgens must also be considered.

Despite the wide range of phenotypes that may be bracketed under the PCOS, their origin is thought to lie in a small number of key genes interacting with so-called environmental factors (Fig. X.1). An autosomal dominant model of transmission has been proposed, enabling the condition to be passed on through either sex (Crosignani & Nicolosi, 2001). A current view is that specific gene mutations affecting androgen synthesis, insulin secretion and insulin activity could explain many of the endocrine and metabolic symptoms. Environmental risk factors may act to convert the PCOS into a clinically manifest condition. Polycystic follicles are reported to have a primary lesion in cells of the theca interna underlying excess synthesis and secretion of progesterone and androgen (Gilling-Smith *et al.*, 1994, 1997). In addition, the cholesterol side chain cleavage gene (*CYP11a*) was revealed as an important locus for susceptibility to polycystic ovarian syndrome hyperandrogenism (Gharani *et al.*, 1997), as was the insulin gene minisatellite (Franks *et al.*, 1997).

Over and above the strong focus on endocrine abnormalities arising from mutations in specific genes (for a review, see Crosignani & Nicolosi, 2001), it is worth giving serious consideration also to the morphological condition of tissues. Under the influence of trophic hormones from the anterior pituitary gland, ovarian tissues are being constantly remodelled. The extracellular matrix of the ovary plays a vital part in this activity (Luck, 1994). Defects arising in the form and function of the extracellular matrix could inhibit normal surface changes in developing follicles, leading ultimately to inhibition of ovulation, even though such follicles would remain hormonally active. Extracellular matrix defects might also explain why cystic follicles in this syndrome seem not to embark on characteristic atresia and shrinkage into islands of tissue; the condition seems not to resolve spontaneously, except in the sense that it may once post-menopausal age is reached (Legro, 1999). Perturbations of ovarian temperature gradients might even be involved (see Chapter III).

Putting to one side the common sense need for obese women to lose weight, the only hopeful treatment for many years appeared to be wedge resection – removing a wedge-shaped portion of ovarian tissue. The objective was to expose an area of fresh ovarian surface, thereby hoping that growth and maturation of healthy follicles could lead to ovulation unimpeded by degenerative surface tissues. Not to be undertaken lightly, since this traditionally involved major abdominal surgery, there was frequently a spontaneous re-establishment of cystic surfaces. In addition, subsequent problems with ovarian adhesions could arise from seepage at the surface of the transected tissue. Contemporary approaches involve drilling of ovarian tissues at laparoscopy to create a pathway or laser- or thermo-cautery to destroy ovarian stroma. No consistent outcome to these approaches has been reported, nor is this surprising. Ovarian tissues tend to respond more predictably to a sensitive and subtle stimulation, less so to gross physical approaches. In this regard, the reviews of Franks (1995, 1997) are especially thoughtful.

A quite distinct means of overcoming infertility as a consequence of poly-cystic ovaries is to aspirate follicular oocytes and subject them to procedures of *in vitro* maturation and IVF (see Trounson *et al.*, 2001). Embryos gener-ated in this manner may give rise to full-term pregnancies when restored to the genetic mother, although the individual will probably require endocrine (pro-gestagen) therapy. Alternatively, embryos arising from aspirated oocytes and *in vitro* procedures may be transplanted to a surrogate mother in appropriate circumstances.

Ovarian dysgenesis

A significant topic under the title of failure of ovulation necessarily con-cerns ovarian dysgenesis. There may be a congenital absence of oocytes and therefore of follicles or there may be an accelerated loss of ovarian oocytes, as in Turner's syndrome. A genetic disorder in women that characteristically produces this syndrome is the 45,XO condition of dysgenesis, with the absence of one X chromosome per cell (Ford *et al.*, 1959). This is viewed as the most common chromosomal abnormality in our own species, even though the inci-dence falls dramatically from about 1 in 100 at conception to 1 in 10 000 females at birth, or 1 in 2500 live births (Barad, 1999). Some 98% or 99% of embryos or fetuses with an XO sex chromosome constitution will die during prenatal development and be aborted (Bishop, 1972; Short, 1982). Typically, females with Turner's syndrome have short stature, skin folds over the neck ('webbing'), narrowing of the aorta, and abnormalities of the renal and lym-phatic systems – the last resulting in oedema. The short stature and possibly abnormal bone growth may, in part, be a reflection of cardiovascular and lym-phatic abnormalities.

In individuals with the XO condition, the ovaries invariably undergo pre-mature germinal wastage. This leads to conspicuously reduced sexual de-velopment, expressed as a juvenile reproductive tract, an accelerated rate of oocyte degeneration, and a premature onset of sterility. In such XO women, primary amenorrhoea is the most frequent consequence, affecting 97% of cases (Simpson, 1976), although menstruation (Weiss, 1971) and even pregnancy have been reported (Dewhurst, 1978; King, Magenis & Bennett, 1978). The ovaries of XO human fetuses aborted spontaneously during the first 10 weeks appear normal, with an almost characteristic complement of germ cells (Singh & Carr, 1966). However, deficiencies of germinal follicular tissue are noted in the ovaries of mid-term XO fetuses aborted as a sequel to amniocentesis, and the expected steps of primordial follicle formation had rarely occurred.

As a consequence of an enhanced rate of atresia during the second half of pregnancy, few if any oocytes remain in the fetus by birth, the ovaries of XO infants being severely affected (Carr, Haggar & Hart, 1968; Bove, 1970; Weiss, 1971). The loss of XO germ cells in the human XO gonad occurs throughout the period when oogonia are embarking upon meiosis (Carr, 1972). In at least three XO human fetuses, Speed (1986) found that arrest in oogenesis can occur at the very earliest stage of meiotic prophase before chromosome pairing is even established. In the monosomic condition for the X chromosome, the ovaries degenerate into streaks or rudiments of fibrous connective tissue usually referred to as streak gonads (Simpson, 1976), resulting in a stroma unable to generate follicles or to secrete hormones. Although affected individuals are born with morphologically normal female external genitalia, there is little or no development of secondary sexual characteristics in the absence of ovarian hormone synthesis.

The preceding paragraphs have stressed the abnormal, but it is worth recalling that Turner's syndrome is not an all-or-none condition as regards ovarian oocytes. In rare instances, a few oocytes may apparently survive until the time of puberty, thereby permitting the formation of functional ovaries, and thus a tiny proportion of women presumed to have Turner's syndrome can be fertile. The invariable assumption in such cases is that the surviving oocytes would be of XX sex chromosome constitution (Burgoyne & Baker, 1985), although these authors concede that there may occasionally be XO oocytes remaining at puberty. They underline that the chances of a successful pregnancy in XO or XO mosaic mothers are low, some 33 of 48 reported pregnancies ending in spontaneous miscarriage, stillbirth, neonatal death or congenital abnormality (King *et al.*, 1978). This would be in line with the anticipated outcome for XO oocytes, for any XO fetuses would almost certainly be aborted.

The aetiology of the condition has been reviewed at some length (Hunter, 1995; Barad, 1999), including details of specific molecular insights. Despite the ever-accumulating molecular information, the way forward in a context of fertility therapy is by no means clear, not least since Turner's syndrome embraces a variety of clinical anomalies stemming from disparate molecular defects. As written in an earlier review, one cannot escape the conclusion that intellectual insights gained by molecular analysis have not yet provided the physician with a ready means of mitigating most clinical problems, and this remains especially so in the emotive context of infertility and sterility (Hunter, 1995). In fact, approaches such as the transplantation of fetal ovarian tissue to women with dysgenic gonads (Gosden, 1992, 2000) or embryo transplantation to suitably prepared (steroid-primed) individuals may offer more hope in the short term than a specific molecular therapy.

Kallmann's syndrome

Occasionally described under the heading of olfactogenital dysplasia, Kallmann's syndrome is a condition with major consequences for the reproductive system (Kallmann, Schoenfeld & Barrera, 1944). It is frequently associated with arrested development of the anterior part of the brain, and is a genetic disorder of both the gonadal and olfactory functions. It may affect 1 in 10 000 males but some five to seven times fewer females (Hardelin *et al.*, 1992). Kallmann's syndrome results in a high incidence of cryptorchidism in males, whereas there is amenorrhoea in females associated with the persistence of primary follicles or the death of follicles at an early stage of growth. Characteristically, there are low levels of circulating gonadotrophins and oestrogens, small ovaries, and primordial follicles assembled in clusters close to a thin tunica albuginea (Goldenberg *et al.*, 1976). A minor degree of follicular growth may be detected. Soon after its recognition, the defect was surmised to reside in the hypothalamo-pituitary axis rather than in the ovary, primarily because ovulation could be induced in patients exhibiting the syndrome as a response to injected gonadotrophins. Recent laboratory investigations have moved to the molecular level and required some modification of the preceding conclusion.

The hypogonadotrophic hypogonadism characteristic of Kallmann's syndrome is a consequence of defective GnRH secretion. The associated defect of anosmia is due to hypoplasia or the complete absence of olfactory bulbs and tracts. Why should these two conditions be linked? Immunochemical studies in mice indicated that neurons producing GnRH share a common embryonic origin and migration pathway to the hypothalamus with the vomeronasal and terminalis nerves, suggesting that a human cell migration defect specifically affecting such neurons in the olfactory placode of the developing brain could be involved in the syndrome (Schwanzel-Fukuda & Pfaff, 1989; Schwanzel-Fukuda, Bick & Pfaff, 1989; Wray, Grant & Gainer, 1989; Rugarli, 1999). This is now accepted to be so, and a gene has been isolated that maps to the Kallmann's syndrome critical region on the distal part of the short arm of the human X chromosome where it escapes X-inactivation (Franco *et al.*, 1991). This particular gene (*KALIG-1*) shares homology with molecules involved in cell adhesion and axonal pathfinding, suggesting that a defect in this gene – a partial or complete deletion – causes the neuronal migration defect underlying Kallmann's syndrome. A contribution from pioneer cells (a form of glial cell) from the olfactory placode was invoked, since such cells leave a neural cell adhesion molecule (NCAM) trail or scaffold that is then followed by vomeronasal and GnRH neurons (Pfaff & Schwanzel-Fukuda, 1993).

However, it is fully accepted that more than one gene will be involved in the different genetic forms of the syndrome, for segregation analysis has revealed three modes of transmission: X-linked, autosomal recessive and autosomal dominant (Hardelin *et al.*, 1992). It is now also appreciated that the *KAL* gene responsible for X-linked Kallmann's syndrome, the most frequent form, has a rôle in other neuronal pathways and in kidney organogenesis (Hardelin *et al.*, 1992, 1993). Further molecular details were summarised by Hunter (1995), with comments on a possible involvement of the KAL protein as an extracellular matrix component exhibiting antiprotease and adhesion functions (Petit, 1993). Evidence suggests that KAL is a glycosylated membrane protein that cleaves to yield a 45 kDa diffusible component that may function as a chemoattractant. In addition, recent embryological studies have detected mRNA encoding the KAL protein in the developing olfactory bulb, kidney, ocular muscles and cerebellum, offering an underlying explanation for the range of disorders noted in Kallmann's patients (Halvorson, 1999). As a further aspect, mutations in the *DAX-1* gene have been noted, and *DAX-1* expression may give rise to hypogonadotrophic hypogonadism (Muscatelli *et al.*, 1994; Swain *et al.*, 1996, 1998).

Returning to the clinical scene in humans, and for reasons already described, the syndrome usually presents as a delayed or incomplete onset of puberty, with patients lacking a sense of smell – even though they may not have appreciated this disorder. Because the gonads remain potentially responsive to gonadotrophic hormones, for the primary lesion involves GnRH, there are at least two routes to promoting functional ovarian activity. Gonadotrophic stimulation of the ovaries can be direct by means of daily injection of appropriate recombinant FSH and LH preparations or purified urinary preparations. Alternatively, the physiological backcloth of pulsatile GnRH can be obtained using a subcutaneous pump programmed to mimic cyclic events in the delivery of such hormone. Viable pregnancies have been obtained by both approaches (e.g. Kousta *et al.*, 1996).

Ovotestis and ovulation failure

Because mammalian gonads are paired structures, there is a widespread assumption that they will be morphologically and functionally similar. Extensive studies in laboratory rodents, domestic animals and humans have demonstrated that this is not always the case. Putting minor differences of size to one side, there are clearly documented instances of putative females having one ovary and a contralateral ovotestis, usually on the right-hand side, or there may be paired ovotestes or, exceptionally, two testis-like structures. These conditions have been reviewed in depth (Hunter, 1995). Worth stressing straight away is that the testis-like tissue in these putative females invariably reveals normal interstitial cells of Leydig but never a germ cell line in the seminiferous tubules. In other words, there is no possibility of spermatogenesis. Such animals usually have an XX sex chromosome constitution, occasionally carrying a translocated portion of a Y chromosome or exhibiting a specific mutation. Even at the level of the gonads, these individuals cannot be considered as hermaphrodites; instead, they are best termed intersexes.

The incidence of their occurrence is uncertain, since a unilateral anomaly may not compromise fertility. A value of 3% has been reported for an intersex condition in mice among 1439 inbred BALB/cWt 15-day fetuses (Beamer, Whitten & Eicher, 1978) compared with 0.4% among 3310 animals of the same strain found at weaning (Whitten, Beamer & Byskov, 1979). The inference was that a proportion of intersex mice may have been resorbed during late fetal development or died during early postnatal development. In certain inbred strains of domestic pig, the intersex condition may reach an incidence of 3–5% (Hunter, Cook & Baker, 1985). Indeed, Bäckström & Henricson (1971) noted that one Landrace boar generated 12 intersex piglets in a total of 286 offspring, a frequency of 4.2% in 25 litters. Although incidental observations on intersexuality in humans have recorded any left–right combination of ovary, ovotestis and testis, the incidence of such anomalies is generally considered to be low, seldom exceeding 1 in 10000. However, extrapolating from observations in animals suggests the possibility of a higher value in isolated populations with some degree of inbreeding, as may still be found even today on islands in the Pacific Ocean or among the indigenous peoples of the Upper Amazon basin or perhaps even in the Eskimo (Inuit) populations of the Arctic.

The relevance to the present chapter of animals bearing ovotestes is that there may be problems of ovulation seen as a reduced number of ovulations or even complete failure of ovulation. As a rule-of-thumb statement, and on the basis of observations in domestic farm animals, the greater the proportion of testicular-like tissue within an ovotestis, the lower the chances of spontaneous ovulation from that gonad. A simplistic observation here would be that there is inappropriate gonadal steroid feedback from the ovotestis to the hypothalamus and anterior pituitary gland to prompt a pre-ovulatory discharge of LH. Accepting that testicular androgens will cause endocrine anomalies, nevertheless inadequate positive feedback cannot be a sufficient explanation for the failure of ovulation within ovarian tissue of an ovotestis if there is a functional contralateral ovary. Indeed, the contralateral ovary may well show normal ovulation as a response to positive oestradiol feedback from its mature

Graafian follicles and the gonadotrophin peak would be available systemically. Accordingly, ovarian tissue in immediate proximity to testicular tissue may not support growth of Graafian follicles to a pre-ovulatory size and stage of maturity or such follicles may be unable to respond to a pre-ovulatory surge of gonadotrophic hormones. There is evidence from domestic animals for both of these propositions. Graafian follicles within an ovotestis frequently show an inability to respond to injected gonadotrophic hormones (Hunter *et al.*, 1985). One specific conclusion here is that the follicular cells have not developed appropriate gonadotrophin receptors. Alternatively, androgen and/or protein secretions from the ovotestis have inhibited the expression of such receptors, and follicular peptide hormone synthesis may have been perturbed.

As noted above, the incidence of ovotestis formation in humans is low, at least in developed societies, suggesting that the contribution of this condition to failure of ovulation will be insignificant; it may even be negligible in most populations today.

Ovulation failure as a response to stress

Putting to one side considerations raised elsewhere in this chapter, anovulation is found under conditions that modify or disrupt endocrine feedback from the ovaries themselves. Such failure of ovulation due to inappropriately programmed activity of the hypothalamus, so-called hypothalamic anovulation, may stem from conditions such as anorexia, diverse forms of physical or mental stress, or from starvation. To the extent that these conditions may be reversible and usually should not be accepted as permanent, they impose a phase of infertility rather than sterility.

Disorders of nutritional status linked to disorders of metabolism are frequently coupled with failure of orderly folliculogenesis and/or follicular maturation. At least amongst women in developed countries, weight-related (overweight or obese) amenorrhoea is considered as one of the most common causes of hypogonadotrophic amenorrhoea. The most typical feature here is a reduction in the frequency of LH pulses leading to inappropriate programming of follicular development. Insulin imbalance, as for example in hyperinsulinaemia (referred to above under the PCOS), may disrupt the orderly sequence of cellular events necessary for final maturation of the ovulatory follicle.

Distinct from nutritional problems leading to hypothalamic and/or pituitary dysfunction, evidence from domestic farm animals shows that exposure to acute stress acts to reduce normal pulsatile patterns of GnRH release. This in turn is seen in a reduced frequency and amplitude of LH pulses from the anterior

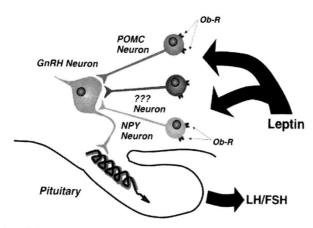

Fig. X.2. A model to suggest the involvement of leptin in the neuroendocrine–reproductive axis by means of influences on specific groups of hypothalamic nuclei and their peptide products – and thereby on gonadotrophin secretion by the anterior pituitary gland. Ob-R, leptin receptor; POMC, proopiomelanocortin; NPY, neuropeptide Y; GnRH, gonadotrophin releasing hormone. (After Cunningham *et al.*, 1999.)

pituitary gland. Examples of stress included transport (mechanical displacement) of animals, administration of high doses of insulin, or chronic administration of ACTH. A consequence of the endocrine disruption is seen as a failure of ovulation or a belated ovulation (Dobson *et al.*, 2001).

Literature appropriate to this field is diverse and extensive, and adequate documentation would appear as a list. Even so, as a useful route into the various topics, readers could consult the reviews of Ashworth & Antipatis (2001), Rhind, Rae & Brooks (2001) and Wallace *et al.* (2001).

Leptin involvement in ovulation failure

As a small peptide molecule (16 kDa) synthesised by white adipocytes, leptin is a hormone that contributes to the regulation of food intake and thereby body weight. It is a product of the obese (*ob*) gene, and mRNAs for both leptin and its receptor have a widespread distribution throughout the body. Circulating concentrations of leptin are correlated with fat mass and, at appropriate values, will act at the hypothalamus to suppress appetite and increase metabolic rate (Granowitz, 1997). In part, this is achieved through reduced production of NPY from the arcuate nucleus and perhaps also an involvement of proopiomelanocortin-containing nuclei and other hypothalamic peptides (Fig. X.2; Cunningham, Clifton & Steiner, 1999; Cowley *et al.*, 2001).

As suggested, leptin is a vital link between nutritional status, somatic energy reserves and neuroendocrine components of the reproductive system. When energy reserves are inadequate, leptin may function as a metabolic gate to inhibit activity within the neuroendocrine reproductive axis (Cunningham *et al.*, 1999). In terms of puberty, individuals may cross the reproductive threshold when sufficient concentrations of leptin are achieved in the circulation (Palmert, Radovick & Boepple, 1998). It may then serve as an important metabolic cue prompting neuronal activation of GnRH synthesis and secretion, in turn leading to pituitary secretion of LH at an appropriate frequency and amplitude. Nonetheless, there continues to be debate as to whether leptin is a primary stimulus of the reproductive system or, more broadly, is a permissive factor within a spectrum of somatic monitors (Clarke & Henry, 1999; Cunningham *et al.*, 1999; Foster & Nagatani, 1999). Among the latest of these is a newly identified protein termed resistin that is also produced by fat cells (Flier, 2001; Steppan *et al.*, 2001).

In addition to an involvement with the hypothalamo-pituitary axis, leptin has been detected in the ovary itself. Moreover, ovarian tissues can seemingly influence leptin availability, since there is a significant reduction in circulating leptin concentrations in 'normal' women after bilateral ovariectomy (Messinis *et al.*, 1999). However, interactions between ovarian leptin and adipocyte leptin appear complex, since ovarian oestradiol has been proposed as a regulator of leptin production by adipocytes (Messinis *et al.*, 1998; Messinis & Milingos, 1999). A key rôle of cytokines such as TNF-α must also be considered.

Extensive studies *in vitro* on different populations of follicular cells (e.g. granulosa and theca) point to an autocrine mechanism involving leptin in the human ovary that may influence pre- and post-ovulatory follicular development. One of the links here could involve the potent influence of leptin on the action of insulin, not least since insulin and IGF are implicated in normal follicular development and maturation (see Chapters V and VI). Despite accumulating experimental observations, it is still not straightforward to invoke anomalous secretion of leptin in instances of ovulation failure in seemingly healthy women. Nor is failure of ovulation in obese women characterised by a leptin deficiency, although there remains the speculation that obesity may represent a state of leptin resistance (Conway & Jacobs, 1997).

Concluding remarks

As clinical colleagues will certainly appreciate, one could write a substantial book on failure of ovulation and its many causes. However, the primary purpose of this concise chapter has not been to be comprehensive but rather to counter-balance the detailed descriptions of the physiological components of ovulation that precede it. The emphasis in the several sections has not been specifically on endocrine anomalies but has included some of the syndromes that may underlie failure of ovulation in a small proportion of patients. Wayward development of the gonads has been amply documented, albeit usually at a low or extremely low incidence. Whether modern forms of industrial pollution, including that from radio-active isotopes and environmental oestrogens, will act to enhance such congenital problems remains to be seen. The level of environmental toxins with noxious embryological influences is certainly on the increase in many countries, particularly so in the developing world.

Very much more could have been written on the relationship between stress and failure of ovulation, not least since there is suggestive evidence that the somewhat pressurised lifestyles in modern urban societies may influence reproductive processes: increasingly sedentary routines set against a backcloth of heat, noise, over-eating and/or consuming inappropriate diets. The much-heralded and widely encouraged march towards computer-based living, working and shopping may not prove ultimately to be the blessing that many would like to imagine. Instead, it may become the basis of serious and diverse problems of health, both social (mental) and somatic, and as such will undoubtedly include problems of reproductive health.

References

Ashworth, C.J. & Antipatis, C. (2001). Micronutrient programming of development throughout gestation. *Reproduction*, **122**, 527–35.

Bäckström, L. & Henricson, B. (1971). Intersexuality in the pig. *Acta Veterinaria Scandinavica*, **12**, 257–73.

Balen, A.H., Conway, G.S., Kaltsas, G. *et al.* (1995). Polycystic ovary syndrome: the spectrum of the disorder in 1741 patients. *Human Reproduction*, **10**, 2107–11.

Barad, D.H. (1999). Turner's syndrome. In *Encyclopedia of Reproduction*, vol. 4, ed. E. Knobil & J.D. Neill, pp. 879–86. San Diego & London: Academic Press.

Beamer, W.G., Whitten, W.K. & Eicher, E.M. (1978). Spontaneous sex mosaicism in BALB/cWt mice. In *Gatlinburg Symposium on Genetic Mosaics and Chimeras in Mammals*, ed. L.B. Russell, pp. 195–208. New York: Plenum Press.

Ben-Shlomo, I., Franks, S. & Adashi, E.Y. (1995). The polycystic ovary syndrome: nature or nurture? *Fertility and Sterility*, **63**, 953–4.

Bishop, M.W.H. (1972). Genetically determined abnormalities of the reproductive system. *Journal of Reproduction and Fertility, Supplement*, **15**, 51–78.

Borsberry, S. & Dobson, H. (1989). Periparturient diseases and their effect on reproductive performance in five dairy herds. *Veterinary Record*, **124**, 217–19.

Bove, K.E. (1970). Gonadal dysgenesis in a newborn with XO karyotype. *American Journal of Diseases of Children*, **120**, 363–6.

Britt, J.H. (1975). Ovulation and endocrine response after LH-RH in domestic animals. *Annales de Biologie Animale, Biochimie, Biophysique*, **15**, 221–31.

Burgoyne, P.S. & Baker, T.G. (1985). Perinatal oocyte loss in XO mice and its implications for the aetiology of gonadal dysgenesis in XO women. *Journal of Reproduction and Fertility*, **75**, 633–45.

Carr, D.H. (1972). Cytogenetic aspects of induced and spontaneous abortions. *Clinics in Obstetrics and Gynaecology*, **15**, 203–19.

Carr, D.H., Haggar, R.A. & Hart, A.G. (1968). Germ cells in the ovaries of XO female infants. *American Journal of Clinical Pathology*, **49**, 521–6.

Casida, L.E. (1976). Ovulation studies with particular reference to the pig. *Biology of Reproduction*, **14**, 97–107.

Clarke, I.J. & Henry, B.A. (1999). Leptin and reproduction. *Reviews of Reproduction*, **4**, 48–55.

Clayton, R.N., Ogden, V., Hodgkinson, J., *et al.* (1992). How common are polycystic ovaries in normal women and what is their significance for fertility in the general population? *Clinical Endocrinology*, **37**, 127–34.

Conway, G.S. & Jacobs, H.S. (1997). Leptin: a hormone of reproduction. *Human Reproduction*, **12**, 633–5.

Cowley, M.A., Smart, J.L., Rubinstein, M., *et al.* (2001). Leptin activates anorexigenic POMC neurons through a neural network in the arcuate nucleus. *Nature (London)*, **411**, 480–4.

Crosignani, P.G. & Nicolosi, A.E. (2001). Polycystic ovarian disease: heritability and heterogeneity. *Human Reproduction Update*, **7**, 3–7.

Cunningham, M.J., Clifton, D.K. & Steiner, R.A. (1999). Leptin's actions on the reproductive axis: perspectives and mechanisms. *Biology of Reproduction*, **60**, 216–22.

Dewhurst, J. (1978). Fertility in 47,XXX and 45,X patients. *Journal of Medical Genetics*, **15**, 132–5.

Dobson, H., Tebble, J.E., Smith, R.F. & Ward, W.R. (2001). Is stress really all that important? *Theriogenology*, **55**, 65–73.

Dunaif, A., Xia, J., Book, C.B., *et al.* (1995). Excessive insulin receptor serine phosphorylation in cultured fibroblasts and in skeletal muscle. A potential mechanism for insulin resistance in the polycystic ovary syndrome. *Journal of Clinical Investigation*, **96**, 801–10.

Dziuk, P.J. (1962). Control of oestrus and ovulation in farm animals with the aid of orally active progestins. *Journal of Endocrinology*, **24**, xxi.

Flier, J.S. (2001). The missing link with obesity? *Nature (London)*, **409**, 292–3.

Ford, C.E., Jones, K.W., Polani, P.E., De Almeida, J.C. & Briggs, J.H. (1959). A sex-chromosome anomaly in a case of gonadal dysgenesis (Turner's syndrome). *Lancet*, **I**, 711–13.

Foster, D.L. & Nagatani, S. (1999). Physiological perspectives on leptin as a regulator of reproduction: role in timing puberty. *Biology of Reproduction*, **60**, 205–15.

Franco, B., Guioli, S., Pragliola, A., *et al.* (1991). A gene deleted in Kallmann's syndrome shares homology with neural cell adhesion and axonal path-finding molecules. *Nature (London)*, **353**, 529–36.

Franks, S. (1989). Polycystic ovary syndrome: a changing perspective. *Clinical Endocrinology*, **31**, 87–120.

Franks, S. (1995). Polycystic ovary syndrome. *New England Journal of Medicine*, **333**, 853–61.

Franks, S. (1997). Polycystic ovary syndrome. *Archives of Disease in Childhood*, **77**, 89–90.

Franks, S., Gharani, N., Waterworth, D., *et al.* (1997). The genetic basis of polycystic ovary syndrome. *Human Reproduction*, **12**, 2641–8.

Garverick, H.A. (1997). Ovarian follicular cysts in dairy cows. *Journal of Dairy Science*, **80**, 995–1004.

Gharani, N., Waterworth, D.M., Batty, S. *et al.* (1997). Association of the steroid synthesis gene *CYP11a* with polycystic ovary syndrome and hyperandrogenism. *Human Molecular Genetics*, **6**, 397–402.

Gilling-Smith, C., Willis, D.S., Beard, R.W. & Franks, S. (1994). Hypersecretion of androstenedione by isolated theca cells from polycystic ovaries. *Journal of Clinical Endocrinology & Metabolism*, **79**, 1158–65.

Gilling-Smith, C., Story, E.H., Rogers, V. & Franks, S. (1997). Evidence for a primary abnormality of thecal cell steroidogenesis in the polycystic ovary syndrome. *Clinical Endocrinology*, **47**, 93–9.

Goldenberg, R.L., Powell, R.D., Rosen, S.W., Marshall, J.R. & Ross, G.T. (1976). Ovarian morphology in women with anosmia and hypogonadotrophic hypogonadism. *American Journal of Obstetrics & Gynecology*, **126**, 91–4.

Gosden, R.G. (1992). Transplantation of fetal germ cells. *Journal of Assisted Reproduction and Genetics*, **9**, 118–23.

Gosden, R.G. (2000). Low temperature storage and grafting of human ovarian tissue. *Molecular and Cellular Endocrinology*, **163**, 125–9.

Granowitz, E.V. (1997). Transforming growth factor-beta enhances and pro-inflammatory cytokines inhibit *ob* gene expression in 3T3-LI adipocytes. *Biochemical and Biophysical Research Communications*, **240**, 382–5.

Halvorson, L.M. (1999). Kallmann's syndrome. In *Encyclopedia of Reproduction*, vol. 2, ed. E. Knobil & J.D. Neill, pp. 921–4. San Diego & London: Academic Press.

Hardelin, J.P., Levilliers, J., Del-Castillo, I., *et al.* (1992). X chromosome-linked Kallmann syndrome: stop mutations validate the candidate gene. *Proceedings of the National Academy of Sciences, USA*, **89**, 8190–4.

Hardelin, J.P., Levilliers, J., Young, J., *et al.* (1993). Xp22.3 deletions in isolated familial Kallmann's syndrome. *Journal of Clinical Endocrinology and Metabolism*, **76**, 827–31.

Hunter, R.H.F. (1995). *Sex Determination, Differentiation and Intersexuality in Placental Mammals.* Cambridge: Cambridge University Press.

Hunter, R.H.F. & Baker, T.G. (1975). Development and fate of porcine Graafian follicles identified at different stages of the oestrous cycle. *Journal of Reproduction and Fertility*, **43**, 193–6.

Hunter, R.H.F., Cook, B. & Baker, T.G. (1985). Intersexuality in five pigs, with particular reference to oestrous cycles, the ovotestis, steroid hormone secretion and potential fertility. *Journal of Endocrinology*, **106**, 233–42.

Isobe, N. & Yoshimura, Y. (2000). Immunocytochemical study of cell proliferation in the cystic ovarian follicles in cows. *Theriogenology*, **54**, 1159–69.

Kallmann, F.J., Schoenfeld, W.A. & Barrera, S.E. (1944). The genetic aspects of primary eunuchoidism. *American Journal of Mental Deficiency*, **48**, 203–36.

Kesler, D.J. & Garverick, H.A. (1982). Ovarian cysts in dairy cattle: a review. *Journal of Animal Science*, **55**, 1147–59.

King, C.R., Magenis, E. & Bennett, S. (1978). Pregnancy and the Turner syndrome. *Obstetrics and Gynaecology*, **52**, 617–24.

Kittok, R.J., Britt, J.H. & Convey, E.M. (1973). Endocrine response after GnRH in luteal phase cows and cows with ovarian follicular cysts. *Journal of Animal Science*, **37**, 985–9.

Kousta, E., White, D.M., Piazza, A., LouMaye, E. & Franks, S. (1996). Successful induced ovulation and completed pregnancy using recombinant luteinising hormone and follicle stimulating hormone in a woman with Kallmann's syndrome. *Human Reproduction*, **11**, 70–1.

Laing, J.A. (ed.) (1970). *Fertility and Infertility in the Domestic Animals*, 2nd edn. London: Baillière, Tindall & Cassell.

Legro, R.S. (1999). Polycystic ovary syndrome. In *Encyclopedia of Reproduction*, vol. 3, ed. E. Knobil & J.D. Neill, pp. 925–9. San Diego & London: Academic Press.

López-Béjar, M.A., López-Gatius, F., Camón, J., *et al.* (1998). Morphological features and effects on reproductive parameters of ovarian cysts of follicular origin in superovulated rabbit does. *Reproduction in Domestic Animals*, **33**, 369–78.

Luck, M.R. (1994). The gonadal extracellular matrix. *Oxford Reviews of Reproductive Biology*, **16**, 33–85.

MacDougall, M.J., Tan, S.-L., Balen, A. & Jacobs, H.S. (1993). A controlled study comparing patients with and without polycystic ovaries undergoing *in vitro* fertilisation. *Human Reproduction*, **8**, 233–7.

Messinis, I.E. & Milingos, S.D. (1999). Leptin in human reproduction. *Human Reproduction Update*, **5**, 52–63.

Messinis, I.E., Milingos, S.D., Alexandris, I.E., *et al.* (1999). Leptin concentrations in normal women following bilateral ovariectomy. *Human Reproduction*, **14**, 913–18.

Messinis, I.E., Milingos, S.D., Zikopoulos, K., *et al.* (1998). Leptin concentrations in the follicular phase of spontaneous cycles and cycles superovulated with follicle stimulating hormone. *Human Reproduction*, **13**, 1152–6.

Muscatelli, F., Strom, T.M., Walker, A.P., *et al.* (1994). Mutations in the *DAX-1* gene give rise to both X-linked adrenal hypoplasia congenita and hypogonadotropic hypogonadism. *Nature (London)*, **372**, 672–6.

Nellor, J.E. (1960). Control of estrus and ovulation in gilts by orally effective progestational compounds. *Journal of Animal Science*, **19**, 412–20.

Osawa, T., Nakao, T., Kimura, M., *et al.*, (1995). Fertirelin and buserelin compared by LH release, milk progesterone and subsequent reproductive performance in dairy cows treated for follicular cysts. *Theriogenology*, **44**, 835–47.

Palmert, M.R., Radovick, S. & Boepple, P.A. (1998). Leptin levels in children with central precocious puberty. *Journal of Clinical Endocrinology and Metabolism*, **83**, 2260–5.

Perry, J.S. & Pomeroy, R.W. (1956). Abnormalities of the reproductive tract of the sow. *Journal of Agricultural Science*, **47**, 238–48.

Petit, C. (1993). Molecular basis of the X-chromosome-linked Kallmann's syndrome. *Trends in Endocrinology and Metabolism*, **4**, 8–13.

Pfaff, D.W. & Schwanzel-Fukuda, M. (1993). Migration of GnRH neurons from olfactory placode to basal forebrain. *Journal of Reproduction and Fertility*, Abstract Series, No. **11**, 5, S5.

Regan, L., Owen, E.J. & Jacobs, H.S. (1990). Hypersecretion of luteinising hormone, infertility and miscarriage. *Lancet*, **336**, 141–4.

Rhind, S.M., Rae, M.T. & Brooks, A.N. (2001). Effects of nutrition and environmental factors on the fetal programming of the reproductive axis. *Reproduction*, **122**, 205–14.

Rugarli, E.I. (1999). Kallmann syndrome and the link between olfactory and reproductive development. *American Journal of Human Genetics*, **65**, 943–8.

Schwanzel-Fukuda, M., Bick, D. & Pfaff, D.W. (1989). Luteinising hormone-releasing hormone (LHRH)-expressing cells do not migrate normally in an inherited hypogonadal (Kallmann) syndrome. *Molecular Brain Research*, **6**, 311–26.

Schwanzel-Fukuda, M. & Pfaff, D.W. (1989). Origin of luteinising hormone-releasing hormone neurons. *Nature (London)*, **338**, 161–4.

Short, R.V. (1982). Sex determination and differentiation. In *Reproduction in Mammals*, 2nd edn, vol. 2, ed. C.R. Austin & R.V. Short, pp. 70–113. Cambridge: Cambridge University Press.

Simpson, J.L. (1976). *Disorders of Sexual Differentiation, Etiology and Clinical Delineation*. New York & London: Academic Press.

Singh, R.P. & Carr, D.H. (1966). The anatomy and histology of XO human embryos and fetuses. *Anatomical Record*, **155**, 369–84.

Speed, R.M. (1986). Oocyte development in XO foetuses of man and mouse: the possible role of heterologous X-chromosome pairing in germ cell survival. *Chromosoma (Berlin)*, **94**, 115–24.

Steppan, C.M., Bailey, S.T., Bhat, S., *et al.* (2001). The hormone resistin links obesity to diabetes. *Nature (London)*, **409**, 307–12.

Sturgis, S.H. (1961). Factors influencing ovulation and atresia of ovarian follicles. In *Control of Ovulation*, ed. C.A. Villee, pp. 213–18. Oxford & New York: Pergamon Press.

Swain, A., Narvaez, V., Burgoyne, P., *et al.* (1998). *Dax*-1 antagonizes *Sry* action in mammalian sex determination. *Nature (London)*, **391**, 761–7.

Swain, A., Zanaria, E., Hacker, A., *et al.* (1996). Mouse *Dax*-1 expression is consistent with a role in sex determination as well as in adrenal and hypothalamus function. *Nature Genetics*, **12**, 404–9.

Thatcher, W.W., Drost, M., Savio, J.D., *et al.* (1993). New clinical uses of GnRH and its analogues in cattle. *Animal Reproduction Science*, **33**, 27–49.

Trounson, A., Anderiesz, C. & Jones, G. (2001). Maturation of human oocytes *in vitro* and their developmental competence. *Reproduction*, **121**, 51–75.

Udoff, L. & Adashi, E.Y. (1995). Polycystic ovarian disease: a new look at an old subject. *Current Opinions in Obstetrics and Gynaecology*, **7**, 340–3.

Ulberg, L.C., Christian, R.E. & Casida, L.E. (1951a). Ovarian response in heifers to progesterone injections. *Journal of Animal Science*, **10**, 752–9.

Ulberg, L.C., Grummer, R.H. & Casida, L.E. (1951b). The effects of progesterone upon ovarian function in gilts. *Journal of Animal Science*, **10**, 665–71.

Wallace, J., Bourke, D., Da Silva, P. & Aitken, R. (2001). Nutrient partitioning during adolescent pregnancy. *Reproduction*, **122**, 347–57.

Webb, R., Gutierrez, C.G., Gong, J.G. & Campbell, B.K. (1998). Dynamics and aetiology of ovarian follicular cysts in post-partum dairy cattle. *Reproduction in Domestic Animals*, **33**, 285–8.

Weiss, L. (1971). Additional evidence of gradual loss of germ cells in the pathogenesis of streak ovaries in Turner's syndrome. *Journal of Medical Genetics*, **8**, 540–4.

Whitten, W.K., Beamer, W.G. & Byskov, A.G. (1979). The morphology of foetal gonads of spontaneous mouse hermaphrodites. *Journal of Embryology and Experimental Morphology*, **52**, 63–78.

Wray, S., Grant, P. & Gainer, H. (1989). Evidence that cells expressing luteinising hormone-releasing hormone mRNA in the mouse are derived from progenitor cells in the olfactory placode. *Proceedings of the National Academy of Sciences, USA*, **86**, 8132–6.

XI

Induction of ovulation in women and domestic animals

Introduction

Having commented briefly in Chapter X on failure of ovulation, a logical means of bringing the text towards a final perspective might now be to discuss induction of ovulation. Under such a heading, there are important contrasts between domestic animals and humans. In many species of farm and laboratory animal, induction of ovulation is not applied primarily in order to overcome an intrinsic problem of ovulation failure. Rather, it is used to control the time of ovulation in conjunction with procedures of artificial insemination and/or embryo transplantation, or perhaps more fundamental research applications, or it may be a means of superovulation – increasing the number of follicles ovulating above that characteristic of the species. Induction of ovulation may also be used as an approach to bring forward the onset of puberty or the onset of the breeding season, or as part of a programme of synchronisation of oestrous cycles. In the last two examples, treatment may follow priming with steroid hormones or their analogues.

Moving the focus to women, and in contrast to the above, gonadotrophin therapy is quite widely applied under circumstances of ovulation failure, stemming from endocrine imbalance. An exception would be in instances of fertility treatment in which the objective is a controlled increase in the number of eggs ovulated – a mild form of superovulation – or available for laparoscopic aspiration shortly before ovulation. Whereas the approach in animals has generally been to give a single injection, administration of gonadotrophic preparations in women has tended to favour multiple injections. Alternatively, a mini-infusion pump worn by the patient may be used to deliver GnRH intravenously on a protocol that mimics the pulsatile release of the hypothalamic peptide and, as a consequence, to generate pulsatile secretion of the corresponding hormones from the pituitary gland. However, clinical approaches are constantly evolving, not least because of the development of new forms of hormone preparation and associated agonists and antagonists.

In one regard, practical objectives in animals and women may be coming closer than is sometimes appreciated. This remark refers to superovulation and embryo transplantation, procedures in animals that could be valuable if applied successfully to individuals of superior genetic merit for specific production traits (i.e. genes). In the case of women, induction of multiple ovulation and recovery of several embryos, or laparoscopic aspiration of several oocytes followed by *in vitro* maturation and *in vitro* fertilisation, may be employed in a wider context than was initially supposed. Not only might embryo transfer be used to overcome infertility in the patient herself, but donated embryos may be transplanted to surrogate mothers. Clearly, this latter practice will have a limited application at the present time but, to the extent that it is already accepted, it may suggest approaches that will be more widely applied in the future. Even in our own species, the march of reproductive technology appears to have no particular limit or boundary, especially when public opinion has been suitably prepared.

To avoid any possible misunderstanding, an important point to mention in this introduction is that human pituitary gonadotrophin (hPG) is no longer prescribed for clinical purposes. This is because of potential risks associated with Creutzfeldt–Jakob disease and an inability to guarantee the absence of prion particles after extraction from cadaver pituitaries.

Treatments for overcoming anovulation in women

Diverse reasons for failure of spontaneous ovulation in women of reproductive age were discussed in Chapter X. In attempting to overcome such causes of infertility, there has been a continuous development and refinement of clinical techniques during the past 30–35 years or so. The topic has been reviewed repeatedly and in depth (e.g. Gemzell, 1961; Franks & Owen, 1992; Yovich, 1995; Mandelbaum *et al.*, 2001). Working primarily with gonadotrophin preparations, substantial progress has been made and many individuals – in reality, couples – have benefited. After a systematic clinical investigation to establish the origin and nature of the anovulatory problem, specific corrective treatment may be applied. The clinical 'work-up' would include ultrasonic examination of the ovaries, serial measurement of hormone concentrations in peripheral plasma, including steroids, gonadotrophins and prolactin, endometrial biopsies, and probably inspection of cervical mucus. The curve of basal body temperature would be followed.

In overcoming ovulatory dysfunction, the objective would be to mimic cyclic physiology to the extent of stimulating final maturation of a single Graafian follicle capable of ovulation, with accompanying changes in the genital tract. The intention would be that the induced ovulation could be followed by normal fertilisation *in vivo* if, in addition to menstrual cycles, actual pregnancy were sought. Assuming that the ovaries are potentially functional, and depending on the anatomical level of the problem – hypothalamus or hypophysis – the following treatments might be considered.

Hypothalamus

In the case of hypothalamic deficiency, frequently associated with anorexia, starvation, or severe physical or mental stress, the anti-oestrogen clomiphene citrate is commonly used at a daily dose of 100 mg for 5 days (e.g. 50–200 mg tablets taken daily for 5 days). Prior to treatment, the systemic concentration of oestradiol-17β would have been established as ≥50 pg/ml. The response to such clomiphene administration is monitored both endocrinologically (Table XI.1) and by ultrasonic scanning and, once a suitably mature follicle has been detected (≥18 mm diameter), ovulation is triggered by a single injection of 5000 i.u. hCG (Fig. XI.1). Administration of human menopausal gonadotrophin (hMG) in conjunction with clomiphene citrate may be required to elicit satisfactory follicular development. Full details of treatment protocols can be found in Yovich (1995), for example, and almost all relevant concerns associated with clomiphene citrate treatment have been assessed by Dickey & Holtkamp (1996).

Pituitary gland

If a problem is diagnosed at the hypophyseal level, then a conventional treatment would be to use GnRH, an approach that is also effective in hypothalamic amenorrhoea that fails to respond to clomiphene citrate. Administration of the GnRH preparation is either subcutaneously or intravenously by means of a mini-pump. Dosage is usually of the order of 5–20 μg introduced at regular intervals, characteristically every 90 minutes to achieve pituitary downregulation. Concomitantly, follicular development may be enhanced by hMG stimulation. Ovulation would be induced 10–20 days after commencing treatment with an injection of hCG, and the GnRH support may then be extended into the luteal phase. An alternative luteal phase therapy is to administer hCG instead of GnRH (Yovich, 1995; Fig. XI.2). Whereas the incidence of pregnancy so generated can achieve 30–35% per induced cycle, the level of spontaneous miscarriage is high (∼25%).

Table XI.1. *Comparison of mean follicle number (±SEM), mean oestradiol concentrations, mean oestradiol per follicle, and implantations per follicle for clomiphene citrate (CC) and human menopausal gonadotrophin (hMG) ovulation induction régimes*

Ovulation induction régime	No. of cycles	No. of follicles per cycle ≥12 mm	No. of patients	Mean oestradiol per cycle on day of hCG (pg/ml)	Mean oestradiol per follicle ≥12 mm (pg/ml)	Implantations per follicle
CC + hMG	119	3.2 ± 0.2	106	962 ± 80	456 ± 76	0.100
CC	524	2.4 ± 0.1	62	613 ± 48	254 ± 18	0.047
hMG	57	4.6 ± 0.6	75	1194 ± 161	278 ± 37	0.054
hMG and CC	79	5.9 ± 0.5	56	1674 ± 739	329 ± 21	0.040

Adapted from Dickey & Holtkamp, 1996.

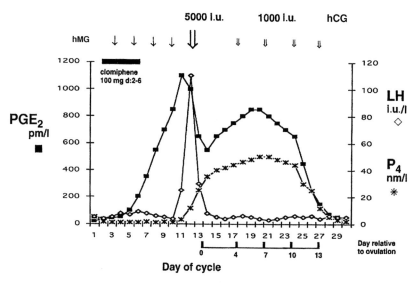

Fig. XI.1. Human ovarian stimulation using clomiphene citrate and human menopausal gonadotrophin (hMG), with human chorionic gonadotrophin (hCG) injected to induce ovulation and support luteal function. LH, luteinising hormone; P_4, progesterone; PGE_2, prostaglandin E_2. (After Yovich, 1995.)

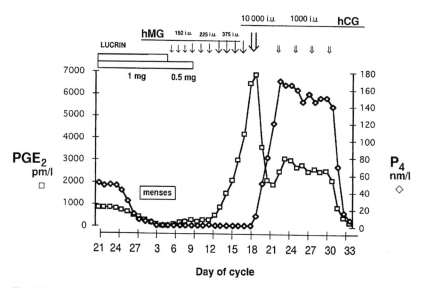

Fig. XI.2. An ovarian stimulation protocol used in *in vitro* fertilisation (IVF) clinics, with pituitary downregulation before using gonadotrophin preparations sequentially. For abbreviations, see Fig. XI.1. (After Yovich, 1995.)

Ovarian level

Targeting the ovaries themselves, treatment will involve administration of a gonadotrophin preparation such as hMG. Daily administration of 75, 150 or 225 i.u. for 5–7 days should produce a detectable response. Maturation of a follicle(s) would be monitored ultrasonically and/or by following plasma oestradiol concentrations (Bomsel-Helmreich *et al.*, 1979). Values for a ripe follicle should have reached 200–400 pg/ml and there would be a marked influence on the viscosity of cervical mucus. Once a suitably mature follicle has been detected, ovulation is induced with 5000 i.u. hCG. Ideally, no more than two Graafian follicles will have reached a pre-ovulatory state, otherwise the risk of multiple embryos – and in due course fetuses – will arise. The timing of the hCG injection requires careful judgement: premature treatment could liberate an immature (primary) oocyte, whereas a belated treatment might be associated with an atretic oocyte. The timing of human ovulation in response to the hCG injection is approximately 38 hours later; and resumption of meiosis would have proceeded during this interval (Edwards, 1962, 1965, 1980). When treatment protocols have been correctly followed, some 80% of patients should ovulate. Perhaps half of these will become pregnant if a healthy sperm population is available at a suitable time.

Induction of multiple ovulation in women

Induction of multiple ovulation – ideally of two or three follicles – is principally used in conjunction with transplantation of embryos, perhaps preceded by a phase of storage. In fact, oocytes may be harvested before the stage of ovulation using procedures based on ultrasound-guided needle puncture and laparoscopic aspiration from selected follicles. Such oocytes would then be subjected to procedures of *in vitro* maturation and *in vitro* fertilisation. The individual ovarian response to a given stimulatory treatment is unpredictable in the sense that it may be poor or, alternatively, may represent hyperstimulation, with serious attendant problems. Nonetheless, the fundamental aim in a superovulation treatment is to over-ride the endogenous feedback of ovarian hormones that regulates pituitary secretion of gonadotrophins during a spontaneous cycle. Within this overall strategy, the ideal tactics would involve adjusting (i.e. titrating) the dose and duration of a trophic treatment to stimulate development of the required number of follicles to the pre-ovulatory stage of maturation.

Should hyperstimulation inadvertently occur as a response to treatment, this is referred to as the ovarian hyperstimulation syndrome (OHSS). Primary evaluation of the number and size of follicles involves trans-vaginal ultrasound.

Women with polycystic ovaries appear especially prone to hyperstimulation, even though – paradoxically – many fail to respond to clomiphene citrate and hMG. The over-response syndrome may be expressed as a massive enlargement of the ovaries, extra-vasation of fluid into the abdominal cavity resulting in ascites, hydrothorax, hypercoagulability, and major electrolyte perturbation. Taken together, these are life-threatening, lead to systematic organ dysfunction, especially renal failure, and require constant clinical surveillance (see Rizk & Smitz, 1992). The incidence of severe OHSS may be approximately 2% (Elchalal & Schenken, 1997). As to underlying causes, a perturbed ovarian renin–angiotensin system may contribute and likewise VEGF; the latter would be acting to influence angiogenesis and capillary permeability (Ong, Eisen & Rennie, 1991; McClure *et al.*, 1994).

Even putting these risk factors associated with ovarian stimulation to one side, there is still the major concern that should more than one embryo be left *in situ* as a response to treatment and give rise to multiple fetuses, then there is an increased danger of high perinatal mortality and morbidity.

As has long been appreciated from studies in domestic farm animals (Hammond, 1961), the primary objectives of a multiple ovulation (i.e. superovulation) treatment are to supplement and prolong the trophic influence of endogenous FSH, thereby amplifying the ovarian response. Hormone preparations used in the attempted induction of multiple ovulation have included hMG, purified FSH (e.g. Metrodin, Serono), and pulsatile GnRH to stimulate endogenous gonadotrophins. Recombinant gonadotrophin preparations have become available in recent years. Blockade with a GnRH agonist during the phase of stimulation with hMG or FSH offers the specific advantage that, when applied on an appropriate schedule, a premature peak of plasma LH can be avoided. This enables a more controlled ovulatory response and, as a consequence, usually a better yield of oocytes or embryos. Premature elevation of systemic LH values appears to compromise normal development and maturation of Graafian follicles.

Treatment with hMG or FSH alone has generally been less satisfactory, even though attempts to adjust the dose and mimic endogenous patterns of secretion have been made by monitoring both development of follicles and secretion of oestradiol. Such attempts have been rendered less effective by the long half-life of hMG, which may be expressed for 48–60 hours or more in the circulation. Surprisingly, highly purified preparations of FSH have offered little advantage so far in obtaining a more precise ovarian response. The anti-oestrogenic feedback influences of clomiphene citrate have been used as a means of manipulating endogenous gonadotrophin levels, especially augmenting the secretion of FSH but, once again, with lack of predictability in individual responses.

As a refinement of the above general approaches to induction of multiple ovulation, growth factors such as IGF-1 have also been included in the protocols. A justification here is that IGF-1 has been found to potentiate the action of FSH on granulosa cells (see Chapter V), thereby improving the intra-follicular environment and hopefully the quality of the oocyte. IGF-1 mediates the physiological effects of growth hormone, so this pituitary trophin has likewise been examined in some treatments as a supplement to the gonadotrophin preparations. Recombinant growth hormone is now widely available and its use may reduce the dose of gonadotrophin needed for inducing multiple ovulation in some hypogonadotrophic patients.

However, as of writing, one method of choice remains the use of long-acting GnRH agonists for management of follicular stimulation in infertile women (Fleming *et al.*, 1982). This means of suppressing plasma peaks of endogenous gonadotrophin prevents full follicular development with elevated secretion of oestradiol-17β and the risk of positive feedback triggering a premature LH surge, not least in patients with a polycystic ovarian syndrome. Treatment with a GnRH agonist is best commenced in the preceding cycle to permit down-regulation of the anterior pituitary gland before promoting follicular growth with injected gonadotrophin. In this relatively controlled situation using hMG stimulation of follicular growth in conjunction with trans-vaginal ultrasonic scanning to monitor follicle maturity, ovulation can be induced at a predetermined time with hCG. No longer restricted to the infertile patient, this approach is established as an attractive method for inducing multiple ovulation (Porter *et al.*, 1984). Even so, this is not to suggest that individual ovarian responses have become predictable, nor does a GnRH agonist in combination with highly purified FSH yield greater precision than hMG (Franks & Owen, 1992).

Short-term protocols with GnRH agonists have also been examined, and their attractions are considerable. For example, the simplicity, efficacy and safety of an intranasal route of administration (insufflation) of a GnRH agonist such as Nafarelin (nafarelin acetate) are not to be overlooked. Doses of only 400–800 μg per day will consistently give an adequate blockade to prevent LH surges during the phase of stimulation. Reduction in the number of hMG/FSH ampoules per treatment cycle may produce more cost-effective outcomes in the long term (for a review, see Wong *et al.*, 2001). One of the intriguing features of GnRH agonists still under examination is the extent to which they may influence the sensitivity of the ovary itself to gonadotrophin stimulation. Speculation as to a mode of action is premature. Unequivocal evidence for the putative effect must first be presented.

An alternative and more recent strategy employs a GnRH antagonist (Hall, 1993), such as a single dose of 3 or 5 mg Cetrorelix given subcutaneously

SINGLE DOSE

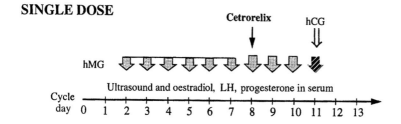

MULTIPLE DOSES

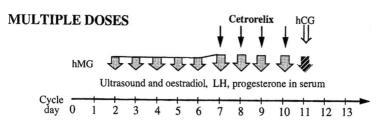

Fig. XI.3. A protocol for human ovarian stimulation involving a GnRH antagonist, as well as classical sources of gonadotrophin. For abbreviations, see Fig. XI.1. (After Olivennes *et al.*, 2000.)

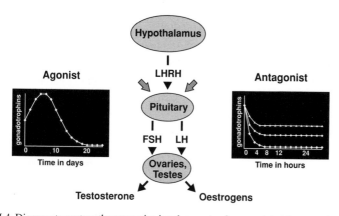

Fig. XI.4. Diagram to portray the respective involvements of an agonist and antagonist used in systems of gonadal stimulation. LH, luteinising hormone; FSH, follicle stimulating hormone; LHRH, LH releasing hormone. (After Reissmann *et al.*, 2000.)

(Fig. XI.3). Administered during the late follicular phase, this decapeptide should prevent an LH surge for 5–15 days or more. Some features of the agonist versus antagonist approaches are presented in Fig. XI.4. Agonists are seen to produce an initial stimulation of gonadotrophins, the so-called 'flare-up' effect, which may be a clinical disadvantage. Used mainly as part of an *in vitro* fertilisation protocol, the GnRH antagonist ensures rapid suppression of pituitary gonadotrophin secretion and, when appropriately administered, can be used to prevent a premature LH surge during stimulation of follicular growth and development (Olivennes *et al.*, 1995; Albano *et al.*, 1996). This is because the antagonist binds competitively to the receptors and thereby prevents endogenous GnRH from stimulating the pituitary gland. However, the responsiveness of the gland is preserved.

In a context of *in vitro* fertilisation, the clinical approach entails early follicular phase recruitment of a group of developing follicles within a normal menstrual cycle. And, as noted above, an administered FSH preparation will be supplementing endogenous FSH, although the optimum time of administration remains to be established. Although not usually suggested in print, this could vary importantly between patients. Most studies reported until 2000 with a GnRH antagonist approach have commenced FSH treatment on days 2 or 3 of the menstrual cycle, but this fact needs to be set against endogenous FSH levels usually 'peaking' on day 6, with dominance being identified on a median day 7 in normally cycling women (Macklon & Fauser, 2000). Thus, there is some justification for starting FSH treatment on day 5 with daily doses as low as 100 i.u. recombinant FSH (rFSH), still with the objective of stimulating a group of follicles for *in vitro* fertilisation procedures (De Jong, Macklon & Fauser, 2000).

It is not necessary to wait for ovulation. Oocytes may be aspirated from suitable follicles (Chang, 1955). In terms of resumption and progression of meiotic maturation in oocytes within stimulated follicles, maintenance of a GnRH antagonist during the late follicular phase should permit control using recombinant LH (rLH) or GnRH agonist instead of hCG. Suitable clinical trials are under way, with the further objective of reducing the risk of ovarian hyperstimulation (De Jong *et al.*, 1998; Macklon & Fauser, 2000).

As to the future, various modifications and novel approaches are discussed by Macklon & Fauser (2000). One stimulating development involves modification of recombinant gonadotrophins to alter their biopotency and half-life. This might be achieved, for example, by altering the glycosylation pattern of FSH to give short-acting but potent rFSH fractions and long-acting but less potent fractions (i.e. basic versus acidic fractions, respectively). Moreover, addition of the carboxy-terminal peptide (CTP) of the hCG β subunit, responsible for the long half-life of the molecule, to FSHβ will extend the half-life of the resultant preparation. Clinical trials of such FSH-CTP are in progress, for its enhanced potency *in vivo* may simplify still further possible routes to ovarian stimulation in women (Macklon & Fauser, 2000), assuming that oocyte quality is not compromised. The advantages of minimal stimulation protocols are attractive: less treatment in the sense of a lowered number of hMG ampoules, less monitoring, lowered risk of hyperstimulation and other side effects, and lowered costs. Together, these will need to be weighed against a loss of clinical flexibility, in the sense that stimulation would be applied in a spontaneous (natural) cycle rather than after a phase of endogenous gonadotrophin suppression with GnRH agonists.

Maintenance of a functional corpus luteum

hCG synthesised and secreted by the developing embryo – as the name suggests – functions to support and stimulate the developing corpus luteum from the seventh or eighth day after fertilisation, or even slightly earlier (for a review, see Sauer, 1979). However, the notion is now widespread that many of the protocols hitherto used in *in vitro* fertilisation clinics to induce ovulation may have compromised subsequent luteal activity. In some undetermined manner, the reasonably intensive phase of ovarian stimulation may have had a negative influence on subsequent corpus luteum function (Macklon & Fauser, 2000). Accordingly, it has become conventional practice after treatments for the stimulation of follicular growth followed by induction of ovulation to extend the period of gonadotrophin treatment. This is reasoned to assist and support establishment, maintenance and full secretory function of the induced corpora lutea. Treatment takes the form of repeated administration of hCG, which may be combined with supplementary progesterone therapy, together ensuring an adequate post-treatment luteal phase. Somewhat surprisingly, clomiphene citrate has also been used to increase progesterone concentrations in patients with luteal insufficiency (Dickey & Holtkamp, 1996).

Having presented the argument from a clinical angle, it may be helpful now to mention the viewpoint of a physiologist. In circumstances of spontaneous ovulation during a normal menstrual cycle, an initial phase of corpus luteum development and progesterone secretion would follow more or less automatically from the sequence of events at ovulation. Support and enhanced function of such a spontaneous corpus luteum 6 or 7 days after ovulation would depend in part on hCG. An inadequate embryo, arising perhaps as a consequence of a defective oocyte or belated fertilisation and possibly having chromosome anomalies (i.e. aneuploidy), might not produce sufficient chorionic support for full luteal maintenance. This in turn could lead to premature luteal failure and loss of a potential pregnancy – a seemingly prudent outcome on nature's part. Under clinical circumstances of induced ovulation, the extent to which luteal structures may be intrinsically defective rather than failing due to inadequate embryonic secretion of hCG has not been resolved. Both aspects may be involved. At a molecular level, induced corpora lutea would doubtless vary in their competence between patients and from cycle to cycle within the same patient.

Control of ovulation time in animals

As indicated in the introduction to this chapter, control of ovulation may be applied in conjunction with procedures of artificial insemination, it being important to deposit a sperm suspension in the female tract some hours before ovulation. This enables a timely progression of spermatozoa to the site of fertilisation and completion of the process of capacitation – the final maturation of spermatozoa that enables penetration of the zona pellucida. Induction of ovulation has also been a valuable technique in fundamental research, for example in examining post-ovulatory ageing of oocytes or the biochemical composition of fluids in the Fallopian tube lumen at specific intervals before and after ovulation.

The approach is simple and straightforward. Late in the follicular phase of a spontaneous oestrous cycle or during a rebound follicular phase after treatment for synchronising cycles, when a follicle or follicles would be close to maturity and approaching pre-ovulatory diameter, a single injection of chorionic gonadotrophin is administered by an intravenous or, more usually, intramuscular route. The twin objectives are to anticipate and imitate the endogenous surge of LH. The preparation most widely used is hCG, which is an LH-rich placental gonadotrophin, and an intramuscular or intravenous route ensures a peak of hormone in the circulation shortly after injection. As suggested, it is important that this peak of exogenous origin is presented shortly (hours, not days) before the endogenous LH surge and this involves careful judgement of the late follicular phase of the oestrous cycle. A useful rule is that if the animal shows behavioural signs of late pro-oestrus (i.e. impending oestrus) together with vulval swelling, coloration in appropriate species or breeds, and copious secretion of mucus but is not yet in standing oestrus, then an injection will be correctly timed to influence a mature follicle or follicles. The dose of hCG commonly used would be 50 i.u. for rabbits, 250–500 i.u. for sheep, 500 i.u. for pigs and 1500–2000 i.u. for cattle.

Extensive evidence indicates a remarkably consistent time interval between such an injection and ovulation, for example 25–26 hours in sheep (Dziuk, 1965) or 40–42 hours in pigs (Hunter & Polge, 1966). Not only is the time of ovulation predictable after injecting a suitable LH preparation (Table VII.1), but resumption of meiosis in pre-ovulatory oocytes proceeds along a well-defined time-course – a finding of value in various research procedures (Table VII.2). Until quite recently, the time-course of resumption of meiosis in ovarian oocytes up to the stage of first polar body extrusion was presumed to be inflexible, but the study of Sakaguchi *et al.* (2000) indicates that this may not always be so, at least under *in vitro* conditions. A combination of two growth factors (EGF and IGF-1) added to the culture medium could accelerate progression of meiosis in bovine follicular oocytes if surrounded by cumulus cells, although their developmental competence was not improved (Sakaguchi *et al.*, 2000).

As an alternative to a late pro-oestrous injection of hCG for inducing ovulation, with the requirement for careful clinical judgement, gonadotrophin may be injected at a fixed time after a protocol for synchronising the oestrous cycle. This could have involved treatment with progestogens (Dziuk & Polge, 1962) or an orally active non-steroidal compound such as methallibure (Polge, 1965). The hCG would be injected a precise number of days after cessation of treatment, corresponding closely to the duration of the follicular phase of an oestrous cycle in that species. Mature and responsive Graafian follicles would be expected at the time of injection. By contrast, premature injection with hCG, for example in the mid rather than late follicular phase, leads to serious problems. Follicular diameter and hormone secretion are inadequate, the animals do not show behavioural oestrus, and a high proportion of primary oocytes will be shed (Polge & Dziuk, 1965; Hunter, Cook & Baker, 1976).

Induction of superovulation in animals

Whereas induction of ovulation employs an LH-rich preparation injected shortly before the onset of oestrus, a traditional approach to superovulation has been to inject an FSH-rich preparation at the beginning of the follicular phase of an oestrous cycle. Such an injection would be made immediately upon regression of the corpus luteum or corpora lutea, and should be in step with the emergence of a new wave of growing follicles (Fig. XI.5). The objective is to enhance the number of follicles that progress to ovulation by reducing the extent of atresia amongst the larger developing Graafian follicles. As a general rule, the later the treatment is given during the follicular phase of the oestrous cycle, the poorer the response in terms of the number of ovulations (Hunter, 1979, 1984; Table XI.2). Superovulation is achieved by providing additional (i.e. exogenous) gonadotrophic support for the wave of ripening follicles emerging at the beginning of a follicular phase or arising upon cessation of treatment for the synchronisation of oestrus. As examples, a spontaneous follicular phase has a duration of approximately 3 days in sheep, 4 days in cattle and 5–6 days in pigs, but the injected gonadotrophin – whether a pituitary or placental preparation – tends to accelerate follicular growth and slightly shorten the follicular phase (Hunter, 1964).

Table XI.2. *The influence of a single injection of 1000 i.u. PMSG on day 17,*
18 or 19 of the oestrous cycle on the number of ovulations at the subsequent
oestrus in a unilaterally ovariectomised model

Day of unilateral ovariectomy and PMSG injection	No. of animals	Expected in intact animal	Number of ovulations: Observed on single remaining ovary	
			Mean	Range
17	8	16.0	16.8	15–18
18	8	15.8	15.5	15–16
19	8	14.8	7.8[a]	7–9

PMSG, pregnant mare serum gonadotrophin.
The mean ovarian response in eight intact control gilts injected with 1000 i.u. of the
same PMSG preparation was 26.3 (range 22–44).
[a]Plus many cystic follicles of 8–12 mm diameter in three of the pigs.
After Hunter & Nichol, 1982.

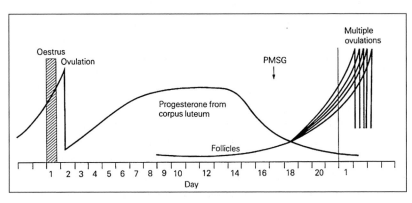

Fig. XI.5. The optimum time at which to inject pregnant mare serum gonadotrophin (PMSG)
in domestic animals for a superovulatory response is at the beginning of the follicular phase
of an oestrous cycle. (After Hunter, 1982.)

Gonadotrophic stimulation of enhanced numbers of maturing follicles
throughout the follicular phase requires relatively long-acting support and this
is traditionally provided by subcutaneous injection of the placental gonadotrophin PMSG, nowadays more commonly referred to as equine chorionic
gonadotrophin (eCG). Such a preparation has a long half-life and is rich in
FSH activity when harvested from mares between days 60 and 90 of gestation (Cole & Hart, 1930; Day & Rowlands, 1940; Allen, 1969). Examples of

Table XI.3. *Examples of some doses of PMSG and hCG*
commonly administered to induce superovulation and/or
control the time of ovulation

Species	Dose of PMSG injected subcutaneously (or intramuscularly)[a] (i.u.)	Dose of hCG injected intramuscularly (or intravenously) (i.u.)
Cow	2000–3000	500–2500
Sheep	500–800	250–500
Pig	750–1500	500–1000

PMSG, pregnant mare serum gonadotrophin; hCG, human chorionic gonadotrophin.

[a] PMSG is administered as a single dose. Because of its relatively long half-life, there appears to be little advantage in dividing the dose into several injections.

After Hunter, 1980.

doses of PMSG used to increase the number of follicles ovulating are shown in Table XI.3. Use of pituitary gland sources of FSH generally involves more than one injection for induction of superovulation. After such an approach, the increased number of follicles will usually ovulate spontaneously in response to the endogenous gonadotrophin surge. Even so, there may be some spread in the timing of ovulation of all the stimulated follicles, so injection of an LH-rich preparation at the onset of oestrus may be imposed as a sequel to superovulation treatment in domestic animals.

An overall point of specific importance is that, although the mean dose–response relationship can be described when superovulating a group of animals, the response of an individual to a given dose of PMSG remains unpredictable in farm animals. This stands in marked contrast to findings in laboratory species such as hamsters, in which an extremely precise response to superovulatory procedures employing PMSG has been reported (Greenwald, 1962). As a partial explanation of this difference, it is probable that the kinetics of follicular maturation are more closely defined during the 4–5-day oestrous cycle of the golden hamster than in the 16–21-day cycles of the larger species.

A quite distinct and more recent approach to inducing superovulation, and used with notable success in sheep, is to interfere immunologically with negative feedback signals from maturing Graafian follicles, thereby augmenting secretion of endogenous gonadotrophic hormones. Antibodies raised, for example, against androstenedione complexed with a macromolecule to make the small steroid molecule immunogenic, or antibodies raised against the peptide inhibin and examined at suitable titres on an appropriate protocol, should permit a greater number of follicles to mature than would do so spontaneously (Scaramuzzi & Hoskinson, 1984). Indeed, the technique can be sufficiently successful to be used under commercial conditions as a means of increasing lambing rate. Large-scale programmes of immunisation against androstene-dione, using active immunisation against the drug fecundin for this purpose, have led to overall improved fertility and fecundity (Stubbings & Maund, 1988). However, the ranges in response to treatment were considerable.

Another method of using PMSG has been to inject animals during the mid-luteal phase of the oestrous cycle, or under the influence of progestogen block-ade, and then to remove the blockade and/or to induce premature regression of the corpus luteum with a $PGF_{2\alpha}$ analogue administered 1 or 2 days after the gonadotrophin treatment. The underlying consideration here is that there may be a more uniform, although smaller-sized, cohort of follicles available for stimulation earlier in the oestrous cycle, and treatment may therefore produce a more consistent response. As an example of this approach to superovula-tion in cattle, a single injection of 2000 i.u. PMSG is administered on day 11 or 12 of the oestrous cycle followed by 750 μg cloprostenol (Estrumate, ICI) 48 hours later. Ovulation will occur without further hormonal treatment, al-though the period of time required for all the stimulated follicles to ovulate may be protracted. A further variation here is to use a GnRH agonist (desloren implant) over a period of days to block endogenous LH surges during FSH stimulation of follicular growth. Control of the time of ovulation in the several (or multiple) mature follicles is then obtained with a single injection of LH. This approach of chronic treatment with a GnRH agonist in conjunction with a purified FSH injection can be used in cattle and sheep (Evans *et al.*, 1994; Twagiramungu, Guilbault & Dufour, 1995; D'Occhio, Jillella & Lindsey, 1999).

Finally, because of the involvement of growth factors in follicular develop-ment (see Chapter V), it is worth noting that administration of growth hormone as recombinant bovine somatotrophin increases serum IGF-1 concentrations in cattle and improves the mean response to a superovulation treatment, tending to reduce variability in response between animals (Gong *et al.*, 1993).

Manipulation of prepuberal animals

Injections of gonadotrophic hormones have been used to induce ovulation in farm animals before the spontaneous occurrence of puberty. A rationale is best offered in a context of the rate of breeding and genetic improvement. If one could breed from livestock several months or more ahead of the time of puberty, this might enable an experimental shortening of the generation interval and an acceleration of genetic progress, especially if used in conjunction with the technique of embryo transplantation to a series of recipients. Because the full complement of oocytes is present in mammalian ovaries by the time of birth (in fact maximum numbers may have been exceeded and atresia will already have taken a significant toll: see Chapter VI), and responsive Graafian follicles have formed within a few weeks or months of birth, a technique that stimulates precocious development of follicles should enable competent oocytes to be shed. Specific treatments applicable to farm animals have been described in a number of texts (e.g. Betteridge, 1977; Cole & Cupps, 1977; Hunter, 1980; Gordon, 1996; Hafez & Hafez, 2000). Although artificial insemination may enable fertilisation of ovulated oocytes *in situ*, embryonic and fetal development are unlikely to occur unless animals have been treated close to the time of puberty. Accordingly, a technique has been to recover the embryos from such immature donors and transfer them to suitable mature recipients.

A current approach may or may not involve pre-treatment with gonadotrophic hormones but would not seek induction of ovulation. Rather, it would require physical removal of oocytes from Graafian follicles by needle puncture and aspiration during laparoscopy, followed by procedures of *in vitro* maturation and *in vitro* fertilisation. An alternative contemporary approach is to take fragments of ovarian tissue and attempt to promote growth *in vitro* of dissected young follicles in the hope eventually of yielding functional oocytes competent to undergo *in vitro* fertilisation. In reality, the extent to which such techniques will be used in animal breeding seems rather minor. There would be questions of cost and efficiency: the greater the number of steps needed to generate a supposedly functional oocyte, the lower the overall fruitfulness. Even so, harvesting ovarian oocytes from prepuberal animals may find application in a context of cloning, for an adequate supply of modified oocytes would be required for the procedure of nuclear fusion (see Campbell *et al.*, 1996; Wilmut *et al.*, 1997).

Concluding remarks

In comparison with studies in farm animals, and specifically on a bodyweight basis, one cannot avoid being struck by the relatively high or even massive doses of gonadotrophic hormones, releasing factors and their respective ana-logues that have been administered to women in so-called fertility clinics. In the light of some of the doses employed, it is perhaps not surprising that there have been instances of ovarian hyperstimulation with all the attendant problems and concerns. Writing as one who has applied gonadotrophin treatments extensively to rabbits, sheep, pigs and cows within veterinary clinical departments, com-plications comparable to those described under the ovarian hyperstimulation syndrome have not been witnessed. Only relatively recently, and in the light of misgivings expressed by reflective medical colleagues, have more sensitive and gentler stimulation protocols been examined in women. A turning point here may have been the thought-provoking paper of Edwards, Lobo & Bouchard (1996). To be fair, however, new protocols have to some extent awaited the development of new hormone preparations.

Even so, if a *raison d'être* for so much of the fundamental work on folliculogenesis and follicular development is to enable design of safer and more satisfactory treatments for ovarian stimulation in human patients and farm animals, then one is forced to wonder if those working in hospitals and fertility clinics have had sufficient time to read much that has already been discovered in the laboratory. Régimes involving gonadotrophin treatments for induction of ovulation or superovulation need to simulate more precisely the endocrine events of a normal spontaneous cycle. These thoughts prompt one to remark, in constructive rather than critical vein, that perhaps busy clinicians should be encouraged to take a sabbatical year on the implied timescale during which their energies could be devoted principally to laboratory work, reading and discussion. The same considerations apply to overworked university lecturers.

References

Albano, C., Snuitz, J., Camus, M. *et al*. (1996). Hormonal profile during follicular phase in cycles stimulated with a combination of human menopausal gonadotrophin and gonadotrophin-releasing hormone antagonist (Cetrorelix). *Human Reproduction*, **11**, 2114–18.

Allen, W.R. (1969). The immunological measurement of pregnant mare serum gonadotrophin. *Journal of Endocrinology*, **43**, 593–8.

Betteridge, K.J. (ed.) (1977). *Embryo Transfer in Farm Animals*. Canada Department of Agriculture Monograph No. 16, Ottawa.

Bomsel-Helmreich, O., Gougeon, A., Thebault, A., *et al*. (1979). Healthy and atretic human follicles in the preovulatory phase: differences in evolution of follicular morphology and steroid content of follicular fluid. *Journal of Clinical Endocrinology and Metabolism*, **48**, 686–94.

Campbell, K.H., McWhir, J., Ritchie, W.A. & Wilmut, I. (1996). Sheep cloned by nuclear transfer from a cultured cell line. *Nature (London)*, **380**, 64–6.

Chang, M.C. (1955). The maturation of rabbit oocytes in culture and their maturation, activation, fertilisation and subsequent development in the Fallopian tubes. *Journal of Experimental Zoology*, **128**, 379–405.

Cole, H.H. & Cupps, P.T. (eds.) (1977). *Reproduction in Domestic Animals*, 3rd edn. New York & London: Academic Press.

Cole, H.H. & Hart, G.H. (1930). The potency of blood serum of mares in progressive stages of pregnancy in effecting the sexual maturity of the immature rat. *American Journal of Physiology*, **93**, 57–68.

Day, F.T. & Rowlands, I.W. (1940). The time and rate of appearance of gonadotrophin in the serum of pregnant mares. *Journal of Endocrinology*, **2**, 255–61.

De Jong, D., Macklon, N.S. & Fauser, B.C. (2000). A pilot study involving minimal ovarian stimulation for *in vitro* fertilisation by extending the 'FSH window' without luteal support. Initiation of low-dose recombinant FSH in the mid follicular phase combined with late follicular phase administration of the gonadotrophin-releasing hormone antagonist Cetrorelix. *Fertility and Sterility*, **73**, 1051–4.

De Jong, D., Macklon, N.S., Mannaerts, B.M., *et al.* (1998). High dose gonadotrophin-releasing hormone antagonist (Ganerelix) may prevent ovarian hyperstimulation syndrome caused by ovarian stimulation for *in vitro* fertilisation. *Human Reproduction*, **13**, 573–5.

Dickey, R.P. & Holtkamp, D.E. (1996). Development, pharmacology and clinical experience with clomiphene citrate. *Human Reproduction Update*, **2**, 483–506.

D'Occhio, M.J., Jillella, D. & Lindsey, B.R. (1999). Factors that influence follicle recruitment, growth and ovulation during ovarian superstimulation in heifers: opportunities to increase ovulation rate and embryo recovery by delaying the exposure of follicles to LH. *Theriogenology*, **51**, 9–35.

Dziuk, P.J. (1965). Timing of maturation and fertilisation of the sheep egg. *Anatomical Record*, **153**, 211–24.

Dziuk, P.J. & Polge, C. (1962). Fertility in swine after induced ovulation. *Journal of Reproduction and Fertility*, **4**, 207–8.

Edwards, R.G. (1962). Meiosis in ovarian oocytes of adult mammals. *Nature (London)*, **196**, 446–50.

Edwards, R.G. (1965). Maturation *in vitro* of mouse, sheep, cow, pig. Rhesus monkey and human ovarian oocytes. *Nature (London)*, **208**, 349–51.

Edwards, R.G. (1980). *Conception in the Human Female*. London & New York: Academic Press.

Edwards, R., Lobo, R. & Bouchard, P. (1996). Time to revolutionise ovarian stimulation. *Human Reproduction*, **11**, 917–19.

Elchalal, U. & Schenken, J.G. (1997). The pathophysiology of ovarian hyperstimulation syndrome – views and ideas. *Human Reproduction*, **12**, 1129–37.

Evans, G., Brooks, J., Struthers, W. & McNeilly, A.S. (1994). Superovulation and embryo recovery in ewes treated with gonadotropin-releasing hormone agonist and purified follicle stimulating hormone. *Reproduction, Fertility and Development*, **6**, 247–52.

Fleming, R., Adams, A.H., Barlow, D.H., Black, W.P., McNaughton, M.C. & Coutts, J.R.T. (1982). A new systematic treatment for infertile women with abnormal hormone profiles. *British Journal of Obstetrics and Gynaecology*, **80**, 80–5.

Franks, S. & Owen, E.J. (1992). Superovulation strategy. In *Gonadal Development and Function*, vol. 94, ed. S.G. Hillier, Serono Symposia Publications, pp. 227–36. New York: Raven Press.

Gemzell, C.A. (1961). The induction of ovulation in the human by human pituitary gonadotropin. In *Control of Ovulation*, ed. C.A. Villee, Symposium Publications Division, pp. 192–209. Oxford & New York: Pergamon Press.

Gong, J.G., Bramley, T.A., Wilmut, I. & Webb, R. (1993). Effect of recombinant bovine somatotropin on the superovulatory response to pregnant mare serum gonadotropin in heifers. *Biology of Reproduction*, **48**, 1141–9.

Gordon, I. (1996). *Controlled Reproduction in Cattle and Buffaloes*. Wallingford, Oxford: CAB International.

Greenwald, G.S. (1962). Analysis of superovulation in the adult hamster. *Endocrinology*, **71**, 378–9.

Hafez, E.S.E. & Hafez, B. (2000). *Reproduction in Farm Animals*, 7th edn. Philadelphia, New York & London: Lippincott, Williams and Wilkins.

Hall, J.E. (1993). Gonadotropin-releasing hormone antagonists: effects on the ovarian follicle and corpus luteum. *Clinics in Obstetrics and Gynecology*, **36**, 744–52.

Hammond, J. Jr (1961). Hormonal augmentation of fertility in sheep and cattle. In *Control of Ovulation*, ed. C.A. Villee, Symposium Publications Division, pp. 163–76. Oxford & New York: Pergamon Press.

Hunter, R.H.F. (1964). Superovulation and fertility in the pig. *Animal Production*, **6**, 189–94.

Hunter, R.H.F. (1979). Ovarian follicular responsiveness and oocyte quality after gonadotrophic stimulation of mature pigs. *Annales de Biologie Animale, Biochimie, Biophysique*, **19**, 1511–20.

Hunter, R.H.F. (1980). *Physiology and Technology of Reproduction in Female Domestic Animals*. London & New York: Academic Press.

Hunter, R.H.F. (1982). *Reproduction of Farm Animals*. London & New York: Longman.

Hunter, R.H.F. (1984). Ovarian follicular development, maturation and atresia in pigs. In *Période Péri-Ovulatoire, Colloque de la Société Française pour l'Étude de la Fertilité*, pp. 47–54. Paris: Masson.

Hunter, R.H.F., Cook, B. & Baker, T.G. (1976). Dissociation of response to injected gonadotrophin between the Graafian follicle and oocyte in pigs. *Nature (London)*, **260**, 156–8.

Hunter, R.H.F. & Nichol, R. (1982). Ovarian follicular dynamics in the pig. In *Proceedings of the 2nd International Congress on Embryo Transfer in Mammals*, Annecy, pp. 307–8.

Hunter, R.H.F. & Polge, C. (1966). Maturation of follicular oocytes in the pig after injection of human chorionic gonadotrophin. *Journal of Reproduction and Fertility*, **12**, 525–31.

Macklon, N.S. & Fauser, B.C.J.M. (2000). Regulation of follicle development and novel approaches to ovarian stimulation for IVF. *Human Reproduction Update*, **6**, 307–12.

Mandelbaum, J., Antoine, J.M., Izard, V. & Salat-Baroux, J. (2001). Maîtrise de la fertilité humaine. In *La Reproduction chez les Mammifères et l'Homme*, ed. C. Thibault & M.C. Levasseur, pp. 770–91. Paris: Ellipses.

McClure, N., Healy, D.L., Rogers, P.A.W., *et al.* (1994). Vascular endothelial growth factor as capillary permeability agent in ovarian hyperstimulation syndrome. *Lancet*, **344**, 235–6.

Olivennes, F., Ayoubi, J.M., Fanchin, R., *et al.* (2000). GnRH antagonist in single-dose applications. *Human Reproduction Update*, **6**, 313–17.

Olivennes, F., Fanchin, R., Bouchard, P., *et al.* (1995). Scheduled administration of a gonadotrophin-releasing hormone antagonist (Cetrorelix) on day 8 of in vitro fertilisation cycles – a pilot study. *Human Reproduction*, **10**, 1382–6.

Ong, A., Eisen, V. & Rennie, P. (1991). The pathogenesis of the ovarian stimulation syndrome (OHS): a possible role for ovarian renin. *Clinical Endocrinology (Oxford)*, **34**, 39–43.

Polge, C. (1965). Effective synchronisation of oestrus in pigs after treatment with ICI compound 33828. *Veterinary Record*, **77**, 232–6.

Polge, C. & Dziuk, P. (1965). Recovery of immature eggs penetrated by spermatozoa following induced ovulation in the pig. *Journal of Reproduction and Fertility*, **9**, 357–8.

Porter, R.N., Smith, W., Craft, I.L., Abdulwahid, N.A. & Jacobs, H.S. (1984). Induction of ovulation for *in vitro* fertilisation using buserelin and gonadotropins. *Lancet*, **2**, 1284–5.

Reissmann, T., Schally, A.V., Bouchard, P., Riethmüller & Engel, J. (2000). The LHRH antagonist Cetrorelix: a review. *Human Reproduction Update*, **6**, 322–31.

Rizk, B. & Smitz, J. (1992). Ovarian hyperstimulation syndrome after superovulation using GnRH agonists for IVF and related procedures. *Human Reproduction*, **7**, 320–7.

Sakaguchi, M., Dominko, T., Leibfried-Rutledge, M.L., Nagai, T. & First, N.L. (2000). A combination of EGF and IGF-1 accelerates the progression of meiosis in bovine follicular oocytes in vitro and fetal calf serum neutralises the acceleration effect. *Theriogenology*, **54**, 1327–42.

Sauer, M.J. (1979). Review. Hormone involvement in the establishment of pregnancy. *Journal of Reproduction and Fertility*, **56**, 725–43.

Scaramuzzi, R.J. & Hoskinson, R.M. (1984). Active immunisation against steroid hormones for increasing fecundity. In *Immunological Aspects of Reproduction in Mammals*, ed. D.B. Crighton, pp. 445–74. London: Butterworths.

Stubbings, L.A. & Maund, B.A. (1988). Effects on the fecundity of sheep of immunisation against androstenedione. *Veterinary Record*, **123**, 489–92.

Twagiramungu, H., Guilbault, L.A. & Dufour, J.J. (1995). Synchronisation of ovarian follicular waves with a gonadotropin-releasing hormone agonist to increase the precision of estrus in cattle: a review. *Journal of Animal Science*, **73**, 3141–51.

Wilmut, I., Schnieke, A.E., McWhir, J., Kind, A.J. & Campbell, K.H.S. (1997). Viable offspring derived from fetal and adult mammalian cells. *Nature (London)*, **385**, 811–13.

Wong, J.M., Forrest, K.A., Snabes, M.C., Zhao, S.Z., Gersh, G.E. & Kennedy, S.H. (2001). Efficacy of nafarelin in assisted reproductive technology: a meta-analysis. *Human Reproduction Update*, **7**, 92–101.

Yovich, J.L. (1995). Ovarian stimulation for assisted conception. In *Gametes – the Oocyte*, ed. J.G. Grudzinskas & J.L. Yovich, Cambridge Reviews in Human Reproduction, pp. 329–56. Cambridge: Cambridge University Press.

XII

Concluding thoughts and a current perspective

Introduction

The principal objective of this final chapter will be to highlight some of the key themes in the text and to try to integrate diverse lines of thought concerning formation, selection and ovulation of a Graafian follicle. Although clearly containing a subjective element, various points judged to be of special significance have cropped up more than once or indeed repeatedly in succeeding chapters, and attention is drawn to this fact in Table XII.1. Even though the text could have contained much more detail in almost all sections, there remains a feeling that the contents of Table XII.1 would still have risen to the surface.

Of course, the dilemma in preparing a scientific monograph in the twenty-first century is the balance to be struck between overwhelming information on the one hand and some form of perspective on the other: in words all too frequently used, attempting to see the wood without it being obscured by the trees.

Historical landmarks

Lessons from history are always valuable and yet often overlooked or even disregarded. In a scientific context, this may be at considerable intellectual cost and certainly so in terms of time. As to knowledge concerning the ovaries, major landmarks have been noted, dating from the views of the Ancient Greeks. However, within a historical sweep up to the twentieth century, no observation was of greater significance to an understanding of ovarian physiology than that reported by John Hunter in 1787 (see Chapter I). The results of his classical experiment involving unilateral ovariectomy in pigs demonstrated the remarkable reserves of follicles and oocytes in a single gonad, but the manner in which follicles are selected and oocytes released remain topics of importance more

Table XII.1. *To highlight key themes concerned with the developmental physiology of a Graafian follicle that recur in the text*

Themes	Chapters
Involvement of specific angiogenic factors and blood flow in follicular development and regression	II, III, V, VI, VII, VIII, IX, XI
Contributions of autonomic innervation and catecholamines to follicular development and selection	I, II, III, V, VI, VII, VIII
Potentiating influences of intra-follicular peptides and peptide growth factors	II, III, IV, V, VI, VIII
Oocyte as centrally located computer programming and integrating conversations with ovarian somatic cells	II, V, VI, VIII
Diverse involvements of locally released nitric oxide in follicular physiology	III, V, VI, VIII
Ovulation as inflammatory-like process with contributions from infiltrating white cells	I, V, VI, VIII
Pre-ovulatory secretion of progesterone by granulosa cells, thereby acting as incipient corpus luteum	V, VI, VIII, IX
Vascular (blood and lymph) counter-current feedback systems to regulate follicular development	III, IV, VIII
Contribution of apoptosis to oocyte and follicle selection	II, VI, VII
Precise means of inter-ovarian communication, including vectors in peritoneal fluid	III, IV, VI
Ovulated cumulus (granulosa) cells as vital paracrine tissue in vicinity of oocyte and embryo	IV, IX

than 200 years later. Chapter I also contains reference to cystic ovarian disease, a condition arising eventually as a consequence of that unilateral ovariectomy in 1787. There are passing comments, too, on the involvement of ovarian follicular fluid in pre- and post-ovulatory reproductive events.

As is an author's prerogative, a few kites have been flown, not least the requirement for more rigorous and focused studies on involvement of the autonomic nervous system and its molecular products in normal and abnormal ovarian function; this view is developed in appropriate chapters. More practically, reproductive technologies based on ovarian tissues are seen as an important way forward in diverse clinical situations, possibly having more to contribute in the short term than so-called gene therapy to initiate or restore potential fertility. A note of caution is sounded in the last regard, for molecular modification of the genome may lead to consequences little imagined even with a clearly defined gene construct or precisely targeted gene deletion as the starting point. In a phrase, interactions between individual components of a genome may be far reaching and subtle.

Formation of ovarian follicles

Stimulating thoughts from Chapter II concern the manner in which a germ cell seemingly acts to attract and organise somatic cells in the embryonic (fetal) gonad, and how the two cell types together generate a functional follicle. Molecules synthesised in the far larger but still growing oocyte would seem of critical importance here, although there can be little doubt that a conversation between the two compartments soon develops and becomes essential: somatic cell factors prompt oocyte responses whilst germ cell molecules programme somatic cell activity. The latter would include structural organisation and differentiation of both theca and granulosa cell layers and the extracellular matrix. Undoubtedly of special significance is elaboration of the cumulus oophorus, a multilayered cellular investment around the oocyte that contributes in a major way before ovulation, during ovulation, and subsequently in a modified state within the Fallopian tube. One of the many functions of the cumulus oophorus in a mature follicle may be as a means of transducing and/or amplifying messages emanating from the oocyte itself. Such messages would act not only within a given follicle but also in neighbouring follicles and perhaps more widely.

Briefly touched upon is the perennial question of why so many oocytes are generated when so few will achieve a fruitful outcome. To suggest that the vast number of oocytes formed in a mammalian ovary is principally an evolutionary vestige may be too simplistic a response and miss a vital physiological point. More probably, formation of large numbers of oocytes enables the generation of very substantial numbers of follicles, within which population waves of development and regression may facilitate selection of a suitable follicle(s) at an optimum time for ovulation. In other words, large numbers of follicles would bestow flexibility on cyclic events, enabling emergence of an appropriate follicle during the wave of developmental changes culminating in shedding of a viable oocyte. Still to be resolved across a range of species is the extent to which the follicular somatic cells contribute to the formation of the zona pellucida. To view this accessory coat as uniquely a product of the oocyte may well be in error in species other than the mouse. Limitation on the passage of macromolecules through the zona pellucida, of both oocytes and fertilised eggs, also requires reassessment.

Blood vessels, lymphatics, innervation and temperature

Although the ovarian vasculature was examined from various angles, with some emphasis on the system of follicular lymphatics, there was one point of special intrigue. This was the long recognised fact that the capillary network does not penetrate inwards beyond the theca interna of mature follicles until progressive dissolution of the basement membrane shortly before ovulation. In other words, whereas the theca cell layers are richly vascularised, those of the granulosa are not. Possible reasons for this distinction were discussed in Chapter III, not least in the context of avoiding premature luteinisation with secretion of progesterone.

A classical rôle for the vascular bed in providing nutrients and substrate to growing follicles was noted, and a proposal made that regulation of blood flow to individual follicles would have an incisive influence on their developmental potential. The vascular and lymphatic networks in the theca of healthy follicles are especially impressive, and doubtless reflect the activities of locally synthesised angiogenic factors together with the influence of relatively high concentrations of oestradiol-17β. An increased rate of capillary blood flow shortly before ovulation would not only support the substrate requirements of the follicle, but also pave the way to establishment of fully functional luteal cells. The vascular and lymphatic networks were viewed conventionally, and also in terms of counter-current systems of exchange within and between follicular microcirculations. These would facilitate inter-follicular communication, and contribute to the establishment of follicular hierarchies and the process of selection.

Turning to innervation of Graafian follicles, the review ranged beyond the involvement of catecholamines in nerve cell function and the control of steroidogenesis, and considered a local release of peptides, cytokines and nitric oxide. Evolving patterns of ovarian innervation would enable sensitive communication between individual follicles and, with the involvement of the preceding molecules, could influence follicular differentiation, development and regression. Moreover, an involvement of the nervous system in integrating physiological events between the two ovaries should not be overlooked.

The existence of intra-ovarian temperature gradients was also touched on, albeit in a conservative way, and examined in the setting of mature Graafian follicles with their relatively high ratio of antral fluid volume to that of the encircling granulosa cells. Different metabolic activities in different ovarian compartments need to be considered, and the possibility of endothermic reactions within maturing follicles was highlighted. If their existence can be substantiated under physiological conditions and placed on a specific molecular footing, then there would seem to be important implications for procedures of *in vitro* maturation of oocytes in fertility clinics and in animal technology laboratories.

Contribution of antral fluid

Follicular fluid has been viewed both conventionally in terms of its formation and composition and more imaginatively as to its potentially diverse functions. Concerning its composition, the point was stressed in Chapter IV that the fluid would not be homogeneous throughout the antrum. Rather, distinct microenvironments should be considered, with different compositions close to the mural granulosa cell surfaces and similarly at the cumulus cell surface. The death or degeneration of follicle cells would act locally to alter fluid composition, and this composition would also alter with time, not least during the pre-ovulatory interval, with modifications to the basement membrane, the vasculature and the nature of GAGs synthesised by the cumulus cells during mucification. Intrafollicular migration of white blood cells and their secretory products could also have an important influence on follicular fluid composition.

Antral fluid acts as a primary reservoir for the diverse molecules secreted by the follicle wall, and this is especially so for steroid hormones and prostaglandins, which subsequently enter the vascular bed and lymphatics. The kinetics of ovarian follicular steroid secretion are complex but will be regulated in part by the activity of binding proteins in the antral fluid that, in a sense, perform a buffering rôle and influence the available local concentrations of a hormone. Whilst follicular hormones are considered primarily to act systemically in a classical endocrine manner, it is more and more appreciated that they also have a critical paracrine and autocrine function and doubtless contribute to patterns of follicular development and establishment of a follicular hierarchy.

Two other lines of thought may place the presence of follicular fluid in a slightly different perspective. Follicular fluid molecules appear to be involved in suppressing resumption of meiosis in the antral oocyte and then in facilitating progression from first meiotic prophase after the pre-ovulatory surge of gonadotrophic hormones. And, when considering the destiny of somatic cells, follicular fluid molecules may be acting to prevent premature luteinisation of the granulosa layers. Once again, only after the pre-ovulatory gonadotrophin surge would this transition be initiated, and at a time when thecal capillaries are beginning to penetrate the basement membrane and colonise the mural granulosa.

In summary, all aspects of follicular fluid composition and function must be seen in a dynamic state. These will be related not least to the activities of a supposedly quiescent oocyte – which in reality is not quiescent at all. In fact, molecules secreted by the oocyte will have a profound and continuing influence on the function of a Graafian follicle and the composition of its antral fluid. Nor is the orbit of follicular fluid restricted to the ovary, for close to the time of ovulation and soon thereafter, this gonadal fluid may be exerting powerful influences on and within the neighbouring Fallopian tube, not least on the microenvironment at the site of fertilisation and on the activity and physiological maturity of spermatozoa therein. Follicular fluid will also be contributing to peritoneal fluid as and when refluxed from the tubal lumen.

Endocrine potential of Graafian follicle

Much of the interest in Chapter V concerned the manner in which cells of the theca interna interacted with those of the stratum granulosum. This was primarily in a context of steroid hormone modification and one that is frequently described under the title of the two-cell theory of oestradiol synthesis. Although aromatisation of theca-derived androstenedione substrate within granulosa cells was accepted as a principal source of ovarian oestradiol, the potential of highly vascularised theca interna cells was questioned shortly before ovulation, as was the existence of different molecular species (isoforms) of aromatase in such theca cells. Oestrogens derived from these cells would have ready access to the capillary bed. Even so, the nature of gonadotrophin receptors on the two somatic cell types lends support to the current dogma for follicular oestradiol synthesis: LH-programmed theca cells and FSH-programmed granulosa cells. However, the nature of gonadotrophin receptors changes during the pre-ovulatory interval, particularly those on granulosa cells, as do the steroid hormones synthesised. Development of LH receptors on the granulosa is associated with secretion of progesterone rather than oestradiol.

Diverse follicular proteins were examined, as were various growth factors, and their local and systemic influences assessed. Interactions between gonadotrophins, proteins and peptide growth factors occurred during follicular development. Recent work on cytokines was linked to the presence of cells from the immune system (white blood cell lineage) in the follicular tissues. An involvement of endorphins, enkephalins and nitric oxide was also considered in follicular physiology, and all of these activities were seen as a sensitive means of coordinating development of a healthy Graafian follicle. Levels of fine-tuning would originate in the different cell layers, and a central rôle of the oocyte in this regard was also raised. Moreover, cells of the cumulus oophorus investment might be acting to amplify messages originating in the oocyte itself, especially in the final hours before ovulation and shortly thereafter. In a reciprocal sense, protein and peptide molecules from the somatic cells would be influencing development and maturation of the oocyte, and peptide growth factors could further provide a detailed system of communication within the ovary. Such a web would integrate the activities of different cell compartments and provide local modulation of follicular development.

Follicular selection, dominance, atresia

The preceding discussion leads on to the question of precisely what cues identify the follicle(s) that will develop and mature to ovulation. Because >99% of the initial complement of ovarian germ cells will be lost, do those that survive represent a privileged population in terms of the time sequence of their formation or the quality of their nuclear and cytoplasmic organelles? And are follicles emerging from the ovarian pool in some manner superior in that they avoid being diverted into atresia? All these questions are responded to in Chapter VI, with the earlier caveat that the autonomic nervous system and its generation of diverse molecules cannot be overlooked when one considers the developmental potential of an individual follicle and similarly the nature and extent of the capillary bed.

Over and above these specific remarks, endocrine considerations are seen to be of prime importance. The interplay of pituitary gonadotrophins and gonadal steroid and peptide hormones exerts a powerful system of control, with the presence of FSH and LH receptors on the granulosa and theca cells, respectively, taking centre stage, together with follicular secretion of oestradiol and associated regulatory peptides. A divergence arises between growth of the dominant follicle and that of subordinate follicles, with oestradiol and inhibin acting to suppress circulating FSH concentrations. Currently considered a critical feature of follicle selection and dominance is the ability of a maturing follicle to withstand the fall in circulating concentrations of FSH. However, the extent to which the transfer of a dominant follicle's gonadotrophin requirement from FSH to LH represents its principal survival strategy requires further examination. Such an interpretation raises complications in terms of granulosa cell aromatase activity, and may divert attention away from many other key factors influencing ovulatory potential, not least the two mentioned at the end of the previous paragraph. The involvement of gonadotrophin receptors or their absence was further discussed in a context of follicular atresia, whereas the extent to which apoptosis represents a means of selection during differentiation of ovarian tissues has yet to be resolved. Once again, the point was made that the oocyte constitutes a centralised system of influence over follicular destiny, coordinating and regulating diverse molecular and cellular events culminating in selection of an ovulatory follicle.

Responses to the gonadotrophin surge

A perspective introduced in Chapter VII was that, at least in one sense, the maturity of a Graafian follicle regulates its own time of ovulation. Only when sufficiently developed and only when an appropriate level – threshold – of oestradiol secretion is achieved will positive feedback from a follicle(s) prompt the pituitary outpouring of gonadotrophic hormones that dictates changes in the somatic cells of the follicle and in the oocyte. At this time, increasing follicular secretion of oestradiol plays diverse rôles of vital importance, including acting within the Graafian follicle and neighbouring ovarian tissues, preparing the reproductive tract for the preliminaries to fertilisation, programming the hypothalamus in terms of both sexual behaviour and GnRH, and influencing the pituitary gland itself.

As to the pre-ovulatory surge of gonadotrophic hormones, this is a consequence of episodes of neurosecretion in the basal region of the hypothalamus acting on pituitary gonadotrophs of enhanced sensitivity to GnRH. Multiple GnRH mRNA transcripts were noted to be differentially regulated, permitting sophisticated levels of control and flexibility. Within the hypophysis, too, there is divergent regulation of subunit mRNAs, enabling LH synthesis and secretion to be upregulated, thereby paving the way to its predominance in the gonadotrophin surge. Whether or not a unique non-steroidal GnSAF or GnSIF attenuates the surge of LH was also touched upon, but new data are eagerly awaited in this regard.

As a response to the gonadotrophin surge, noteworthy steps within the preovulatory follicle concerned a rapid decrease in granulosa cell aromatase, expressed as a progressive change from oestradiol to progesterone secretion, with local and systemic influences. Remodelling of the germ cell layers and basement membrane was closely examined, with mucification and expansion of the cumulus oophorus assuming centre stage. Here, the involvement of specific GAGs in reorganisation of the cumulus was emphasised, and likewise remodelling of the basal lamina and other components of the extracellular matrix in specific regions of the follicle wall, for ultimately an aperture is required to enable liberation of the oocyte and its cellular investment. Throughout growth of a Graafian follicle, a balance must be achieved between breakdown and synthesis of the extracellular matrix.

Factors enabling resumption of meiosis were described in detail, both from the angle of interrupted gap junctional contacts between corona cell processes and oocyte and in terms of modulation of cAMP levels in the germ and neighbouring somatic cells. A resumed meiotic division becomes apparent as follicular steroid secretion shifts from oestradiol to progesterone predominance, and consideration was given to MPF and a specific MAS. Nonetheless, molecular control of meiotic resumption was presumed to be complex, with the interaction and modification of diverse components under physiological conditions. Mention was made of an experimental divergence between follicular somatic cell responses and those of the oocyte when ovulation was induced prematurely with LH or hCG at a time when intra-follicular oestradiol concentrations were still relatively low: in fact, shedding of primary oocytes could be demonstrated at a high incidence. Such oocytes failed to show a block to polyspermy if exposed to competent spermatozoa, nor did sperm heads within the vitellus undergo significant chromatin decondensation. In other words, nuclear and cytoplasmic immaturity could be demonstrated in precociously liberated oocytes, a point not without clinical significance. As proposed almost 40 years ago, a potential block to polyspermy develops in step with acquisition of meiotic competence.

Components of ovulation and shedding of the oocyte

Chapter VIII sought to understand the manner whereby the oocyte is released from a mature Graafian follicle in response to the gonadotrophin surge. Attention was drawn to the highly active phase of angiogenesis shortly before ovulation, with the follicle undergoing dramatic changes in colour in various species during modification of its vascular bed and seepage from distended capillaries. Attention also focused on intra-follicular pressure and the physical condition of the follicle wall, with the overall structure becoming progressively more flaccid and pendulous shortly before ovulation in a number of animals. Rupture of a Graafian follicle is clearly not an expression that can be applied across species, even if it is characteristic of events in rats, rabbits and cats.

As to dissociation of somatic cells and their matrix that is a preliminary to shedding of the oocyte, contributions from numerous molecules were reviewed, progressing from a major involvement of local progesterone and prostaglandin synthesis through diverse proteolytic enzymes and components of an inflammatory cascade: contributions from fibroblasts were highlighted. Cytokines in particular may act as specific vectors and triggers, and are certainly regulatory products of white blood cell traffic in an ovulatory follicle. The exact timescale of such involvements requires further characterisation for, although the oocyte has been shown to respond to the gonadotrophin surge on a predictable timescale, molecular events in its associated somatic cells have not yet been so defined. On matters chronological, it must be clarified whether a migrating population of white cells initiates changes in somatic tissues of the follicle or is a response to changes already under way. Components of the ovulatory programme downstream from the gonadotrophin surge can be compromised or blocked experimentally and yet ovulation may still occur, albeit less efficiently. This would suggest a multi-factorial system with a built-in degree of redundancy: in other words, a series of fail-safe mechanisms.

Factors within tissues of a pre-ovulatory follicle clearly exert the major influence in precipitating ovulation, but a possible contribution from the massaging action of the fimbriated extremity of the Fallopian tube was also considered. Although the conventional distinction between spontaneous and induced ovulation was noted, there was a call for flexibility in this classification. Spontaneous ovulators such as women might very well become induced ovulators under appropriate – if exceptional – circumstances.

Post-ovulatory fate of follicle and contents

Having discussed maturation of a Graafian follicle and ovulation in considerable detail, it seemed important in Chapter IX to comment on the fate of the structure and that of its released oocyte, somatic cells and antral fluid. A question that has received little attention to date is the extent to which residual fluid – that portion remaining in the collapsed follicle – contributes to the formation of a competent corpus luteum. Because of its potent content of hormones and growth factors, it could have a powerful local influence on the function of a developing corpus luteum. Circumstances in which inadequate corpora lutea were formed were recalled, with the comment that these might have been derived from follicles that were less than mature at induced ovulation. Part of such putative immaturity could be reflected in the composition of the residual follicular fluid.

As to the fate of follicular fluid after ovulation, the bulk was accepted to contact the Fallopian tube ampulla and then to be refluxed into the peritoneal cavity rather than passing down into the isthmus. Even so, transient contact with the endosalpinx could initiate specific responses in the underlying vascular bed and in the composition of secretions. Similarly, the population of granulosa cells shed with the oocytes, for these would act as a potent paracrine tissue in the vicinity of the oocyte. One incisive rôle of this follicular cell suspension could be to amplify signals from the oocyte or a newly fertilised egg that would be transmitted to the adjacent ovary. Whilst the oocyte might undergo a final phase of post-ovulatory maturation during progression along the Fallopian tube ampulla, there could also be an influence of the oocyte and its associated cell suspension on the fluid microenvironment within the Fallopian tube. The viable lifespan of the oocyte may depend not only on its maturity at ovulation but also on the fluids bathing it in the lumen of the tube.

Ovulatory failure and ovarian anomalies

Chapter X opens with what appear as gross instances of ovulatory perturbation seen as cystic follicles or PCOS. In reality, quite subtle anomalies of an endocrine or biochemical nature may lead to these tissue disorders, and these in turn may be associated with genetic and/or environmental factors. The point is made that ovulation cannot be taken for granted, even when follicular growth and development are demonstrable. Indeed, although unable to ovulate, such structures can enlarge far beyond the size at which the oocyte would normally be shed, and especially so in domestic farm animals. By contrast, in women suffering from PCOS, a proportion of patients may have almost normal reproductive health. More commonly, however, there are disturbed menstrual cycles and chronic anovulation, set against a backcloth of elevated concentrations of plasma androgens and a decreased sensitivity to insulin (i.e. insulin resistance): obesity and hirsutism are characteristic. Despite its diverse modes of expression, the syndrome may be prompted by a small number of key genes interacting with environmental factors. The causes leading to such wayward gene activity have yet to be clarified, but a contribution from mutations would be anticipated.

Examples were also provided of conditions in which follicular growth is usually not possible. These included an absence of trophic hormones, albeit at the hypothalamic level, as in Kallmann's syndrome, or a dearth or almost complete absence of ovarian germ cells, as in Turner's syndrome. Both conditions are found at a low incidence and both would appear open to experimental amelioration: by hormone injection in the former and by transplantation of appropriate groups of cells in the latter. Unfolding molecular descriptions may eventually lead to the design of alternative therapies, but this is looking deep into the crystal ball. Moreover, molecular therapies would seem to depend on early fetal diagnosis of impending problems.

A more bizarre scenario that may be associated with failure of ovulation is the development of one or paired ovotestes. The local activity of Leydig cells in the testicular tissue can inhibit normal development of adjoining ovarian follicular tissue, i.e. in the same gonad. Although the aetiology is intellectually challenging, and especially so at a molecular level, ovotestis tissue offers a unique model – or rather series of models – for examining the diverse components of follicular support and development, and the manner in which these can be disorganised by high local concentrations of androgens and also by Sertoli cell factors.

A final section touches on stress as a cause of ovulatory failure, including a link between nutritional and metabolic disorders expressed in our own species as hypogonadotrophic amenorrhoea. Inappropriate pulsatile patterns of GnRH release may be the central mode of expression. Leptin was seen as a vital link between nutritional status, somatic energy reserves and neuroendocrine components of the reproductive system, functioning as a metabolic gate when energy reserves are inadequate and also as a key to the onset of puberty. In the latter regard, however, leptin should be seen as but one factor within a spectrum of somatic monitors. So critical a threshold as that of puberty will doubtless be linked to a wide-ranging assessment of somatic and social status, if not in the conventional meaning of the latter term.

Induction of ovulation and control of ovulation time

In the penultimate chapter, reference is made to various means of inducing ovulation in women and experimental animals. In order to prompt follicular growth, gonadotrophins or their releasing hormones are generally administered as a series of injections in women, or via an infusion pump, whereas a single bolus injection is favoured in animals. In overcoming ovulatory dysfunction in women, the objective would be to stimulate maturation of a single Graafian follicle capable of ovulation, and treatments aim to overcome problems at the hypothalamic, pituitary or ovarian level. As a further contrast to procedures in domestic animals, responses during treatment in women are monitored in the concentrations of steroid hormones and perhaps also those of peptide hormones. Premature induction of ovulation must be avoided, not least since shedding of primary oocytes might occur. In a setting of so-called fertility treatment, stimulation of several follicles would be the usual objective and oocytes aspirated some hours before ovulation would be matured *in vitro*, prior to procedures of IVF. The serious syndrome of ovarian hyperstimulation must be avoided, and ultrasonic monitoring of follicular development is a useful management tool in this regard.

Overcoming problems of anovulation is not the usual reason for gonadotrophin treatment in animals. The two specific objectives are either to induce superovulation, frequently in conjunction with a programme of embryo transplantation, or to control the time of ovulation, invariably in conjunction with artificial insemination. Whereas ovulation time can be regulated with precision, the response of an individual animal to a given superovulation treatment remains unpredictable, although mean dose–response curves can be plotted for a population of animals.

* * *

As to the future, a greater understanding of follicular development and selection, in conjunction with hormone preparations rendered more specific at a molecular level – 'designer gonadotrophins' – may produce a more predictable response to a given treatment. However, at least as rewarding as a continuing search for somatic cell factors influencing folliculogenesis would be the profiling of oocyte molecular contributions at succeeding stages of follicular development. The impact of such molecules could be far more profound than hitherto appreciated, not least with amplification of oocyte signals by cells of the cumulus oophorus. And in this quest, exploiting the model in which precocious induction of ovulation leads to shedding of primary oocytes might be especially rewarding.

Despite the efforts that have gone into the above text, one can usefully leave the last remark to lines learnt at school (taken from Hamlet's chastening of Horatio):

> There are more things in heaven and earth, Horatio,
> Than are dreamt of in your philosophy.
>
> W. Shakespeare, *Hamlet*, Act I, sc. 5

Index

abortion 331, 332
acrosome reaction 41, 129
ACTH 160, 337
actin 80, 245
actin filaments 37, 234
active transport 115, 116
activin 50, 144, 150–7, 199
activin receptor II 50
adenosine 237
adenylyl cyclase 169, 197, 238, 297
adipocytes 337, 338
adipose tissue 329
adrenal androgens 329
adrenal cortex 147, 148, 150, 296
adrenal glands 24, 78
adrenergic nerves 77, 80
adrenergic receptors 82, 311
adrenocorticotrophic hormone, *see*
   ACTH
ageing 313
albumin 119, 147, 230
Alexandria 2, 3, 18
alpaca 263
amenorrhoea 331, 333, 336, 347
Amh gene 26
amniocentesis 331
amphibia 239
ampulla 126, 127, 229, 301, 306, 307
ampullary epithelium 107
ampullary–isthmic junction 126,
   128–30, 232, 262, 283, 299, 307,
   316, 328
anaesthesia 84, 85, 88, 124
anaphase 35
anastomosis 7, 61, 63, 64, 70
anatomy 4

androgens 82, 117, 141, 143, 144, 146, 147,
   154, 155, 167, 170, 197, 199, 200–2, 208,
   327, 329, 330, 335
androgen-binding protein 113
androstenedione 117, 118, 139, 141, 143–5,
   148, 197, 208, 329, 359, 360
aneuploid 316, 355
angiogenesis 61, 72, 88, 139, 163, 171, 251,
   271, 277, 302, 351
angiogenic factors 52, 84, 98, 271, 297
angiotensin 114
animalcules 13, 14
anoestrus 80, 87, 262, 296, 325
anoikis 210
anorexia 336, 347
anosmia 333, 334
anovulation 325, 329, 336, 346
anticoagulant 112
antral cavity 43, 45–7, 270
antral follicles 24, 36, 46, 48, 68, 69, 83, 107,
   159, 170, 205, 210, 248
antrum 36, 43, 46, 47, 49, 68, 70, 81, 85, 90,
   96, 106–31, 224, 270, 272, 284
aorta 61, 69
aortico-renal ganglion 77
aplastic ovaries 19
apoptosis 32–4, 47, 170, 206, 208–15, 316
apoptotic bodies 211–15
apoptotic cells 211–15
apoptotic programme 30, 209–15
aquatic ancestors 33, 187, 215
aqueous humour 108
arachidonic acid 168, 280
Aristotle 2, 3, 11, 15
aromatase activity 82, 143, 144, 147, 154, 161,
   191, 195, 197, 198, 201, 208, 244, 269, 296

aromatisation 139, 141, 143, 146, 199, 204, 208
aromatisation potential 139, 141, 191, 195
arterioles 68, 70, 77
arterio-venous anastomosis 64
arterio-venous contacts 68
artificial insemination 345
ascites 351
asymmetry 65
atresia 19, 33, 36–8, 42, 47, 52, 80, 117, 150, 167, 186, 190, 191, 195, 201, 202, 204–9, 212, 215, 230, 248, 325, 326, 329, 330, 332, 357, 361, 372, 373
atretic follicles 85, 86, 150, 156, 191, 195, 204–9, 210–12
autocrine action 144, 151, 169, 170
autocrine factors 144, 277, 338
autocrine influences 48, 49, 154, 157, 163, 188, 203, 277
autonomic nervous system 19, 60, 80–2, 84, 172, 187, 192, 204, 215, 225, 264, 265, 276, 286
autopsy 65
autosomal dominant 329, 334
autosomal recessive 148, 334
autotransplantation 19
avian gonads 7
avian studies 11
axons 78

Baer, K.E. von 14
basal body temperature 346
basal lamina 206, 248, 250–2
basement membrane 45–7, 73, 112, 113, 115, 117, 119, 124, 138, 139, 141, 160, 163, 206, 249, 250, 270, 271, 279, 326
basic fibroblast growth factor 33, 162
bats 128, 313
Bax protein 214
Bcl-2 family 213, 214
bicarbonate 115, 301
bicornuate uterus 3, 8
bifurcated uterus 2
binding protein 1, 106, 113, 119, 156, 198, 309
binding sites 143, 145, 224, 277
bioassays 16, 230
biosynthetic ability 138
biosynthetic pathways 94, 138–40, 143
birds 50
birth 42, 47, 61, 204, 205, 230
bladder 2, 3

blastocysts 11, 33, 310
blastomeres 41
block to polyspermy 38, 41, 241, 264, 310, 315
blood circulation 7
blood components 111, 112
blood clotting 96, 111, 326
blood flow 65, 68, 77, 80, 83–9, 108, 171, 214, 225, 284, 297
blood plasma 106, 116, 277
blood pressure 65
blood serum 116
blood supply 61, 68, 282, 296
blood velocity 87
blood vessel dilatation 84
blood vessel morphology 68
blood vessels 4, 8, 12, 61, 70, 75, 77, 80, 108, 113, 326
bloodstream 3, 15
Bmp4 gene 29
body weight 97, 337
bone morphogenetic protein 29, 203
bovine ovaries 49, 93
bradykinin 275
Brännström, M. 269
breeding season 345
broad ligament 61, 65, 78
Brussels 3
bulk flow 126, 127

C terminus 148
cadaver pituitaries 346
cadherins 210, 211
calcitonin gene-related peptide 80
calves 42
Cambridge 7
camelids 263, 265
cancer 19, 214
canine family 34
capacitation 107, 129, 130, 356
capillaries 68, 124, 277
capillary bed 46, 68, 77, 88, 108, 109, 113, 141, 143, 251, 271, 277, 284, 296
capillary blood flow 84, 86–8
capillary network 45, 46, 61, 64, 65–8, 279, 326
capillary permeability 271, 277, 351
capillary portal system 226, 228
caspases 212, 213
castration 11, 15
catalytic domain 31
catamenia 2

catecholamines 82–4, 129, 187, 192, 204
cathepsins 278
cats 263, 272
cattle 64, 90, 156
cavernous sinus blood 227
cell adhesion molecules 210, 248, 333
cell cycle 239
cell death 158, 209–11, 213, 214, 327
cell division 46
cell migration 29, 296
cell numbers 45, 107, 204
cell organelles 96
cell surface 124
cellular components 14
cellular interactions 128
central nervous system 79, 169
centrally placed computer 215, 268
centrioles 235, 237
cervical mucus 346, 350
cervix 2, 166
Cetrorelix 352, 353
checkpoint 51, 239
chemical constituents 110
chemokines 165, 280
chemotaxis 275, 278
chicken egg 6, 7, 11
chicken embryology 7
chicken reproductive tract 10
chimera 29
cholesterol 143, 144, 147–9, 240, 330
cholesterol side chain cleavage 82, 148, 149,
   330
cholinergic nerves 77, 80
chondroitin sulphate 250
chondroitin sulphuric acid 115
chorionic gonadotrophin 262, 297, 312
chromatin 38, 44, 45, 231, 236, 326
chromatin decondensation 38
chromatography 117
chromosome abnormality 331, 355
chromosome banding 26
chromosome condensation 239
chromosome duplication 27
chromosomes 41, 232, 315, 316
cilia 298–301
ciliary activity 15, 126, 128, 281, 299, 300,
   306, 309
circular muscle 311
c-Kit 31, 34
cleavage stages 36, 37, 302
clinical observations 50, 354

clinical techniques 346, 350–6
clomiphene citrate 347–9, 351
cloned embryos 19
cloning 361
cloprostenol 360
clotting potential 111, 112, 326
c-*mos* 239
c-*mos* mRNA 51, 239
coagulability 95, 110–12
coagulation 3
coding sequences 41
coelomic epithelium 29, 30, 46
coiled arteries 68
coitus 48, 129, 235, 262, 263, 313
collagen 248, 250, 251, 274, 277
collagenase 95, 114, 169, 274, 275, 278, 281,
   284
collagenolytic proteins 166
colloid osmotic pressure 110, 111
colony stimulating factors 279–81
computer 215, 268
conception 7
conceptus 18
congenital lipoid adrenal hyperplasia 148
connective tissue 61, 274, 277, 279
connexin 248
connexin 43 238, 248
constitutive gene pathway 25
contractile activity 15, 128, 263, 275, 279, 306
contractile elements 46, 80, 82, 263
contractile proteins 80
contralateral ovary 204, 207
cooling 94, 97
cooling rates 92, 93
Copenhagen 90, 110
copulation 11
Corner, George 16
corona cell processes 39, 40, 47, 207, 233,
   234, 238
corona cells 47, 233, 234, 238, 250, 252, 267,
   271, 301, 302, 314
corona radiata 47, 128, 233, 234, 238, 246,
   250, 267, 271, 302
corpora lutea 7, 8, 11, 15, 87, 156, 169, 296,
   357
corpus luteum 11, 15–18, 47, 74, 82, 84–6, 97,
   131, 138, 139, 153, 167, 192, 194, 197,
   268, 295–7, 303, 310, 316, 355, 357
cortex 30, 31, 45, 61, 64, 68–70
cortical granules 38, 41, 241, 242, 314–16
cortical tissue 45

corticosteroid-binding globulin 119
cortisol 160
cotyledonary placenta 4
counter-current exchange 67, 76, 94, 95, 98, 308, 309
counter-current feedback 68
cow 111, 116, 119, 124, 147, 149, 150, 153, 160, 166, 168, 188, 191, 192, 194, 195, 198, 204, 208, 212, 238, 247, 267, 271, 304, 328, 356
cow ovaries 11, 49, 93
cow placenta 3, 4
Creutzfeldt–Jakob disease 346
Cruikshank, William 14
cryopreservation 19
cryptorchidism 333
cuboidal cells 45, 49
culture medium 51
culture temperature 95, 96
cumulus cell mass 14, 42, 43, 46, 47, 110, 126, 127, 130, 153, 161, 172, 234, 237–40, 244–7, 270, 273, 274, 284, 285, 299, 302, 303, 305, 314, 357
cumulus expansion 93, 161, 233, 244–7, 270
cumulus investment 119, 122, 126, 161, 233–5, 245, 284, 285, 299, 301
cumulus–oocyte complex 82, 153, 155–7, 166, 233, 237, 240, 245, 246, 272, 273, 299, 301
cumulus oophorus 43, 46, 47, 93, 106, 107, 112, 113, 127, 188, 233, 244, 246, 252, 267, 317, 326
cyclic AMP 49, 50, 82, 119, 121, 122, 146–50, 157, 159, 161, 162, 167, 203, 238, 240, 246, 252, 297
cyclical changes 70
cyclins 239, 240
CYP11a gene 330
cystic follicles 116, 122, 268, 283, 326–30, 358
cytochrome *c* 213
cytochrome P450 143, 148, 161, 191
cytokine networks 278
cytokines 18, 98, 138, 163–8, 170, 171, 211, 212, 263, 265, 275–82, 284, 306, 325, 338
cytoplasm 36, 38
cytoplasmic maturation 38, 95, 119, 241
cytoplasmic organelles 34, 36, 211
cytoplasmic proteins 150
cytoplasmic volume 235
cytoskeleton 37, 95, 245, 246, 315

cytostatic factor 51, 239
cytotoxic signals 206

DAX-1 gene 26, 27, 334
De Formato Foetu 7, 8
De Generatione Animalium 7, 9
De Graaf, R. 7, 10–14
deadenylation 41
decondensation of sperm head 242, 243
deep rectal temperature 90, 91
deer 7
defective gonads 19
degeneration 326, 328
degradation 41
Delft 7, 13
depolymerisation 95, 110, 115, 124, 234
dermatan sulphate 246, 250
desloren 360
developmental errors 27
developmental programme 37
developmental strategy 172
diakinesis 230
dichotomy 241, 243, 264, 286
dictyate nucleus 34, 52, 204, 230, 231
differential display 284
differential regulation 229
differentiation 7, 36
diffusion 124, 206
dihydrotestosterone 148
dimeric proteins 151, 154
diplotene 34, 37, 42, 51, 52, 204, 230
disulphide bonds 151, 154
DNA 36, 212
DNA bands 212
DNA degradation 210
DNA synthesis 162
dog 14, 88, 235
domestic animals 3, 11, 88, 90, 110, 113, 122, 124, 189, 210, 264, 299, 345, 351
dominant follicle 48, 88, 147, 151, 153, 156, 186–93, 195, 198–202, 204, 205, 210, 215, 372, 373
Dominant White Spotting gene (*W*) 30, 31, 51
Doppler ultrasound 85, 86, 88, 89
downregulation 26, 27
dwarfs 159
dynorphin family 169

early pregnancy factors 303, 313
ectopic pregnancy 11, 13, 311, 312, 317
egg 5, 6, 7, 9, 11, 13–15, 128–30, 272, 299

egg activation 242
egg ageing 313, 314
egg capture 97, 234, 301
egg cells 5
egg environment 299
egg microvilli 34
egg release 11, 262, 263, 266, 272
egg surface 299, 301
egg transport 11, 14, 15, 299, 309, 311–13
eicosanoids 163, 168, 169, 278
electrical lesions 79
electrical properties 109
electrical stimulation 78
electrolytes 116
electron microscopy 80
elemental composition 306, 307
embryo 2, 5, 7, 15, 18, 19, 25–7, 30, 37, 76, 130, 302, 303, 309
embryo transplantation 332, 345, 346, 350, 361
embryonic death 316
embryonic development 32, 36, 41, 42, 295, 302, 310
embryonic factors 17, 303
embryonic genome 37, 51
embryonic gonad 27, 30, 33
embryonic nutrition 302
embryonic programme 36, 37
embryonic signals 303
embryonic transport 311
embryonic viability 106
endocrine activity 16, 83, 197, 371, 372
endocrine changes 70, 90
endocrine cues 42
endocrine functions 97
endocrine imbalance 327, 329, 337, 345
endocrine influences 48, 212
endocrine information 70, 197
endocrine programming 60, 198, 212
endocrine studies 83
endocrinology 15, 16
endometrial biopsies 346
endometrium 18, 312
endonuclease 212
endoplasmic reticulum 143
endorphins 169, 170
endosalpingeal secretion 302
endosalpinx 130, 302, 306
endoscope 90, 92
endothelium 111, 170, 280
endothermic reactions 93, 94, 97, 98, 106

enkephalins 169
environmental factors 325, 329, 330
environmental influences 25
environmental oestrogens 339
enzymatic activity 124
enzymatic conversion 138, 139, 143
enzymatic treatment 299
eosinophils 164
epiblast 29
epidermal growth factor 34, 36, 45, 144, 158, 160–2, 197, 246, 357
epididymal tissues 6
epigenesis 7
epithelial cells 109, 165, 250
epithelium 45, 46, 109, 115
equids 45, 312, 313
equine chorionic gonadotrophin 149, 358
equine follicle 61, 150
erectile organ 65
Espey, L.L. 269
eutherian mammals 25, 204, 314
evolution 25, 33
exocytosis 38, 242
explosive process 65
extracellular fluid 107
extracellular matrix 158, 160, 163, 210, 245, 246, 248–50, 252, 270, 274, 280, 330, 334
exudate 112

Fabrica 3–6
Fabricius 5, 7, 8
Falck, B. 138, 139, 141
Fallopian tube 2, 3, 5–7, 11–15, 34, 36, 41, 42, 63, 64, 70, 75, 77, 88, 96, 97, 106, 107, 124, 126, 128–30, 232, 234, 235, 252, 262, 264, 265, 268, 277, 281–3, 295, 297, 299–303, 306, 310–12, 314, 316, 317, 327, 356
Fallopian tube contractions 15
Fallopian tube fluids 41, 127, 356
Fallopian tube lymphatics 69
Fallopius 5
farm animals 16, 36, 49, 90, 112, 115, 335
Fas antigen 212
Fas ligand 212
fat cells 338
fat mass 337
fecundin 360
feedback 36, 166, 192, 199, 335, 336
feedback loop 139, 151, 159, 199, 270
female 30

female fluids 3
female gamete 36
female genital tract 48
female germ line 26
female gonads 2, 26
female seed 2
female semen 2, 5, 7
female testes 2, 5, 11
female testicles 3
ferrets 263
fertilisation 5, 14, 24, 25, 33, 38, 42, 48, 51, 106, 127, 129, 130, 232, 235, 262, 263, 295, 302, 309, 310, 313
fertilising spermatozoon 25, 27, 36, 38, 41, 51, 232, 235, 242, 314, 316
fertility 15, 25, 51, 213, 268, 276
fertility clinics 188, 229, 266
fertility treatment 345
fetal life 33, 42, 163
fetal ovaries 33, 42, 163
fetus 2, 3, 7, 11, 13, 33, 61, 96, 157, 188, 205
fibrillar material 38, 39
fibrin 111
fibrinogen 112
fibroblast growth factor 34, 36, 45, 162
fibroblasts 162, 274, 277, 280, 296
fibronectin 248, 250, 251
Fick principle 88
filamentous processes 33
filaments 39
filtration 3
filtration pressure 111
fimbriated epithelium 298, 299
fimbriated extremity 5, 126, 127, 281, 282, 286, 297, 299, 304
fimbriated infundibulum 97, 281, 282, 297, 299, 301
fine tuning 188
first meiotic division 34, 35, 42, 51, 235
first polar body 34, 35, 237
fish 14
'flare-up' effect 353
FLICE 214
FLIPs 214
fluid viscosity 110, 115
follicle 14, 34, 37, 38, 42, 43, 47, 48, 50, 96, 124, 147, 150, 171, 186–9, 192, 264, 271–3
follicle apex 88, 169, 271, 273, 274, 281, 282, 284, 299
follicle cells 29, 39, 40, 42, 49, 141, 166, 277, 279, 299, 302, 307, 310

follicle cohort 187, 191
follicle deviation 191, 201
follicle diameter 46, 48, 61, 93, 96–8, 125, 153, 171, 189–92, 194, 196, 197, 264, 268, 282, 284, 297, 347
follicle formation 34, 38, 42, 43, 368
follicle growth inhibitory factors 204
follicle inhibition 50
follicle pool 48, 186–9, 201
follicle recruitment 43, 47, 50, 60, 68, 186, 187, 195, 198, 203, 215
follicle selection 48, 68, 80, 98, 159, 163, 186, 187, 191, 194, 195, 198–200, 203, 215, 372, 373
follicle stimulating hormone, *see* FSH
follicle wall 76, 93, 96, 124, 245, 252, 263, 272–4, 277, 278
follicular antrum 36, 47, 68, 90, 106–31, 138, 245, 268
follicular cell processes 34, 40
follicular collapse 106, 124, 128, 267, 272, 295, 296, 304
follicular cysts 326–8
follicular degeneration 209
follicular development 37, 38, 42, 43, 47–50, 52, 60, 61, 65, 68, 80, 82, 83, 98, 117, 119, 131, 144, 150, 151, 154, 156, 158, 161–3, 165, 167, 168, 186–8, 193–5, 197, 200, 202, 205, 250, 336, 338, 347, 360
follicular epithelium 45, 109, 124
follicular fluid 14, 15, 46, 47, 82, 85, 88, 90, 93, 95, 96, 106–31, 138, 143, 147, 148, 151, 153, 154, 156, 159, 162, 165–9, 171, 195, 197, 198, 204, 208, 210, 224, 230, 235, 237, 240, 244, 245, 248–51, 268, 270, 272–5, 278, 284, 295, 297, 301, 304, 305–7, 316, 317, 326, 370, 371
follicular fluid fate 304, 305, 307
follicular fluid volume 304, 305
follicular growth 24, 34, 37, 42, 43, 45–8, 50, 65, 68, 107, 117, 119, 124, 150, 162, 186, 189, 194–6, 200–2, 204, 248, 276, 326, 333, 357
follicular hierarchy 131, 139, 189, 190, 195, 201
follicular maturation 61, 87, 88, 131, 138, 144, 147, 153, 156, 168, 188, 189, 195, 200, 263, 264, 268, 270, 276, 296, 336, 338
follicular phase 48, 70, 144, 154, 157, 190, 194, 195, 199, 243, 263, 264, 357
follicular pressure 122

follicular regression 189
follicular rupture 11, 65, 128, 267, 272
follicular size 96, 125
follicular surface 82, 272, 273
follicular temperature 90–7
follicular waves 47, 187–92, 205, 215, 357
follicular wreaths 68
folliculogenesis 42, 45, 49, 50, 52, 161, 163,
    188, 198, 248, 250, 336
follistatin 150, 153, 155–7
follistatin gene 156, 157
food intake 337
forskolin 238, 240
fox 235
frogs 14
frozen–thawed tissue 60
FSH 143, 147, 151, 154, 159, 161, 170, 191–3,
    195, 201, 203, 205, 208, 224, 225, 226,
    229, 244, 351, 354
FSH concentrations 119, 121, 122, 190–5,
    198–200
FSH dependence 48, 192
FSH receptors 49–51, 82, 144–6, 154, 192,
    197, 200
FSH secretion 156, 196
FSH stimulation 190, 197
FSH synthesis 151
FSH treatment 149
FSHR gene 50
functional sperm reservoir 128, 129

G protein 49, 229
galactosyl transferase 113
Galen 2, 3, 5, 7, 11
gamete maturation 90
gametes 5, 29, 36, 41, 48, 90, 107, 204, 215,
    262, 283, 306
gap junctions 34, 39, 40, 47, 207, 234, 238,
    240, 246, 270
gastrin releasing peptide 80
gel electrophoresis 38
gene action 38
gene activity 209
gene constructs 26
gene deletion 157, 158
gene expression 25, 27, 41, 157, 277
gene knockouts 50, 157–9
gene pathway 25, 26
gene product 25
gene programmes 295
gene sequence 1
gene transcription 51, 150

generation 3, 7
generation interval 360
genetic backcloth 325
genetic constitution 25, 30
genetic error 27
genetic selection 186
genital duct 24, 131, 277, 281, 347
genital ridge 25–7, 29, 30
genital tissues 29
germ cell lineage 29, 33, 95
germ cell loss 30
germ cells 24–7, 29–34, 37, 38, 40, 42, 45, 95,
    109, 114, 158, 205, 206, 212, 215, 264,
    326, 331
germinal epithelium 31, 46
germinal unit 186
germinal vesicle 34–6, 207, 230, 231, 239,
    247, 326
gestation 8, 262
glands of internal secretion 15
glucocorticoids 160
glycolytic enzymes 113
glycoprotein 38, 39, 41, 42, 93, 94, 128, 151,
    164, 168, 230, 246, 277, 301, 304, 306,
    314
glycosaminoglycans 110, 112, 114, 115, 122,
    129, 245, 248, 250
glycosylation 41, 156
goat 188
golden hamster 126, 267, 302, 304, 313, 359
Golgi apparatus 36, 39
gonad 6, 7, 25–7, 30, 31, 48, 62, 131, 147,
    212, 234, 281, 286, 317, 325, 335, 339
gonadal anlagen 29
gonadal function 90, 215
gonadal hormones 139, 172, 200
gonadal organisation 30, 31
gonadal primordia 29, 30
gonadal tissues 60
gonadal tumour 157
gonadotrophic activity 16
gonadotrophic hormones 46, 49, 51, 82, 84,
    109, 115, 138, 144, 147, 151, 156, 158,
    164, 172, 188, 224, 225, 229, 240, 327,
    336, 345, 359, 361
gonadotrophic preparations 116, 334, 345
gonadotrophic stimulation 49, 143, 188, 215,
    250, 326, 345
gonadotrophic support 192, 205
gonadotrophin binding 49, 200
gonadotrophin receptor activity 49, 187, 200,
    201, 208, 336

GnRH 87, 224, 226–9, 252, 270, 328, 333, 334, 336, 338, 345, 347, 351, 352, 354, 360
GnRH agonists 352–5
GnRH antagonist 352–4
gonadotrophin releasing hormone, *see* GnRH
gonadotrophin secretion 224–7, 264
gonadotrophin surge 46, 97, 106, 108, 113, 148, 373, 374
gonadotrophin surge attenuating factor 229, 230, 252
gonadotrophin treatment 122, 149, 215, 230, 326, 362
gonadotrophs 226, 228, 229
Graafian follicles 11, 13, 14, 16, 18, 19, 24, 34, 38, 42, 43, 45–8, 60, 61, 65–70, 72–4, 79–83, 87, 89–95, 106, 108–10, 115–17, 122, 124, 126, 127, 130, 131, 138, 139, 142–5, 150, 158, 165–7, 172, 187, 194, 196, 201, 208, 224, 225, 230, 235, 248, 250, 262, 264, 265, 267, 268, 270, 272, 273, 276–8, 281, 283–5, 295, 297, 301, 308, 310, 316, 317, 326, 328, 336, 347, 350, 351, 357, 359, 361
Graeco-Roman period 3
granulocytes 277–9
granulosa cell tumours 157
granulosa cells 29, 34, 36, 39, 40, 42–7, 49, 50, 65, 68, 73, 76, 82, 88, 97, 106, 107, 109, 110, 112, 113, 115–17, 119, 122, 124, 126, 130, 131, 138, 141, 143–7, 149–51, 153, 154, 156, 158–60, 163, 166–70, 172, 189–91, 195, 197–9, 201, 202, 204–7, 210, 215, 224, 237, 243, 245, 247, 248, 250, 251, 268–72, 274, 296–9, 301–3, 305, 317, 326, 338, 352
granulosa lutein cells 47, 106, 166
gravid uterus 3, 4, 7
Greek philosophers 1
Greek physicians 2, 3
growth differentiation factor (GDF) 9, 50, 161, 162, 203, 247
GDF-9B 203
growth factor 1, 18, 36, 50, 52, 158, 159, 163, 187, 198, 200, 204, 249, 276, 306, 316, 325
growth hormone 158
growth hormone resistance 160
growth phase 34, 36, 38, 42, 48, 49
growth-regulated oncogene 165
guinea-pigs 64
gynaecology 2

haemopoietic cells 31
haemorrhage 82, 96, 111, 312, 326
haploid chromosomes 25
Harris, G.W. 16
Harvey, William 7, 9, 11
heat transfer 94, 95
heparan sulphate 156, 250, 251
heparin 111, 279
hepatocyte growth factor, *see* HFG
hermaphrodite 335
Herophilus 2, 3
heterogametic sex 25
HGF 163
hilus 61, 64, 65, 68–70, 75, 77, 278
hind gut 30
Hippocrates 2
hirsutism 329
histamine 84, 271, 275, 279, 281
histamine-like substances 69
homeostasis 114, 211
homogametic sex 27
hormone concentrations 106
hormone exchange 68, 94
hormone treatment 296
horse 61, 111, 124, 126, 188, 194, 268, 296, 304, 312
human cadavers 3
human chorionic gonadotrophin 84, 97, 120, 123, 150, 166, 168, 169, 230, 238, 243, 263, 282, 347, 350, 352, 355–9
human egg 299
human embryos 30, 70
human follicle 61, 69, 93, 118, 119, 159, 161, 197, 204, 208, 279, 280, 347
human menopausal gonadotrophin 347–52, 354
human ovaries 4, 5, 11, 32, 46, 49, 50, 64, 66, 71, 77, 78, 148, 149, 165, 169, 188–90, 198, 204, 279, 338, 347
human pituitary gonadotrophin 346
human uterus 3, 4, 12, 71
humans 3, 26, 27, 30, 34, 36, 40, 41, 46–8, 50, 64, 77, 89, 90, 93, 96, 110–13, 117–9, 124, 126, 127, 153, 162, 166, 170, 171, 188, 212, 239, 267, 279, 304, 335, 347
humoral interaction 17
Hunter, John 15, 366
hyaluronic acid 115, 234, 245–7, 250, 270, 271
hyaluronidase 115, 234, 245, 302
hydration 93
hydrolases 274

hyperactivation 130
hyperaemia 69, 84, 271
hyperandrogenism 330
hyperplasia 296
hyperstimulation 350, 351
hypertrophy 296
hypogastric 71
hypogastric nerve 78
hypogonadism 333, 334
hypophysectomy 16, 49
hypophyseal portal blood 226, 228
hypophysis 42
hypothalamic anovulation 336
hypothalamic control 16, 228
hypothalamic function 79, 228
hypothalamic neurons 79, 228
hypothalamic releasing factors 326, 345
hypothalamic secretion 226, 337, 345
hypothalamus 16, 18, 48, 78, 79, 139, 224,
    228, 251, 263, 270, 333, 335, 337,
    347
hypoxanthine 237
hysterectomy 16, 18

iliac nodes 71
immature eggs 264
immature follicles 50, 264
immune cells 164, 165
immune mediator 165
immune system 163, 164, 209, 276
immunofluorescence 130, 170
immunoglobulins 113
immunostaining 150, 156, 163, 279
in situ hybridisation 82, 149
in vitro culture 206
in vitro fertilisation 19, 33, 38, 96, 98, 106,
    117, 129, 156, 165, 208, 283, 296, 329,
    331, 346, 349, 350, 354, 355, 361; *see
    also* IVF clinics
in vitro maturation 33, 95, 96, 98, 155, 160,
    161, 208, 246, 331, 350
inbreeding 28
indifferent gonads 26
indomethacin 118, 121, 169, 275
induced ovulator 11, 122, 241
inductive signal 29
inert gases 95
inferior mesenteric ganglion 77
infertility 15, 33, 325, 329, 331, 332, 346
inflammation 171, 210, 211, 246, 263, 270,
    271, 274, 275, 277, 278, 280, 284
inflammatory cells 206, 246, 277

inflammatory mediator 165, 171
inflammatory response 263, 270, 271, 274,
    275, 277, 278
infra-red scanning 90
infundibulum 75
infusion pump 345, 347
inhibin 114, 119, 121, 144, 147, 150–7, 160,
    192, 198–200, 202, 204, 208, 229
innervation 46, 60, 77, 78, 80, 98, 144
insemination 129
insulin 147, 159, 160, 167, 329, 330, 337, 338
insulin imbalance 336
insulin insensitivity 329
insulin-like growth factor (IGF) 50, 147, 154,
    158–60, 167, 197, 198, 199, 249, 338
    IGF-1 144, 154, 155, 158–60, 167, 352,
       357, 360
    IGF-2 158–60
IGF binding proteins 158, 160, 198, 208
integrin receptors 249
intercellular bridges 33
intercellular communication 109
intercellular coupling 122
intercellular junctions 41, 109
intercellular spaces 46, 111, 246
interferon 18, 212
inter-follicular communication 60, 70
interleukins (IL) 6, 32, 165–7
    IL-1 system 165, 166, 168, 171, 278–80
    IL-2 166
    IL-6 166
    IL-8 165, 166
    IL-10 166
intermediary follicle 45, 47
inter-ovarian communication 188
intersexes 27, 335
interstitial cells 169, 207
intra-follicular pressure 75, 122, 124, 271
intranasal administration 352
intra-ovarian factors 49, 50
intra-ovarian regulation 158, 163
intra-peritoneal insemination 266
ipsilateral changes 79
irradiation 19
ischaemia 206, 271, 274, 282
isoforms 161, 170, 171, 251, 283
isomeric forms 141
isthmus 126–9, 304, 306, 307, 310
IVF clinics 188

jugular blood 193
jugular temperature 90

junctional complex 109, 234
junctional contacts 39, 40, 109

kallikrein 275, 281
Kallmann's syndrome 333, 334
karyotype 27
keratin 248
kidney 26
kinetochores 315
kinin formation 275
Kit-ligand 31, 34, 36, 50

lactation 327
lactational anoestrus 80, 325
laminin 248, 250, 251
laparoscope 90
laparoscopy 106, 112, 117, 312, 330, 345, 346, 350, 361
laparotomy 90–2
Leeuwenhoek, Antonie van 13, 14
Leiden 7
Leonardo da Vinci 3, 4
leptin 337, 338
leptin resistance 338
leptotene 34
leucocytes 164–6, 275–9, 284
leukaemia inhibitory factor 32, 33, 168
leukotrienes 169
Leydig cells 26, 27, 335
LH 84, 143, 147, 154, 155, 170, 192, 193, 197, 203, 224–30, 297, 329, 336, 338
LH concentrations 119, 121
LH injection 263, 266, 275
LH pulses 144, 193–5, 200, 336
LH receptors 49, 51, 144, 145, 146, 161, 167, 195, 197, 200, 202, 238, 243, 296, 297
LH secretion 151, 225–7, 229, 230
LH surge 84, 89, 118, 119, 121, 122, 124, 196, 244, 296, 356
lipid droplets 148
lipids 34
lipoprotein 143
liquor folliculi 108
liver 159
llama 263
local communication 68, 192
local modulation 70, 112, 192
local recirculation 68
local signals 49, 192
lordosis 48
Louvain 3

lower vertebrates 25
lumbar nodes 70
lumbo-ovarian ligament 61
lumbo-sacral region 77
luminal fluid 126, 130
luteal cells 87, 97, 138, 277, 283, 295–8, 310, 316, 327
luteal failure 355
luteal function 16, 87, 355
luteal insufficiency 296, 355
luteal lifespan 17, 355
luteal phase 48, 84, 89, 153, 154, 189, 264, 347, 355
luteal regression 84
luteal tissue 11, 17, 93
lutein cells 138, 296
luteinisation 13, 47, 88, 131, 138, 139, 149, 150, 163, 251, 268, 271, 275, 295, 296, 328
luteinised follicles 327
luteinising hormone, *see* LH
luteinising hormone-releasing hormone 79
luteolysins 84
luteolysis 18, 76, 167, 194
luteotrophic complex 297
luteotrophic factors 17, 18
lymph nodes 69, 76
lymphatic bed 69, 131, 138, 141, 271
lymphatic capillaries 69, 70, 72, 73
lymphatic drainage 69, 71, 74, 75
lymphatic flow 75, 164
lymphatic network 46, 60, 69, 70, 72, 95
lymphatic pathways 69, 283
lymphatic system 70, 77, 308
lymphatic trunks 69, 75
lymphatic valves 70, 71
lymphatic vessels 60, 70, 73–5, 164
lymphocytes 164, 167
lysosomal enzymes 210
lysosomes 162
lysozyme 278

macromolecular synthesis 36, 51
macromolecules 37, 38, 51, 95, 246
macrophage colony stimulating factor 168
macrophage secretions 110
macrophages 18, 110, 164–8, 171, 278–80
male 25, 30
male-determining genes 25
male dominance 25
male ducts 5

male fluids 3
male gametes 41
male gonads 3, 6
male precocity 25
male pronucleus growth factor 38, 120, 242
Malpighi, Marcello 11, 13
mammalian egg 13–15
mammalian ovaries 11, 52
mammals 5, 7, 14, 25, 27, 41, 45, 50, 65, 122, 131
marsupials 29
mast cells 164, 279
maternal genome 51
maternal legacy 37, 41
maternal transcripts 37, 235
materno-fetal interface 3
mating 7, 129, 235, 262, 264
matrix metalloproteinases 249
mediaeval thought 3
medical school 1, 3, 7
medical students 3
medicine 2
Mediterranean 2, 3
medulla 30, 31, 45, 61, 64, 68, 69, 165, 278
medullary tissue 45
meiosis 30, 33, 37, 38, 50, 51, 95, 112, 131, 170, 204, 230–2, 235, 237, 266, 332
meiosis-activating substance 240, 241, 252
meiosis-inducing substance 37
meiosis-inhibiting substance 37
meiotic arrest 34, 51, 131, 207, 235, 237–9
meiotic competence 241
meiotic inhibitor 237–9
meiotic maturation 34, 35, 37, 51, 95, 155, 170, 231, 232, 235, 240, 266
meiotic prophase 34, 37, 38, 42, 231, 232, 235
meiotic spindle 95, 235, 236, 315
membrana propria 109
membrane 36, 109
membrane folding 36
membrane remodelling 129
membrane vesiculation 41, 129
menopause 42, 205, 286
menstrual blood 2, 7
menstrual coagulum 2
menstrual cycle 48, 61, 62, 65, 83, 88–90, 127, 139, 144, 147, 150, 153, 154, 167, 171, 186, 189, 190, 207, 264, 314, 329, 347, 354
menstruation 15
mesenchymal cells 29, 33

mesenchymatous cells 30
mesenchyme 26, 30, 250
mesenteries 15, 30, 299
mesoderm 162
mesometrium 75
mesonephric cords 30
mesonephric precursor cells 29
mesonephros 26, 29, 30, 61
mesovarium 60, 61
metabolic cooperation 246
metabolic coupling 39
metabolic cue 338
metabolic disorder 336
metabolic gate 338
metabolic insult 206
metabolic rate 337
metabolic relationship 206, 246
metalloproteases 249, 276, 284
metaphase 51, 230, 232, 315
metaphase promoting factor 239, 240
methallibure 357
microcirculation 68, 70, 84, 98, 131
microelectrodes 90
microenvironment 41, 107, 109, 124, 128, 130
microfilaments 80, 248
micropuncture 306
microscopy 11, 13
microsphere technique 62, 85, 86
microspheres 85, 86
microsurgery 312
microtubules 95, 234–6, 239, 248, 315
microvasculature 60, 69, 98
microvilli 39
midwifery 2
miscarriage 329, 347
mitochondria 36, 45, 143, 148, 205, 213, 297
mitochondrial genome 205
mitochondrial membrane 148–50
mitogen 162
mitosis 30, 33, 167, 204, 206, 211
mitotic cycles 107, 167
molecular access 109
molecular exchange 41, 42
molecular gradient 168
molecular influence 42, 43, 45, 203
molecular inhibitor 237
molecular orientation 1
molecular programmes 27
molecular prompting 45, 203
molecular signals 317
molecular studies 25, 50

molecular techniques 26
molecular weight 41
monkey 11, 40, 110, 198, 229, 230
monocytes 278, 280
monocyte chemoattractant protein 165
monomeric protein 155
morulae 311
Mos protein 51
mother 13
mouse 26, 29–34, 37, 39–41, 45, 49, 50, 51, 60, 109, 126, 157, 159, 161, 163, 166, 168, 170, 171, 210, 212, 237–9, 245, 246, 279, 302, 335
mRNA 34, 82, 141, 149, 150, 152, 156, 157, 159–61, 165–9, 191, 197, 198, 200, 208, 210, 212, 226, 275, 279, 280, 334, 337
mRNA transcripts 51
mucification 93, 106, 110, 233, 244–6, 267, 317
mucin layer 14
mucus 128, 245, 356
Müllerian ducts 214
Müllerian inhibiting substance 214
multiple ovulation 190
multiple transcripts 229
mural granulosa 43, 47
muscular movements 15
mutant embryos 26
mutants 30
mutation 1, 31, 50, 95, 148, 159, 186, 205, 330, 334, 335
myosalpingeal contraction 126, 128, 130, 299, 301, 304, 306, 309, 311
myosalpinx 310, 311, 313
myosin 80

Nafarelin 352
necrosis 208–11
negative feedback 229
neonatal ovary 80
neonate 33
nerve fibres 80
nerve growth factor 52, 83, 276
nerve plexus 80
nervous programming 60, 204
nervous system 52, 187, 188, 204
neural control 16
neural tissue 170
neuroendocrine loop 263
neuromuscular complex 80
neuronal activation 338

neurons 79, 252
neuropeptide Y 80, 337
neuropeptides 80
neurosecretion 251
neurotransmission 170
neurotransmitters 80, 82
neutrophils 164–6, 278, 279
nitric oxide 84, 98, 166, 170–2, 204, 214, 215, 278, 280, 284
nitric oxide synthase 170, 171
non-myelinated fibres 77
noradrenaline 82
norepinephrine 82
nuclear condensation 207, 211
nuclear envelope 211
nuclear fusion 361
nuclear maturation 38, 119, 235, 241
nuclear membrane 236, 326
nuclear pores 36
nuclear proteins 212
nucleolus 34, 44, 45, 231, 235
nucleus 39, 40, 44, 45, 231
nutrition 160, 206, 297
nutritional exchange 47
nutritional status 325, 338
nymphomania 327

obese gene 337
obesity 329, 330, 336, 338
Observationes Anatomicae 5
oedema 271, 284, 310, 331
oestradiol 47, 82–4, 87, 88, 98, 112, 117–19, 131, 139, 141, 143–5, 147, 154, 159, 162, 166–8, 171, 188, 191–5, 197, 199, 200, 202, 208, 224, 226, 229, 243, 244, 262, 263, 270, 281, 296, 297, 299, 308, 310, 311, 335, 338, 347, 348, 350–2
oestradiol threshold 225
oestriol 147
oestrogen administration 69
oestrogen-binding proteins 119
oestrogen secretion 139
oestrogen synthesis 139, 141, 143, 154
oestrogens 147, 208, 327
oestrone 117, 119, 147
oestrous behaviour 139
oestrous cycle 61, 64, 83–5, 87, 90, 120, 122, 127, 139, 144, 150, 156, 165, 168, 186, 189, 191, 192, 194, 196, 200, 226, 227, 241, 243, 265, 314, 327, 328, 345, 356, 357, 360

oestrus 15, 48, 84, 87, 120, 123, 127, 243, 265, 327, 356
offspring 15
olfactory bulbs 333
olfactory function 333
olfactory placode 333
oocyte 5, 24, 26, 33–47, 49, 50, 52, 61, 68, 90, 93, 95, 106, 107, 109, 112, 113, 117, 119, 120, 124, 126, 127, 130, 131, 139, 147, 162, 166, 167, 171, 172, 186, 188, 198, 203, 205, 206, 210, 215, 230, 231, 233, 235, 244, 246, 247, 250, 251, 262, 267, 268, 271–3, 277, 281, 285, 295, 298, 301, 305, 307, 310, 314, 316, 317, 326
oocyte ageing 313, 314
oocyte competence 52
oocyte diameters 34, 49, 96
oocyte growth 33, 34, 36–9, 41, 45, 46, 51, 235
oocyte macromolecules 51
oocyte maturation 51, 122, 151, 154–7, 161, 237, 243, 244, 275, 354
oocyte maturation inhibitor 237, 239
oocyte metabolism 301
oocyte quality 52, 155
oocyte release 88, 96, 97, 262, 263, 266, 268, 272–7, 286, 301, 374, 375
oocyte retention 317
oocyte retrieval 33, 117, 165, 188, 205, 331, 345, 354
oocyte volume 36, 41
oocyte wastage 205
oogenesis 33, 332
oogonia 30, 33, 37, 42, 204, 205, 214, 332
oolemma 39, 47, 207, 238
ooplasm 40
opioid peptides 169, 170
organ culture 282
organelles 45
osmotic gradients 109
osmotic potential 109
osmotic pressure 110
ostium 5, 126, 298, 299
ovarian activity 15, 60, 163
ovarian ageing 210
ovarian arteries 7, 61–6, 76, 78, 87, 94, 308
ovarian asymmetry 188
ovarian blood supply 61–3, 74, 282
ovarian bursa 126, 170, 281, 304
ovarian compartments 68, 158
ovarian control 15
ovarian cyclicity 18, 262

ovarian cysts 15, 239
ovarian differentiation 24–7, 29, 37, 210
ovarian drainage 65, 75
ovarian drilling 330
ovarian dysgenesis 50, 331, 332
ovarian endocrine activity 33
ovarian extracts 15
ovarian follicles 7, 10, 11, 33, 37, 42, 49, 60
ovarian formation 27, 29
ovarian function 11, 18, 19, 60, 68, 83, 87, 163
ovarian grafting 15, 60
ovarian hormone 18
ovarian hyperstimulation syndrome 350, 351, 355, 361
ovarian innervation 60, 77, 80, 82, 264
ovarian lymphatics 69–71, 75, 76
ovarian morphology 3, 45, 265, 328
ovarian palpation 188
ovarian pedicle 66, 94
ovarian physiology 18, 61, 76, 77, 166, 169, 186
ovarian plexus 75, 78, 94
ovarian relocation 60
ovarian response 120, 350–2
ovarian size 65
ovarian stimulation 50, 347–9
ovarian stroma 90–2, 168
ovarian temperatures 90–4, 330
ovarian tissues 18, 19, 24, 27–30, 46, 60, 65, 83, 169, 224, 262, 330, 336, 338
ovarian transplantation 83, 187, 195
ovarian tumour 157
ovarian vasculature 61, 88
ovarian vein 62, 65, 66, 94, 124, 308
ovarian venous plasma 115, 124
ovarian vesicles 7
ovariectomy 15, 79, 149, 282, 309, 338
ovariectomy, unilateral 15, 327
ovaries 2, 3, 5–8, 10, 11, 13–16, 18, 24, 25–7, 29–31, 33, 37, 45, 47, 64, 65, 69, 70, 75, 204, 338, 347
ovaries, human 2
ovary-determining genes 24–6, 37
oviductin 42
ovotestis 27, 28, 334–6
ovulation 5, 11, 13, 15, 16, 18, 24, 33, 34, 38, 41, 45–8, 51, 60, 65, 68, 80, 82, 83, 88, 89, 94, 97, 98, 106, 107, 110, 117, 118, 121–4, 126–9, 147, 163, 166, 168–71, 190, 201, 209, 230, 235, 245, 249, 262–86, 296, 337, 345–62, 374, 375

Index

ovulation advancement 265, 266
ovulation cascade 282–4
ovulation delay 337
ovulation failure 200, 325–39, 345–7, 376, 377
ovulation fossa 45
ovulation, induced 16, 123, 263–6, 282, 284, 296, 345–62, 377, 378
ovulation inhibition 169
ovulation rate 98, 187, 278
ovulation, spontaneous 16, 186, 263–6, 282, 284
ovulation time 19, 34, 52, 84, 87, 107, 113, 115, 139, 171, 225, 226, 264–7, 277, 279, 282, 299, 359
ovulatory follicle 190, 263, 336
ovum 11, 14
oxygen 206, 207
oxygen concentration 124
oxygen diffusion 207
oxytocin 18, 79, 308

p53 214
pachytene 34
Padua 3, 5, 7
pampiniform plexus 65, 66
paracrine factors 45, 49, 52, 203, 247, 277
paracrine influences 48, 130, 144, 151, 153, 154, 160, 163, 169, 170, 188, 205, 277, 308, 310
paracrine tissue 130, 302, 303, 310, 311, 317
parasympathetic nerves 77, 78
paraventricular nucleus 79
Paris 3
parthenogenesis 51
patterns of innervation 77
pCO$_2$ 124, 125
PDGF 163
peak-systolic velocity 88, 89
pelvic nerves 78
penile intromission 263, 265
penis 26
peptide factors 138, 203
peptide fraction 265
peptide growth factors 32, 45, 61, 106, 158, 162, 193, 197, 302
peptide hormones 42, 68, 106, 116, 119, 121, 131, 172, 308, 336
peptidergic nerves 77
peptides 16, 18, 93, 112, 116, 121, 150, 151, 156, 160, 162, 169, 172, 188, 302
perfused ovaries 168, 171, 279, 280

peripheral blood 117, 188
peripheral circulation 113, 224, 226
peripheral plasma 121
peritoneal cavity 15, 24, 126, 127, 304, 307, 308
peritoneal fluid 124, 127, 188, 304
peritoneal membrane 127
peritoneum 46, 127
perivitelline fluid 41
perivitelline space 41, 241, 315
Pflüger's cords 30
pH 124, 125
phagocytosis 114, 211
pharmacological regulation 327
phosphatidylserine 211
phosphoprotein 148
phosphorylation 150, 239, 240
physiological syncytium 238
pig 8, 15, 28, 40, 42, 61, 68, 70–2, 90, 93, 95, 109, 111, 116, 120, 122–5, 143, 156, 157, 168, 188, 194–7, 204, 208, 210, 230, 241, 243, 245–7, 249, 265, 267, 268, 275, 278, 282, 283, 302, 305, 327, 335, 336, 356
pituitary, anterior 16, 49, 138, 151, 156–8, 224, 228, 229, 270, 335
pituitary dysfunction 336
pituitary gland 16, 48, 49, 51, 60, 78, 138, 139, 154, 170, 172, 224, 228, 263, 327, 337, 345, 347
pituitary hormones 119, 164, 169, 226, 244, 262, 297
pituitary, posterior 16, 228
pituitary trophic hormones 60
placenta 139, 150, 296, 310
placental cotyledons 3, 4
placental gonadotrophins 169
placental proteins 113
placentation 3
plasma membrane 39, 40, 47, 49, 160, 210, 228, 233, 234, 310, 314, 315
plasmin 111, 274, 284
plasminogen 114
plasminogen activator 114, 274, 275, 278, 280, 284
platelet activating factor 271, 276, 284
platelet-derived growth factor, *see* PDGF
plexus 64, 70, 75
pluripotent cells 33
pO$_2$ 124, 125
point mutation 50
polar body 232, 235, 236, 357

polycystic ovarian disease 19, 328–31, 351, 352
polycystic ovaries 113
polymeric filaments 41
polymorphonuclear leucocytes 18, 130, 266, 278
polypeptides 37, 162
polyspermic penetration 38, 41, 120, 128, 241, 242, 315
polytocous species 126
portal system 16
positive feedback 131, 159, 200, 224–6, 228, 335, 352
postganglionic fibres 77
postnatal life 24, 163
post-ovulatory ageing 313–16
post-ovulatory events 277
post-ovulatory maturation 316
pre-antral follicle 49, 50, 159, 161, 189, 197, 205, 210
pre-antral stage 34, 49
precocious ovulation 264
precursor cell lineage 29
precursor molecules 141
preformation 7
preganglionic fibres 77
pregnancy 14, 61, 62, 64, 65, 205, 215, 262, 264, 296, 297, 331
pregnancy-associated plasma protein 113
pregnant mare serum gonadotrophin 169, 262, 264, 358–60
pregnenolone 143, 148
pre-granulosa cells 29, 34, 42, 45, 49, 167
premature ovarian failure 50
premature ovulation 38
prenatal development 24
prenatal loss 262
pre-ovulatory diameter 90, 93, 263
pre-ovulatory follicle 61, 68, 84, 87, 90, 93, 110, 116, 117, 119, 122, 144, 148, 156, 162, 165, 197, 198, 202, 212, 224, 239, 245, 246, 262, 266, 268, 270, 271, 274, 276–8, 280, 282, 285, 286, 297, 338
pre-ovulatory gonadotrophin surge 38, 84, 110, 117, 122, 124, 131, 139, 153, 165, 169, 194–6, 200, 224, 226, 228–30, 233, 237, 245, 262, 263, 266, 267, 269–71, 275, 277–80, 282, 283, 286, 336
pre-ovulatory interval 122, 229, 233, 234, 241, 266, 275
pre-ovulatory maturation 34, 148, 197, 235

pre-ovulatory period 89, 114, 148
pre-ovulatory progesterone 82
pre-ovulatory stage 40
prepuberal animals 360
presumptive gonads 24, 29, 30, 39
primary fluid 108, 109, 113
primary follicle 44, 45, 47–50, 68, 69, 82, 163, 170, 210, 331
primary oocytes 24, 30, 34, 37, 42, 120, 204, 231, 241, 242, 264, 350, 357
primates 16, 18, 90, 122, 153, 156, 157, 161, 198, 203, 207, 238, 296, 297, 311
primordial follicles 19, 37, 41–5, 47–52, 61, 159, 163, 186, 188, 201, 202, 206, 210, 214, 237
primordial germ cells 24, 27, 29–34, 51
prion particles 346
procreation 2
pro-enkephalin 169
progesterone 16, 47, 82–5, 87, 112, 117, 118, 129, 139, 144, 147, 148, 150, 162, 166, 167, 193, 195, 197, 200, 201, 208, 244, 268, 277, 280, 281, 284, 296, 297, 299, 305, 308, 310, 311, 330, 349
progesterone treatment 88, 327, 355
progestins 143, 327, 357
programmed cell death 32, 209, 214, 316
prolactin 119, 121, 146, 193, 297, 346
prometaphase 35
pronuclear structures 242
pronucleate eggs 27
pronuclei 26
pronucleus 38
pro-oestrus 89, 356
pro-oestrous injection 123
pro-opiomelanocortin 169
prophase 38, 42, 51
prostacyclin 111
prostaglandins 18, 93, 95, 106, 112, 122, 123, 127, 129, 130, 166, 169, 262, 274, 275, 280, 283, 284, 306, 308, 313
prostaglandin analogue 360
prostaglandin concentrations 275, 349
prostaglandin E_2 122, 123, 169, 271, 275, 313, 349
prostaglandin $F_{2\alpha}$ 17, 18, 76, 84, 123, 169, 275, 328
prostaglandin synthase 283
protease inhibitors 113, 246, 275
proteases 274, 280
protein folding 96

protein kinase 51, 147, 150, 167, 228, 238–40, 297
protein phosphorylation 162
protein synthesis 37, 168, 239
proteins 34, 37, 40, 93, 95, 112, 113, 119, 148, 150, 246, 248
proteoglycans 93, 156, 246, 248, 250, 251, 270, 271
proteolytic cascade 212, 284
proteolytic cleavage 156, 162, 168
proteolytic degradation 249
proteolytic enzymes 95, 113, 169, 267, 274, 277, 302
proto-oncogene c-*mos* 51
puberty 16, 27, 61, 90, 163, 204, 268, 286, 334, 338, 345
pulsatile secretion 193, 194, 198, 226–8, 251, 336, 345
pulse frequency 200, 226–8
pulse generator 252
pycnotic nuclei 206
pyometritis 18
pyruvate 302

rabbit 11, 14, 15, 38, 40, 61, 64, 65, 67, 68, 76, 90, 95, 109, 110, 115, 116, 122, 126, 128, 162, 166, 168, 169, 263, 267, 271, 272, 275, 278, 279, 299, 302, 328, 356
radio-active isotopes 339
rat 61, 64, 68, 78, 83, 84, 116, 122, 151, 153, 156, 157, 159, 161, 162, 165–71, 207, 210, 211, 214, 226, 229, 230, 245, 268, 271, 272, 275, 276, 278, 279, 302
recombinant gonadotrophins 354, 355
recruitment 47, 60, 68
rectal palpation 271
reflex ovulator 263
regional environments 306, 307
relaxin 308
releasing factors 16, 326
remodelling 93, 107, 108, 171, 248, 250, 284, 330
renal failure 351
renal vein 65
renin–angiotensin system 284, 351
reproduction 7, 13, 14
reproductive biologists 60
reproductive efficiency 83
reproductive failure 327
reproductive health 339
reproductive lifespan 19, 32

reproductive organs 3
reproductive technology 33, 38, 346
reproductive threshold 338
reproductive tract 24, 77, 97, 139, 141, 262
resistin 338
resting follicles 47, 48
resumption of meiosis 36–9, 41, 51, 95, 106, 112, 131, 205, 207, 230–2, 235, 237–41, 244, 247, 252, 266, 350, 356
retrograde flow 70, 76
rhesus monkey 61, 62, 207, 311
ribosomal RNA 37
RNA 37, 41, 51, 235
RNA content 37, 41
RNA synthesis 37, 235, 239
rodents 170, 188
Rome 2, 3
Royal Society, the 13, 14
ruminants 18, 156, 161, 188
rupture 65, 280, 295, 304
rutting 7

sacral nerves 78
scaling 96, 268
Scandinavian school 77
scar tissue 15, 207, 326
scrotal testes 90, 94
scrotum 26, 90
seasonal breeding 262, 296
second meiotic arrest 51
second meiotic division 235, 236
second messenger 49, 50, 229
secondary fluid 108, 109, 110, 113
secondary follicle 38, 44, 45, 49, 50, 61, 68, 170
secondary oocyte 34, 232, 235, 268, 270
secondary sexual development 50
secretion 5, 108, 130, 194
secretory glands 15
secretory granules 228
semen 2, 3, 5, 7, 13, 14
seminal ducts 5
seminal plasma 129, 265, 266
seminiferous epithelium 26
seminiferous tubule 335
sensory neurons 77
serotonin 279
Sertoli cells 26, 27, 29, 42
sex chromosomes 25–8, 332, 335
sex determination 24–6
sex hormone-binding globulin 119, 147

sex reversal 26, 27
sexual behaviour 225
sexual cycle 16
sexual development 27
sexual differentiation 24–6, 61
sexual dimorphism 26
sheep 60, 64, 74–6, 84–7, 108, 116, 124, 126, 127, 154, 156, 160, 187, 188, 193–5, 198, 200, 204, 207, 208, 211, 226, 227, 267, 278, 296, 356
shrinkage 208
shunts 84
signal transduction pathway 31, 32, 155, 238, 265
silent death 211
site of fertilisation 5, 41, 126–8, 130, 232, 299, 301, 302, 316
site of ovulation 45
skeletal abnormalities 157
smooth muscle 46, 65, 77, 80, 82, 107, 170, 265, 272
sodium pump 115
somatic cells 26, 31, 34, 36–40, 42, 45, 47, 51, 113, 114, 119, 126, 130, 139, 158, 187, 188, 197, 205, 206, 208, 210, 215, 231, 233, 234, 247, 264, 267–9, 295, 298, 299, 303, 317
somatic monitors 338
somatic tissue growth 158
somatomedins 158
somatotrophin 360
Soranus 2
sow 7, 15
Sox-9 gene 26
species differences 41
species specificity 41
spectroscopy 116
sperm activation 96, 129
sperm distribution 128
sperm binding sites 41
sperm–egg interactions 38, 234, 262, 301
sperm–egg ratios 301, 304, 310
sperm flagellum 130
sperm head 38, 120, 128, 241, 242
sperm nucleus 129
sperm penetration 41, 235
sperm receptors 41
sperm release 128, 308
sperm reservoir 128, 264, 308
sperm storage 264
sperm transport 356

sperm viability 264
sperma 2
spermatogenesis 25, 26, 335
spermatozoa 14, 25, 90, 96, 128–30, 264, 310
spinal cord 77, 78
spiral arteries 64, 65
Sry gene 26, 27, 29
SRY gene 26
StAR gene 148–50, 167
Steel gene (*Sl*) 26, 31, 32, 51, 213
Stein–Leventhal syndrome 328
stem cell factor 31–3, 213
stem cells 19, 157
sterility 11, 31, 38, 331, 332
steroid acute regulatory protein 148–50
steroid-binding proteins 119
steroid hormones 16, 42, 48, 61, 68, 75, 78, 80, 82, 83, 87, 95, 97, 106, 112, 116–19, 131, 138, 141, 143, 147, 148, 151, 160, 161, 172, 188, 190, 192, 208, 212, 225, 243, 267, 283, 299, 301, 302, 306, 314
steroid molecules 119
steroid secretion 45, 48, 61, 83, 87, 166, 283
steroid signals 244
steroid synthesis 139–47, 151, 159, 166, 168, 267, 283
steroidogenesis 68, 80, 82, 98, 117, 143, 146, 148, 150, 158, 161, 167, 169, 170, 197, 198, 271, 277, 281, 297
steroidogenic enzymes 143, 147, 149, 191, 267
stigma 273, 281, 286
streak gonads 332
stress 336, 337, 339
stroma 16, 45, 65, 70, 71, 74, 80, 86–8, 90, 93, 94, 278, 280, 328–30
stromal capillaries 68
substrate 36, 40, 68, 87, 206
sub-ovarian plexus 70, 75
subunits 151, 156, 157
subordinate follicles 191, 192, 204, 209
substrate 297
superior ovarian nerve 78, 79
superovulation 19, 149, 205, 264, 309, 345, 346, 350, 351, 357–60, 362
surgery 15
surgical models 282, 283
sympathetic innervation 79, 83
sympathetic nerves 82, 83
sympathetic neurons 77
synapses 78, 80
synchronised cycles 327, 345

syncytium 39
synergism 138, 143, 154, 193, 276
synthetic activity 36, 37
systemic blood 119, 139
systemic circulation 139, 143, 198, 225, 270, 296, 308, 310

T cell 166
targeted deletion 157
telophase 35
temperature differential 90–3, 109, 369
temperature gradients 60, 90–8, 109, 110, 112, 214, 301, 330, 369
tenrecs 107, 297
teratomas 239
tertiary fluid 108, 110
tertiary follicles 7, 42, 43, 45, 46, 156
testes 5, 26, 29, 31, 37, 42, 240
testicular tissue 26–8
testis determination 24–7
testis development 27, 29, 31, 42
testis formation 25–7, 37, 42
testosterone 117, 118, 143, 145, 147, 148, 329
theca capillaries 65, 77, 108, 109, 111, 117
theca cells 42, 43, 45, 46, 82, 97, 119, 138, 141, 143, 146, 150, 154, 159, 160, 163, 167, 168, 170, 172, 190, 199, 210, 249, 251, 280, 283, 298, 326, 327, 338
theca externa 43, 45, 46, 68, 70, 71, 80–2, 271, 272, 278
theca interna 43, 45, 46, 68, 69, 71, 112, 138, 139, 141–3, 145, 149, 150, 155, 158, 163, 167, 197, 200, 207, 210, 224, 248, 264, 270, 271, 297, 330
theca layers 65, 73, 74, 77, 80, 107, 206, 274, 277, 278, 329
thecal hypertrophy 206, 207
thecal tissues 277
thermistor probes 90
thermo-imaging 90
thrombin 111
thromboxanes 169
thymus gland 163
time of birth 30, 33
tissue inhibitors of metalloproteases 249, 284
tissue movements 30
TNF-α 167, 168, 278–80, 338
toxins 206
transcription 37, 41, 50, 235
transcription factors 60
transcriptional repressor 50

transcripts 36, 37, 50, 235
transforming growth factors (TGFs) 50, 158, 160
    TGF-α 160, 162
    TGF-β 50, 151, 152, 161, 211, 248
transgenic mice 157
translation 37
translocation 335
transmembrane growth factor 31
transplantation evidence 60, 82, 83
transplantation model 60, 82
transplantation therapy 19, 332
transudation 108, 109, 111, 130, 277, 302, 306
trophectoderm 29
trophic factors 34
trophoblast 310
trophoblastin 18
trypsin inhibitor 246
Tsafriri, A. 269
tubal fluid 301, 304–6, 311
tumour 157
tumour necrosis factor α, *see* TNF-α
tumour suppressor gene 157
tunica albuginea 30, 31, 274, 278, 333
Turner's syndrome 331, 332
twin ovulations 192
two-cell theory 139, 141, 143
tyrosine kinase 211

ultrasonic scanning 19, 188, 190, 191, 194, 204, 265, 266, 312, 346, 347, 350, 352
ultrasound 85, 88, 204, 350
underdeveloped ovaries 50
ungulates 18, 310
unilateral influences 76, 79
unilateral ovariectomy 79, 358
urinary bladder 2, 3
urine 2
uterine arteries 7, 61–4
uterine atrophy 15
uterine blood supply 63
uterine fundus 69
uterine tissues 15
uterine tubes 5, 63
uterine vein 62, 65
utero-ovarian vessels 65, 75, 76
utero-tubal junction 15, 311
utero-vaginal ganglion 77, 78
uterus 2–5, 7, 8, 11, 13–16, 18, 69, 70, 76, 77, 88, 128, 139, 311

vagus nerve 78
van Horne, J. 7
vascular arcade 63, 283, 308
vascular baskets 68
vascular bed 84, 96, 111, 114, 131, 138, 142, 187, 206, 215, 275, 277, 326
vascular endothelial growth factor 52, 163, 251, 297, 351
vascular innervation 83
vascular network 52, 84, 207
vascular permeability 65, 106, 111, 271, 277
vascular system 52, 64, 77, 206, 251
vasculature 11, 46, 47, 61, 206, 207
vasoactive intestinal peptide 79, 82
vasoactive substances 280, 281
vasoconstriction 84
vasodilation 69, 84, 87, 170, 214, 275, 279
vasodynamics 68
vena cava 65
venous cannulation 84
Vesalius, Andreas 3, 5
vesicular follicles 24, 107, 205
vesicular nucleus 34, 230, 231
viable lifespan 313
visceral peritoneum 46
viscosity 110, 115, 250
viscous fluid 273, 304, 316, 350
vitelline factors 242
vitelline membrane 38, 233, 242
vitelline proteins 95
vitellus 25, 34, 38, 41, 240, 241
vomeronasal nerve 333

wedge resection 330
white blood cells 18, 110, 164, 172, 188, 262, 263, 265, 266, 276–8, 280, 283, 286
Wilm's tumour gene 50
Wnt-4 gene 26
Wnt-6 gene 26
women 16, 18, 50, 61, 64, 69, 88, 118, 119, 121, 122, 125–7, 147, 148, 167, 205, 230, 264, 268, 282, 296, 304, 311, 328–32, 338, 345
Wt gene 26

X-chromosome 25, 27, 37
X-inactivation 25
X-irradiation 38
Xenopus 240

Y-chromosome 25, 335
yeast 239
yolk 10, 11, 34, 235
yolk sac 24

Zeus 9
zinc 115
zona hardening 42
zona matrix 41
zona pellucida 34, 38–41, 47, 130, 234, 299, 302, 310, 313–16, 356
zona pellucida functions 41
zona permeability 42
zona proteins 41, 314
zygote 25, 36, 37, 51, 130, 303, 316, 317
zygotene 34